Lecture Notes in Computer Scier

T0238413

Commenced Publication in 1973
Founding and Former Series Editors:
Gerhard Goos, Juris Hartmanis, and Jan van Leeuwen

Shin-ichi Nakano Md. Saidur Rahman (Eds.)

WALCOM: Algorithms and Computation

Second International Workshop, WALCOM 2008
Dhaka, Bangladesh, February 7-8, 2008
Proceedings

 Springer

Volume Editors

Shin-ichi Nakano
Gunma University
Faculty of Engineering
Department of Computer Science, Gunma 376-8515, Japan
E-mail: nakano@cs.gunma-u.ac.jp

Md. Saidur Rahman
Bangladesh University of Engineering and Technology (BUET)
Department of Computer Science and Engineering
Dhaka 1000, Bangladesh
E-mail: saidurrahman@cse.buet.ac.bd

Library of Congress Control Number: 2007943472

CR Subject Classification (1998): F.2, C.2, G.2.2, I.3.5, G.1.6, J.3

LNCS Sublibrary: SL 1 – Theoretical Computer Science and General Issues

ISSN   0302-9743
ISBN-10   3-540-77890-X Springer Berlin Heidelberg New York
ISBN-13   978-3-540-77890-5 Springer Berlin Heidelberg New York

Springer is a part of Springer Science+Business Media

springer.com

© Springer-Verlag Berlin Heidelberg 2008
Printed in Germany

Typesetting: Camera-ready by author, data conversion by Scientific Publishing Services, Chennai, India
Printed on acid-free paper   SPIN: 12223176   06/3180   5 4 3 2 1 0

Preface

WALCOM 2008, the 2nd Workshop on Algorithms and Computation, held during February 7–8, 2008 in Dhaka, Bangladesh, covered the areas of algorithms and data structures, combinatorial algorithms, graph drawings and graph algorithms, parallel and distributed algorithms, string algorithms, computational geometry, graphs in bioinformatics and computational biology. The workshop was organized jointly by Bangladesh Academy of Sciences (BAS) and Bangladesh University of Engineering and Technology (BUET), and the quality of the workshop was ensured by a Program Committee comprising researchers of international repute from Australia, Bangladesh, Canada, Germany, India, Italy, Japan, Taiwan and UK.

The Program Committee thoroughly reviewed each of the 57 submissions for contributed talks and accepted 19 of them after elaborate discussions on review reports. Three invited talks by Satoshi Fujita, Alejandro Lopez-Ortiz and Ryuhei Uehara were included in the workshop.

We thank the invited speakers for joining us and presenting their talks on recent research areas of computer science from which researchers of this field could benefit immensely. We thank the members of the Program Committee and external reviewers for their wonderful job in reviewing the manuscripts. We thank the Steering Committee members Kyung-Yong Chwa, Costas S. Iliopoulos, M. Kaykobad, Petra Mutzel, Takao Nishizeki and C. Pandu Rangan for their continuous encouragement. We also thank the Advisory Committee Members M. Shamsher Ali, Naiyyum Choudhury and A.M.M. Safiullah for their inspiring support to this workshop. We are indebted to M. Kaykobad for his all-out support throughout the whole process. We thank Md. Shamsul Alam, Masud Hasan and M.A. Mazed for their prompt organizational support. Members of the Organizing Committee worked hard for the success of the workshop; we also thank them.

We would like to acknowledge the EasyChair system—a free conference management system that is flexible, easy to use, and has many features to make it suitable for various conference models. Finally, we thank our sponsors for their assistance and support.

February 2008
Shin-ichi Nakano
Md. Saidur Rahman

WALCOM Organization

WALCOM Steering Committee

Kyung-Yong Chwa	KAIST, Korea
Costas S. Iliopoulos	KCL, UK
M. Kaykobad	BUET, Bangladesh
Petra Mutzel	University of Dortmund, Germany
Takao Nishizeki	Tohoku University, Japan
C. Pandu Rangan	IIT, Madras, India

WALCOM 2008 Committees

Advisory Committee

M. Shamsher Ali	President, BAS
Naiyyum Choudhury	Secretary, BAS
A.M.M. Safiullah	Vice-Chancellor, BUET

Program Committee

Satoshi Fujita	Hiroshima University, Japan
Stanley P.Y. Fung	University of Leicester, UK
Michael Juenger	University of Cologne, Germany
Subir Kumar Ghosh	TIFR, India
Seok-Hee Hong	NICTA, Australia
Giuseppe Liotta	University of Perugia, Italy
Md. Abul Kashem	BUET, Bangladesh
Naoki Katoh	Kyoto University, Japan
Alejandro Lopez-Ortiz	University of Waterloo, Canada
M. Manzur Murshed	Monash University, Australia
Shin-ichi Nakano	Gunma University, Japan (Co-chair)
Subhas Chandra Nandy	Indian Statistical Institute, Kolkata, India
Md. Saidur Rahman	BUET, Bangladesh (Co-chair)
William F. Smyth	McMaster University, Canada
Ryuhei Uehara	JAIST, Japan
Sue Whiteside	McGill University, Canada
Hsu-Chun Yen	National Taiwan University, Taiwan

Organizing Committee

Muhammad Abdullah Adnan
Md. Mostofa Akbar
Md. Ashraful Alam
S.M. Mahbub Alam
Md. Shamsul Alam (Chair)
Muhammad Masroor Ali
Tanveer Awal
A.M. Choudhury
Naiyyum Choudhury
S.M. Farhad
Md. Ehtesamul Haque
Masud Hasan (Secretary)
Mojahedul Hoque Abul Hasnat
Md. Shohrab Hossain
Anindya Iqbal
A.B.M. Alim Al Islam
Md. Mahfuzul Islam
Md. Humayun Kabir
Md. Rezaul Karim
M. Kaykobad
Ahmed Khurshid
M.A. Mazed
Momenul Islam Milton
Mahmuda Naznin
Md. Mostofa Ali Patwary
A.K.M. Ashikur Rahman
Md. Saidur Rahman
M. Sohel Rahman
Md. Muhibur Rasheed
Md. Abul Hassan Samee
Khaled Mahmud Shahriar

External Reviewers

Muhammad Abdullah Adnan
Reaz Ahmed
Toru Araki
Carla Binucci
Arijit Bishnu
Christoph Buchheim
Sandip Das
Emilio Di Giacomo
Walter Didimo
Birgit Engels

Fazle Elahi Faisal
Chun-I Fan
Sasthi C. Ghosh
Francesco Giordano
Luca Grilli
Carsten Gutwenger
Masud Hasan
Md. Rezaul Karim
Akinori Kawachi
Mohammad Kaykobad
Karsten Klein
Frauke Liers
A.K.M. Ashikur Rahman
Hsueh-I Lu
Hiroshi Nagamochi
Wolfgang Paul
Simon Puglisi
Muntasir Rahman
Md. Abul Hassan Samee
Michael Schulz
WF Smyth
Shin-ichi Tanigawa
Shi-Chun Tsai
Yushi UNO
Shu Wang
Roman Wienands
Tian-Li Yu
Munira Yusufu

WALCOM 2008 Organizers

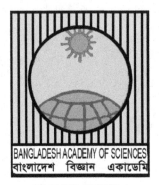

BANGLADESH
ACADEMY OF SCIENCES

BANGLADESH
UNIVERSITY OF
ENGINEERING AND
TECHNOLOGY (BUET)

WALCOM 2008 Sponsors

Table of Contents

Invited Talks

Bioinformatics Algorithms

Computational Geometry and Graph Drawing

Graph Algorithms I

Algorithm Engineering

Graph Algorithms II

Vertex Domination in Dynamic Networks[*]

Satoshi Fujita

Department of Information Engineering
Graduate School of Engineering, Hiroshima University
fujita@se.hiroshima-u.ac.jp

Abstract. This paper studies a vertex domination problem in dynamic networks, which allows dynamic change of the set of vertices, the set of edges, and the set of weights. In particular, we will examine the following two theoretical issues arising in designing an adaptive vertex domination scheme for such networks: 1) How can we transfer a given configuration to a dominating configuration while keeping that any intermediate configuration is safe? 2) How can we reduce the computational complexity of the vertex domination problem by allowing defections in the domination?

Keywords: Dominating set, dynamic network, defection, regular graphs.

1 Introduction

Given a graph $G = (V, E)$ with vertex set V and edge set E, a **dominating set** for graph G is a subset of vertices such that for any $u \in V$, either u or at least one neighbor of u is contained in the subset [15]. Mathematical properties of dominating set have been extensively investigated from various aspects during the past three decades, which include graph theoretic characterization [2,5,6,16,18,24], computational complexity of finding a dominating set with a designated cardinality [13,17,20], and polynomial time algorithms to find minimum dominating set for special graphs such as interval graphs and perfect graphs [1,3,4,19,21]. It has also been pointed out that the notion of dominating set is closely related with several **resource allocation problems** in networks, such as file allocation problem [9], facility allocation problem [12], and path assignment problem in wireless communication networks [7,14,22,23].

In this paper, we consider a vertex domination problem in dynamic networks, which allows dynamic change of the set of vertices, the set of edges, and the set of weights in the given network. A dynamism in network can commonly be observed in actual applications. For example, in distributed systems such as peer-to-peer and grid computers, node arrival and node departure dynamically change the set of vertices; in wireless networks such as mobile ad hoc networks and sensor networks, the node mobility can easily connect and disconnect links in the wireless network; and even when the overall network configuration does

---

[*] This research was partially supported by the Grant-in-Aid for Scientific Research, Priority Areas (B)(2) 16092219.

S.-i. Nakano and Md. S. Rahman (Eds.): WALCOM 2008, LNCS 4921, pp. 1–12, 2008.

not change, the change of traffic load would affect to the set of optimum configurations for the corresponding resource assignment problem. Among them, the
most critical issue we have to settle is that the disconnection of edges and the removal of dominating vertices could generate a vertex which is not dominated by
any vertex in V. In order to apply the notion of dominating set to such dynamic
environments in a safe and efficient manner, we have to realize a procedure to
converge a given configuration to a dominating one by conducting appropriate
reassignments of dominating vertices over the network.

In this paper, we will consider the following concrete scenario to realize such
a recovering procedure: Initially, a given graph is dominated by several (disjoint)
dominating sets, each of which corresponds to a specific service provided by a set
of designated servers. By a dynamic change of the underlying network, some of
those dominating sets are "damaged" and several vertices become undominated
by those damaged sets. The objective of the recovering scheme is to transfer a
given configuration with damages to a normal configuration with no damages,
while supplementing the role of damaged sets by the undamaged ones during the
course of transfer. More specifically, in the following, we will focus our attention
to the following two theoretical issues which should appear in designing *safe* and
efficient recovering scheme:

1. What is a sufficient condition to realize a safe transfer to a normal configuration? To clarify this point, we will examine the number of dominating
   vertices (i.e., the number of mobile servers to be prepared in some applications) which realizes a safe transfer between two dominating configurations,
   while keeping that any intermediate configuration in the transfer is also
   dominating one. It should be stressed here that although we do not directly
   consider a transfer from an undominated configuration to a dominated configuration, in many practical situations, it is very important to move the role
   of dominating vertex from an instable node to a stable one beforehand, in
   order to avoid (unnecessary) damages of the dominating sets.
2. The problem of partitioning the vertex set of a given graph into a maximum
   number of dominating sets is known to be NP-hard [15]. So, it should be
   interesting to ask how can we reduce the computational complexity of such
   partitioning problem by allowing *defections* in the domination. Note that
   this question is also important from practical point of view, because if the
   answer is "yes", then any configuration with some defections could be used
   as a safe state to realize an efficient recovery scheme.

The remainder of this paper is organized as follows. In Section 2, we describe
preliminary results on the safe transfer between dominating configurations. Section 3 describes basic properties of domatic partition with defections. Finally
Section 4 concludes the paper with future problems.

2   Convergence to a Given Configuration

In this section, we assume that each dominating set is a multiset, by technical reasons. Let $\mathcal{D}(G)$ denote an (infinite) set of all dominating (multi)sets for

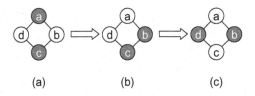

(a)   (b)   (c)

Fig. 1. A sequence of single-step transfers among dominating configurations for a ring with four vertices (dominating vertices are painted gray)

$G = (V, E)$. For any $S_1, S_2 \in \mathcal{D}(G)$, we say that S_1 is **single-step transferable** to S_2 and denote it as $S_1 \rightarrow S_2$, if there are two vertices $u, v \in V$ such that $S_1 - \{u\} = S_2 - \{v\}$ and $\{u, v\} \in E$. In other words, single-step transfer from S_1 to S_2 is realized by moving the "role of dominating vertex" from $u \in S_1$ to its neighbor $v \in S_2$, where each vertex can own more than one roles since each dominating set is assumed to be a multiset. For example, in a ring network consisting of four vertices a, b, c, and d arranged in this order, a dominating configuration $\{a, c\}$ is transferred to configuration $\{b, c\}$ in a single-step by moving the role of dominating vertex from a to b (see Figure 1 for illustration). A transitive closure of the relation of single-step transferability naturally defines the notion of **transferability**, which is denoted as $S_1 \xrightarrow{*} S_2$, in what follows. Note that this definition of transferability requests every subset appearing in a transfer from S_1 to S_2 to be a dominating set for G.

2.1 Mutual Transferability

The notion of mutual transferability among dominating configurations is formally defined as follows. A set $\mathcal{D}' \subseteq \mathcal{D}(G)$ is said to be **mutually transferable** if it holds $S_1 \xrightarrow{*} S_2$ for any $S_1, S_2 \in \mathcal{D}'$, where a sequence of single-step transfers from S_1 to S_2 can contain a dominating configuration contained in $D(G) - \mathcal{D}'$. In [11], we have derived interesting results on the mutual transferability for several classes of graphs. In this subsection, we will review some of those results. The first theorem gives a tight bound for the class of trees [11]:

Theorem 1 (Trees). *For any tree T with n vertices, the set of dominating sets for T consisting of $k \geq \lfloor n/2 \rfloor$ vertices is mutually transferable, and there is a tree T_0 with n vertices such that $\gamma(T_0) = \lfloor n/2 \rfloor$, where $\gamma(T_0)$ denotes the size of a minimum dominating set for T_0.*

This theorem indicates that if we can use $\lfloor n/2 \rfloor$ tokens, each of which represents a dominating vertex, we can always realize a transfer between *any* dominating configurations for a tree represented by the $\lfloor n/2 \rfloor$ tokens, and that there is a tree which cannot be dominated by $\lfloor n/2 \rfloor - 1$ vertices. In other words, $\lfloor n/2 \rfloor$ is a tight bound for the class of trees.

The next theorem provides a lower bound on the number of dominating vertices for the class of Hamiltonian graphs [11].

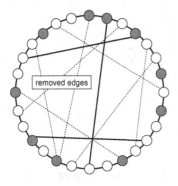

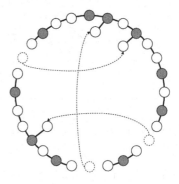

(a) An example of graph G'.   (b) An example of graph G''.

Fig. 2. Explanation of the proposed scheme

Theorem 2 (Lower Bound). *For any $r \geq 2$ and $n \geq 1$, there is a Hamiltonian r-regular graph G with more than n vertices such that the set of dominating sets for G with cardinality at least $\lceil (n+1)/3 \rceil - 1$ is not mutually transferable.*

It should be noted that for any Hamiltonian graph G consisting of n vertices, $\gamma(G) \leq \lceil (n+1)/3 \rceil - 1$, since it has a ring of size n as a subgraph. It is in contrast to the case of trees, since the theorem claims that there is a Hamiltonian r-regular graph such that $\lceil (n+1)/3 \rceil - 1$ tokens are not sufficient to guarantee the mutual transferability among dominating configurations, while $\lceil (n+1)/3 \rceil - 1$ tokens are sufficient to dominate it. By combining Theorem 2 with the following theorem, we could derive a tightness of the bound for the class of Hamiltonian graphs [11].

Theorem 3 (Hamiltonian Graphs). *For any Hamiltonian graph G with n vertices, the set of dominating sets for G consisting of $k \geq \lceil (n+1)/3 \rceil$ vertices is mutually transferable.*

In the next subsection, we provide an outline of the algorithm which connects two given dominating configurations by a sequence of single-step transfers.

2.2 Algorithm

Let $G = (V, E)$ be a Hamiltonian graph with n vertices, and R a Hamiltonian cycle in it. In what follows, edges contained in R will be referred to as *ring edges* and the other edges in G are referred to as *chord edges*. Let $S \subseteq V$ be a dominating set for G consisting of $\lceil (n+1)/3 \rceil$ vertices. The key idea of our scheme is to transfer S to a dominating configuration for R by consecutively removing chord edges and by moving the role of dominating vertices accordingly. Since it is known that the set of dominating sets for R consisting of $\lceil (n+1)/3 \rceil$ vertices is mutually transferable [10], the existence of such transfer implies that the set of dominating sets for G with $\lceil (n+1)/3 \rceil$ vertices is also mutually transferable.

An outline of the scheme is as follows [11]. In the first step of the transfer, we apply the following rule until it could not be applied to the resultant graph:

Rule 1: If the removal of a chord edge does not violate the condition of domination for its end vertices, then remove it.

Let G' be the resultant graph. Figure 2 (a) illustrates an example of G and G'. Note that S is a dominating set for G', and that G' contains at most $n - \lceil (n+1)/3 \rceil - 2$ $(= |V - S| - 2)$ chord edges, since there are at most $|V - S|$ vertices to be dominated by the vertices in S, and at least two of them have already been dominated via ring edges. In addition, for any chord edge in G', 1) exactly one of the end vertices must be a member of S and 2) the other vertex must be connected with exactly one chord edge (otherwise, Rule 1 can be applied to remove a chord edge).

Next, we consider a subgraph G'' of G' which is obtained by removing all ring edges incident to the vertices dominated via chord edges. Figure 2 (b) shows an example of the resultant graph. By construction, G'' is a forest of trees such that every leaf is a member of $V - S$ and every vertex with degree more than two is a member of S. Since $|S| = \lceil (n+1)/3 \rceil$ is assumed, in at lease one of the resultant trees, the number of dominating vertices exceeds one third of the number of vertices. Let T be one of such trees and S_T $(\subseteq S)$ be the set of dominating vertices contained in T. The proof can be completed if we could show that in graph G'', S_T can be transferred to a dominating configuration for T in which at least one leaf is a dominating one. In fact, we can prove the claim by using the fact that $|S_T|$ is greater than one third of the number of vertices in T. See [11] for the details.

3   Domatic Partition with Defections

Let $G = (V, E)$ be an undirected graph with vertex set V and edge set E. For each $u \in V$, let $N[u]$ denote the closed neighbor of u in G; i.e., $N[u] = \{u\} \cup \{v \in V : \{u, v\} \in E\}$. A partition $\mathcal{V}$ of V is said to be a domatic partition (DP) of G if each element in $\mathcal{V}$ is a dominating set for G. In the following, we refer to a DP of size d as d-DP. A DP with maximum cardinality is called a maximum DP, and the cardinality of which is referred to as the domatic number. In this section, we introduce the notion of **defective domatic partition** (DDP) of graph G, and investigate basic properties of that. A partition $\mathcal{V}$ of V is said to be a (d, k)-DDP of G if it satisfies: 1) $|\mathcal{V}| = d$, and 2) for any $u \in V$,

$$|\{W \in \mathcal{V} : N[u] \cap W \neq \emptyset\}| \geq d - k.$$

In other words, in a (d, k)-DDP of G, each vertex is dominated by at least $d - k$ dominating sets, and allows at most k defections among prepared d dominating sets. Note that $(d, 0)$-DDP is equivalent to d-DP.

3.1   Basic Properties

This subsection describes basic properties of (d, k)-DDP. The first lemma is related with the sufficiency on the number of defections.

Lemma 1. *If G has a (d, k)-DDP, then for any $1 \leq i \leq |V| - d$, G has a $(d + i, k + i)$-DDP.*

Proof. We may repeat the following process for i times: 1) select an element W in $\mathcal{V}$ consisting of at least two vertices, 2) remove any vertex u from W, and 3) add a new element $W' = \{u\}$ to $\mathcal{V}$. In each iteration, 1) the number of defections at a vertex not in $N[u]$ increases by one, and 2) that of vertices in $N[u]$ does not increase. Thus the lemma follows.

The above lemma implies that, for example, every graph with no isolated vertex has a $(2 + i, i)$-DDP for any $1 \leq i \leq |V| - d$, since such graph has a 2-DP consisting of a maximal independent set and its complement.

Let $\Delta(G)$ and $\delta(G)$ be the maximum and minimum vertex degree of graph G, respectively. Since every vertex can be dominated by at least $\delta(G) + 1$ different dominating sets in a partition of V of size $|V|$, we have the following claim:

Property 1. Any graph G has a $(|V|, |V| - (\delta(G) + 1))$-DDP.

The above claim can be improved by using the notion of "square" of graphs, which is formally defined as follows:

Definition 1. *A square of graph $G = (V, E)$, denoted by G^2, is defined as follows: 1) $V(G^2) = V$ and 2) $E(G^2) = \{\{u, v\} \in V \times V : \{u, v\} \in E$ or there exists w such that $\{u, w\}, \{w, v\} \in E\}$.*

Since any two vertices contained in an independent set of G^2 are not contained in a closed neighborhood of a vertex in G, we can reduce the size of partition shown in Property 1 from $|V|$ to $\chi(G^2)$, while keeping the gap between d and k, where $\chi(G)$ denotes the chromatic number of G. By the Brooks' theorem, it is known that $\chi(G) \leq \Delta(G)$ unless G is complete, and that such a coloring can be found in a polynomial time. Thus, we have the following theorem.

Theorem 4. *If G is a graph such that a square of G is not complete and consists of at least $\Delta^2 + 1$ vertices, then G has a $(\Delta^2, \Delta^2 - (\delta + 1))$-DDP, where Δ and δ denote the maximum and minimum vertex degree of graph G.*

By this theorem, we can immediately have a corollary such that any 3-regular (i.e., cubic) graph with at least twelve vertices has a $(9, 5)$-DDP, any 4-regular graph with at least 18 vertices has a $(16, 11)$-DDP, and so on. Figure 3 shows a cubic graph consisting of ten vertices which does not have a $(9, 5)$-DDP.

Finally, when $d > \delta(G) + 1$, the number of defections at a vertex with degree $\delta(G)$ is at least $d - (\delta(G) + 1)$. Thus, we have the following negative claim:

Property 2. Any graph G has no $(\delta(G) + 2 + i, i)$-DDP for any $i \geq 0$.

3.2   Regular Graphs

This subsection examines (d, k)-DDP for regular graphs, in more detail. In the following, we first consider the class of rings, and then extend it to cubic graphs.

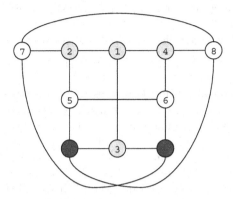

Fig. 3. A cubic graph consisting of ten vertices which does not have a $(9, 5)$-DDP

The domatic number of a ring is at least two and at most three. In addition, a ring has a 3-DP (i.e., $(3, 0)$-DDP) iff the number of vertices is a multiple of three. As for the existence of $(4, 1)$-DDP, the following lemma holds (note that although it is true that a ring has a $(4, 1)$-DDP if the number of vertices is a multiple of three, the reverse is not true):

Lemma 2 (Ring). *Any ring has a $(5, 2)$-DDP, and a ring has a $(4, 1)$-DDP iff the number of vertices in the ring is not equal to five.*

Proof. By Theorem 4, a ring with more than five vertices has a $(4, 1)$-DDP, since a square of such ring is not complete (note that $\Delta = \delta = 2$ for rings). In addition, by Lemma 1 (resp. Property 1), a ring with three (resp. four) vertices has a $(4, 1)$-DDP. Thus, the sufficiency follows.

The necessity could be verified as follows. First, if a ring with five vertices has a $(4, 1)$-DDP, each vertex in $N[u]$ must belong to different dominating sets for any u. Let $v_1, v_2, \ldots, v_5$ be five vertices arranged in this order in the ring, and without loss of generality, let us assume that vertices v_1, v_2, v_3 belong to different dominating sets W_1, W_2, W_3, respectively. Now, we have $v_4 \notin W_2 \cup W_3$ since three vertices in $N[v_3]$ must belong to different dominating sets, and we have $v_4 \notin W_1$ since three vertices in $N[v_5]$ must belong to different dominating sets. Thus, vertex v_4 must belong to a new dominating set, say W_4. However, the same is true for vertex v_5, i.e., $\{v_4, v_5\} \subseteq W_4$ must hold, a contradiction.

Finally, the proof on $(5, 2)$-DDP is trivial. Hence the lemma follows.

Note that by combining the above results with Property 2, we can check the existence of (d, k)-DDP in a ring with a designated number of vertices, for any combination of d and k.

Now, let us proceed to the examination of cubic graphs. The domatic number of a cubic graph is at least two and at most four, and it is known that the complexity of asking if a given cubic graph has a i-DP is NP-hard for $i = 3, 4$. In fact, we can prove the NP-hardness for 3-DP by using a reduction from the

3-dimensional matching, and that for 4-DP by using a reduction from 3-edge-coloring problem.

As for $(4, 1)$-DDP, we have the following positive result.

Theorem 5. *Any cubic graph G has a $(4, 1)$-DDP.*

Proof. First, let us consider bridgeless cases. It is known that the set of edges in a bridgeless cubic graph can be partitioned into a collection of one- and two-factors, where a one-factor is a matching and a two-factor is a collection of rings. Let $\mathcal{C}$ be the resultant collection of rings, where we may assume $|\mathcal{C}| \geq 2$ without loss of generality. Let G' be a connected subgraph of G which is obtained from $\mathcal{C}$ by adding minimum number of edges (i.e., $|\mathcal{C}| - 1$ edges) contained in the one-factor, in such a way that the resultant graph is a tree of rings (i.e., a cactus). By Lemma 2, each element in $\mathcal{C}$ has a $(4, 1)$-DDP if it consists of less than or more than five vertices, and moreover, even if it consists of five vertices, we can construct a $(4, 1)$-DDP for the overall graph G by "supplementing" defective dominating sets between neighboring rings connected by an edge.

If G contains a bridge vertex, on the other hand, we can construct a $(4, 1)$-DDP of G as follows. Let G'' be a collection of connected components obtained by removing bridge vertices from G. In the first step, it constructs a $(4, 1)$-DDP for each connected component in G''. It then associates a dominating set for each bridge vertex, and "relabels" the partition for each component, if necessary. Note that such a relabeling always terminates since there are no cyclic dependence among connected components, and it is clear that such a relabeling correctly terminates the labeling procedure. Hence, the theorem follows.

Thus, by Lemma 1, any cubic graph has a $(4 + i, 1 + i)$-DDP for any $i \geq 0$.

Next, as for $(8, 4)$-DDP, we have the following positive result.

Theorem 6. *Any cubic graph with at least twelve vertices has a $(8, 4)$-DDP.*

Proof. The basic idea of the proof is to introduce the notion of layer of vertices, and to assign labels to the vertices according to the order of layers. Given a subgraph G' of $G = (V, E)$, a chain C is said to be a **chordless chain** with respect to G' if it satisfies the following two conditions:

1. If C is a simple path, then each of its two end vertices is adjacent with at least one and at most three vertices in $V(G')$. Otherwise, the shape of C is a concatenation of a cycle and a simple path, and the end vertex of C in the path part is adjacent with at least one and at most two vertices in $V(G')$.
2. Let $v_1, v_2, \ldots, v_p$ be a sequence of vertices appearing in C in this order, where v_1 is a vertex adjacent with a vertex in $V(G')$ and if the shape of C is a concatenation of a cycle and a simple path, then vertex v_p is adjacent with a vertex in C. For any $2 \leq i \leq p - 1$, there is no vertex $u \in V(G') \cup \{v_1, v_2, \ldots, v_{i-2}\}$ such that $\{v_i, u\} \in E(G)$.

A **layering** of V is a sequence of subsets of V, which is defined as follows:

1. V_0 is a set of vertices which induces a chordless cycle in G.
2. For any $i \geq 1$, V_i is the vertex set of a chordless chain with respect to a subgraph of G induced by $\bigcup_{j=0}^{i-1} V_j$.

Note that any cubic graph has at least one layering, which could be found in a linear time by using depth-first search. Let $V_0, V_1, \ldots, V_q$ be a layering of V such that $\bigcup_{j=0}^{q} V_j = V$. In the following, we will assign eight labels to the vertices in V in such a way that for each $u \in V$, four vertices in $N[u]$ are assigned different labels (in what follows, we refer to it as Condition A, to clarify the exposition).

At first, we label vertices in V_0 with five colors in such a way that Condition A is satisfied for the subgraph induced by V_0. We then try to assign labels to vertices in layers $V_1, V_2, \ldots$ in this order. More concretely, the labeling of vertices in V_i is conducted as follows:

Case 1: First, let us consider cases in which $|V_i| \geq 2$ and the subgraph C of G induced by V_i is a simple path. If each of the end vertices of C is not adjacent with two vertices in $\bigcup_{j=0}^{i-1} V_j$ with the same label, then we can easily label vertices in V_i in such a way that Condition A is satisfied for the subgraph induced by $\bigcup_{j=0}^{i} V_j$. On the other hand, if an end vertex of C is adjacent with two vertices with the same label, then we may label the end vertex to satisfy Condition A after "flipping" the label of one of its adjacent vertices appropriately. Note that eight labels are sufficient not to cause a propagation of such flippings, and that the labeling of intermediate vertices of C can be substituted without causing a propagation of flippings.

Case 2: When the shape of subgraph C is concatenation of a cycle and a path, we can complete the labeling by using the same argument to above.

Case 3: Finally, let us consider the case in which $|V_i| = 1$, i.e., when the unique vertex $u \in V_i$ is adjacent with three vertices $v_1, v_2, v_3 \in \bigcup_{j=0}^{i-1} V_j$. Such case can be reduced to the first case by removing the label assigned to v_1 and considering $V_i' = \{u, v_1\}$ as the chordless chain to be examined in the ith round.

Hence the theorem follows.

On the other hand, we have the following negative results on the existence of (d, k)-DDP. First, by Property 2, any cubic graph has no $(6, 1)$-DDP (thus by Lemma 1, it has no $(5, 0)$-DDP). Figure 4 shows a part of a cubic graph which has no $(7, 3)$-DDP. Since the non-existence of $(7, 3)$-DDP immediately indicates the non-existence of $(6, 2)$-DDP and $(5, 1)$-DDP, the above claim implies that for each $4 \leq d \leq 7$, there exists a cubic graph which does not have a $(d, d-4)$-DDP. On the other hand, we can show that there exists a cubic graph to have a $(d, d-4)$-DDP for any $d \geq 4$ (we may consider a cubic graph which has a perfect domination [15]). Thus, at least in the range of $4 \leq d \leq 7$, the problem of asking if a given instance has a $(d, d-4)$-DDP is non-trivial in the sense that it is neither trivially "yes" nor trivially "no".

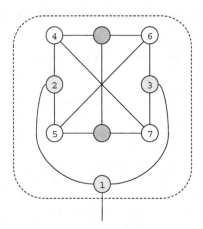

Fig. 4. A component of cubic graph which does not have a $(7,3)$-DDP

In other words, the remaining problem we have to solve is what is the computational complexity of the problem of finding a $(5,1)$-DDP, $(6,2)$-DDP, and $(7,3)$-DDP of cubic graphs. Note that since the problem of asking if a given cubic graph has a 4-DP (i.e., $(4,0)$-DDP) is NP-hard, and the decision problem is in P for $(8,4)$-DDP by the above results, in the course of increasing the size of partition from four to eight by keeping the gap to four, the complexity of the problem changes from NP-hard to P (if the gap is three, the problem is NP-hard for $k = 0$ and in P for $k \geq 1$; and if the gap is five or more, there are no such DDPs in cubic graphs).

4   Concluding Remarks

In this paper, we examined two theoretical issues in designing adaptive vertex domination problem for dynamic networks. There remain several interesting open problems to be investigated:

- We have to clarify the computational complexity of the problem of finding a $(5,1)$-DDP, $(6,2)$-DDP, and $(7,3)$-DDP of cubic graphs. A natural conjecture is that a cubic graph has no $(7,3)$-DDP iff it includes the graph shown in Figure 4 as a subgraph (if the conjecture is true, then the decision problem is in P for $(7,3)$-DDP).
- We should design an efficient distributed algorithm to find a $(8,4)$-DDP for given cubic graphs with at least twelve vertices. Is it difficult to parallelize?
- We should design a *distributed* algorithm to converge a given configuration to a dominating one. A promising approach to realize such an autonomous convergence is to apply the notion of self-stabilization invented by Dijkstra in [8], which has been extensively investigated in the field of distributed algorithms.

- Defections could be introduced to the notion of r-configurations [9]. How can we realize mutual transfers between domatic partitions with defections and r-configurations with defections?

References

1. Atallah, M.J., Manacher, G.K., Urrutia, J.: Finding a minimum independent dominating set in a permutation graph. Discrete Appl. Math. 21, 177–183 (1988)
2. Bange, D.W., Barkauskas, A.E., Slater, P.T.: Efficient dominating sets in graphs. In: Ringeisen, R.D., Roberts, F.S. (eds.) Applications of Discrete Mathematics, pp. 189–199. SIAM (1988)
3. Bertossi, A.A.: On the domatic number of interval graphs. Information Processing Letters 28(6), 275–280 (1988)
4. Chang, G.J., Rangan, C.P., Coorg, S.R.: Weighted independent perfect domination on cocomparability graphs. Technical Report 93-24, DIMACS (April 1993)
5. Cockayne, E.J., Hedetniemi, S.T.: Optimal domination in graphs. IEEE Trans. Circuit and Systems CAS-22, 855–857 (1975)
6. Cockayne, E.J., Hedetniemi, S.T.: Towards a theory of domination in graphs. Networks 7, 247–261 (1977)
7. Dai, F., Wu, J.: An Extended Localized Algorithm for Connected Dominating Set Formation in Ad Hoc Wireless Networks. IEEE Transactions on Parallel and Distributed Systems 53(10), 1343–1354 (2004)
8. Dijkstra, E.W.: Self-stabilization in spite of distributed control. Communications of the ACM 17(11), 643–644 (1974)
9. Fujita, S., Yamashita, M., Kameda, T.: A Study on r-Configurations – A Resource Assignment Problem on Graphs. SIAM J. Discrete Math. 13(2), 227–254 (2000)
10. Fujita, S., Liang, Y.: How to Provide Continuous Services by Mobile Servers in Communication Networks. In: Liew, K.-M., Shen, H., See, S., Cai, W. (eds.) PDCAT 2004. LNCS, vol. 3320, pp. 326–329. Springer, Heidelberg (2004)
11. Fujita, S.: A Tight Bound on the Number of Mobile Servers to Guarantee the Mutual Transferability among Dominating Configurations. In: Deng, X., Du, D.-Z. (eds.) ISAAC 2005. LNCS, vol. 3827, pp. 563–572. Springer, Heidelberg (2005)
12. Fujita, S.: Loose Covering by Graphs. In: Proc. of 2006 Japan-Korea Joint Workshop on Algorithms and Computation, pp. 9–15 (2006)
13. Garey, M.R., Johnson, D.S.: Computers and Intractability: A Guide to the Theory of NP-Completeness. W.H. Freeman and Company, San Francisco (1979)
14. Guha, S., Khuller, S.: Approximation Algorithms for Connected Dominating Sets. In: Proc. European Symposium on Algorithms, pp. 179–193 (1996)
15. Haynes, T.W., Hedetniemi, S.T., Slater, P.J.: Fundamentals of Domination in Graphs. Marcel Dekker, Inc. (1998)
16. Haynes, T.W., Hedetniemi, S.T., Slater, P.J.: Domination in Graphs: Advanced Topics. Marcel Dekker, Inc. (1998)
17. Irving, R.W.: On approximating the minimum independent dominating set. Information Processing Letters 37, 197–200 (1991)
18. Livingston, M., Stout, Q.F.: Perfect dominating sets. Congressus Numerantium 79, 187–203 (1990)

19. Lu, T.L., Ho, P.H., Chang, G.J.: The domatic number problem in interval graphs. SIAM J. Disc. Math. 3, 531–536 (1990)
20. Matheson, L.R., Tarjan, R.E.: Dominating sets in planar graphs. Technical Report TR-461-94, Dept. of Computer Science, Princeton University (May 1994)
21. Rao, A.S., Rangan, C.P.: Linear algorithm for domatic number problem on interval graphs. Information Processing Letters 33(1), 29–33 (1989)
22. Wu, J., Li, H.: Domination and Its Applications in Ad Hoc Wireless Networks with Unidirectional Links. In: Proc. of International Conference on Parallel Processing, pp. 189–200 (2000)
23. Wu, J.: Extended Dominating-Set-Based Routing in Ad Hoc Wireless Networks with Unidirectional Links. IEEE Transactions on Parallel and Distributed Computing 22, 327–340 (2002)
24. Yen, C.C., Lee, R.C.T.: The weighted perfect domination problem. Information Processing Letters 35, 295–299 (1990)

Closing the Gap Between Theory and Practice: New Measures for On-Line Algorithm Analysis

Reza Dorrigiv and Alejandro López-Ortiz

Cheriton School of Computer Science,
University of Waterloo,
Waterloo, ON, N2L 3G1, Canada
{rdorrigiv,alopez-o}@uwaterloo.ca

Abstract. We compare the theory and practice of online algorithms, and show that in certain instances there is a large gap between the predictions from theory and observed practice. In particular, the competitive ratio which is the main technique for analysis of online algorithms is known to produce unrealistic measures of performance in certain settings. Motivated by this we examine first the case of paging. We present a study of the reasons behind this apparent failure of the theoretical model. We then show that a new measure derived from first principles and introduced by [Angelopoulos, Dorrigiv and López-Ortiz, SODA 2007] better corresponds to observed practice. Using these ideas, we derive a new framework termed the cooperative ratio that generalizes to all other online analysis settings and illustrate with examples in list update[1].

1 Introduction

Competitive analysis has long been established as the canonical approach for the analysis of on-line algorithms. Informally, an on-line algorithm processes the input in an on-line manner; that is, the input is a sequence of requests that arrive sequentially in time and the algorithm must make irrevocable decisions with only partial or no knowledge about future requests.

The competitive ratio was formally introduced by Sleator and Tarjan [33], and it has served as a practical framework for studying on-line algorithms. An algorithm (assuming a cost-minimization problem) is said to be α-competitive if the cost of serving any specific request sequence never exceeds α times the optimal cost (up to some additive constant) of an *off-line* algorithm which knows the entire sequence. The competitive ratio has been applied to a variety of problems and settings such as on-line paging, list update, geometric searching/motion planning and on-line approximation of NP-complete problems. Indeed the growth and strength of the field of on-line algorithms is due in no small part to the effectiveness of this measure in the course of practical analysis: the measure is

---

[1] This paper presents the unifying concepts behind a series of papers on on-line algorithm analysis by the authors, which as a whole lead to a new model for on-line algorithm analysis. See [4,5,18,19,20,21] where aspects of this proposal are discussed separately and at length on their own.

S.-i. Nakano and Md. S. Rahman (Eds.): WALCOM 2008, LNCS 4921, pp. 13–24, 2008.
© Springer-Verlag Berlin Heidelberg 2008

relatively simple to define yet powerful enough to quantify, to a large extent, the performance of an on-line algorithm. Furthermore computing the competitive ratio has proven to be effective—even in cases where the exact shape of the off-line optimum OPT is unknown.

On the other hand, there are known applications in which competitive analysis yields unsatisfactory results. In some cases it results in unrealistically pessimistic measures, in others it fails to distinguish between algorithms that have vastly differing performance under any practical characterization. Most notably, for the case of paging and on-line motion planning algorithms, competitive analysis does not reflect observed practice, as first noted by Sleator and Tarjan in their seminal paper [33]. Such anomalies have led to the introduction of many alternatives to the competitive analysis of on-line algorithms [8,10,11,13,14,16,22,24,25,27,28,35].

In this paper we study the reasons behind this disconnect, using paging as a case study. We show then that a newly introduced measure by Angelopoulos, Dorrigiv and López-Ortiz [5] refines the model and resolves most of these issues for paging. We then generalize the ideas to other settings which leads to a general framework termed the *adaptive/cooperative ratio* for the analysis of on-line algorithms. This model gives promising results when applied to three well known on-line problems, paging, list update and motion planning. The idea is to normalize the performance of an on-line algorithm by a measure other than the performance of the off-line optimal algorithm OPT. We show that in many instances the performance of OPT on a sequence is a coarse approximation of the difficulty or complexity of this input. Using a finer, more natural measure we can separate paging and list update algorithms which were otherwise indistinguishable under the classical model. This creates a performance hierarchy of algorithms which better reflects the intuitive relative strengths between them. Surprisingly, certain randomized algorithms for paging and list update which are superior to the deterministic optimum in the classical model are not so in the cooperative model. This confirms that the ability of the on-line adaptive algorithm to ignore pathological worst cases can lead to algorithms that are more efficient in practice.

2   Definitions

Let $\sigma = (\sigma_1, \sigma_2, \ldots)$ be an input sequence. We denote by $\sigma_{1:j} = (\sigma_1, \sigma_2, \ldots, \sigma_j)$ the prefix subsequence of the first j requests in σ. An on-line algorithm $\mathcal{A}$ for an optimization problem takes as input a sequence $\sigma = (s_1, s_2, \ldots, s_n)$. The algorithm $\mathcal{A}$ processes the request sequence in order, from σ_1 onwards and produces a partial solution with cost $\mathcal{A}(\sigma_{1:j})$ after the arrival of the jth request (for convenience of notation we will denote as $\mathcal{A}(\sigma) = \mathrm{cost}_{\mathcal{A}}(\sigma)$).

In general it is assumed that the length of the sequence is unknown beforehand and hence an on-line algorithm performs the same steps on the common prefix of two otherwise distinct input sequences. More formally, if σ' is a prefix of σ then $\mathcal{A}(\sigma') = \mathcal{A}(\sigma_{1:|\sigma'|})$. In contrast, the off-line optimal algorithm, denoted as

OPT has access to the entire sequence at once and hence does not necessarily meet the prefix condition.

Definition 1. *An on-line algorithm $\mathcal{A}$ is said to have competitive ratio $C(n)$ if, for all input sequences σ we have: $\mathcal{A}(\sigma) \leq C(|\sigma|) \cdot \text{OPT}(\sigma)$.*

Equivalently, using the more conventional ratio notation, we have that an algorithm is $C(n)$–competitive iff

$$C(n) = \max_{|\sigma|=n, n \geq N_o} \left\{ \frac{\mathcal{A}(\sigma)}{\text{OPT}(\sigma)} \right\}.$$

3   Paging

Paging is a fundamental problem in the context of the analysis of on-line algorithms. A paging algorithm mediates between a slower and a faster memory. Assuming a cache of size k, it decides which k memory pages to keep in the cache without the benefit of knowing in advance the *sequence* of upcoming page requests. After receiving the ith page request if the page requested is in the cache (known as a *hit*) it is served at no cost; else, in the case of a *fault* the page is served from slow memory at a cost of one unit. In this event the request results in a cache miss and the on-line algorithm must decide irrevocably which page to evict without the benefit of knowing the *sequence* of upcoming page requests. The goal of a paging algorithm is to minimize the number of faults over the entire input sequence, that is, the *cost* of a particular solution.

Three well known paging algorithms are *Least-Recently-Used* (LRU), *First-In-First-Out* (FIFO), and *Flush-When-Full* (FWF). On a fault, if the cache is full, LRU evicts the page that is least recently requested, FIFO evicts the page that is first brought to the cache, and FWF empties the entire cache. All these paging algorithms have competitive ratio k, which is the best among all deterministic on-line paging algorithms [9].

3.1   Theory Versus Practice

As was mentioned earlier, the standard model for paging does not lead to satisfactory conclusions which are replicated in practice. With the goal of closing the gap between theory and practice, we examine the difference in assumptions between the theoretical competitive ratio model and the practical systems research approach to paging. We now discuss in detail the differences which also appear summarized in Table 1.

1. The theoretical model for the study of paging algorithms is the competitive ratio framework, in contrast, the vast majority of systems research on paging uses the fault rate measure, which simply determines the percentage of page requests leading to a page fault. Consider for example a request sequence of 1M pages, such that an on-line algorithm A has 200 page faults while the off-line optimum has twenty faults. This means that A has a competitive

ratio of 10 which is high, while in terms of the fault rate model A has a page fault rate of 0.002% which is very good.

2. In the worst case one can devise highly contrived request sequences with a very high competitive ratio for any paging algorithm. Since these sequences do not occur naturally, measuring the performance of an online algorithm using them does not shed light on the actual relative performance of various algorithms. Practical studies in contrast use an extensive set of real-life request sequences (traces) gathered from diverse set of applications, over which the performance of any online strategy can be measured.

3. Under the competitive ratio all marking algorithms have the same competitive ratio. In other words LRU and FWF are equal under this measure. In contrast, experimental analysis has consistently shown that LRU and/or minor variants thereof are typically the best choices in practice, while FWF is much worse than LRU. The competitive ratio then fails to separate between these algorithms with very different performance in practice.

4. In terms of practice the theoretical model suggests that LRU might be preferable for "practical" heuristic reasons. In actuality, since paging algorithms are executed concurrently with every page access this limits the complexity of any solution, and hence practical heuristic solutions are simplifications and approximations of LRU.

5. Competitive analysis uses an optimal off-line algorithm as a baseline to compare on-line algorithms. While this may be convenient, it is rather indirect: one could argue that in comparing A to B all we need to study is the relative cost of the algorithms on the request sequences. The approach we follow stems from this basic observation. The indirect comparison to an off-line optimal can introduce spurious artifacts due to the comparison of two objects of different types, namely an online and an off-line algorithm. As well the off-line optimum benefits from aspects other than the difficulty of the instances, namely it can take advantage of knowledge of the future, so regardless of the difficulty of servicing a request it might do better as a consequence of this[2]. In contrast the fault rate measure uses a direct comparison of the number of faults per access of paging algorithms to determine which one is preferable.

6. Interestingly, even if algorithms are measured using the competitive ratio, in practice the *worst case* request sequence encountered using LRU has competitive ratio 4, and most sequences measure are well below that with competitive ratio between two and four. Contrast this with the predicted competitive ratio of k under the theoretical model.

7. The off-line optimum model implicitly creates an adversarial model in which the paging algorithm must be able to handle all request sequences, including those maliciously designed to foil the paging algorithm. In contrast, in real life, programmers and compilers purposely avoid bad request sequences and

---

[2] For example consider the decision whether to purchase car insurance or not [8]. If one purchases insurance then the adversary selects the input in which no claim is filed, if alternatively no insurance is bought then the adversary selects the input in which an accident takes place. In real life, however, it is easy to see that the best on-line strategy is to buy insurance so long as it is priced below the expected loss.

try to arrange the data in a way so as to maximize locality of reference in the request sequence (e.g. the I/O model [1], or the cache oblivious model [30]). In game theoretical terms, the theoretical competitive model is a zero sum game in which the adversary benefits from a badly performing paging algorithm, while in practice paging is a positive sum game in which both the user and the paging algorithm can maximize their respective performances by cooperating and coordinating their strategies. Indeed it has been observed that paging algorithms optimize for locality of reference because this was first observed in real life traces, and now compilers optimize code to increase locality of reference because those paging algorithms excel on those sequences.

8. Lastly, we observe that finite lookahead does not help in the theoretical model, as this is a worst case measure (simply repeat each request for as long as the lookahead is) yet in practice instruction schedulers in many cases know the future request sequence for a small finite lookahead and can use this information to improve the fault rate of paging strategies.

Table 1. Contrast of theory versus practice for paging

Theoretical Model	Systems Framework
Competitive ratio framework	Fault rate measure
Worst case analysis	Typical case analysis
Marking algorithms optimal	LRU and variants thereof are best
In practice LRU is best	LRU is impractical
LFD is off-line optimal	No analogous concept
Competitive ratio is k	Comp. ratio over observed sequences is at most 4
User is a malicious adversary	User (compiler/programmer) seeks locality of reference
No benefit from lookahead	Lookahead helps

3.2 Related Work

In this section we overview some alternatives to the competitive ratio. We refer the reader to the survey of Dorrigiv and López-Ortiz [18] for a more comprehensive and detailed exposition.

Loose competitiveness, which was first proposed by Young in [35] and later refined in [38], considers an off-line adversary that is oblivious to the parameter k (the cache size). The adversary must produce a sequence that is bad for most values of k rather than for just a specific value. It also ignores the sequences on which the on-line algorithm incurs a cost less than a certain threshold. This results in a weaker adversary and hence in paging algorithms with constant performance ratios. The *diffuse adversary* model by Koutsoupias and Papadimitriou [28] as well as Young [36,37] refines the competitive ratio by restricting the set of legal request sequences to those derived from a class (family) of probability distributions. This restriction follows from the observation that although a good performance measure could in fact use the actual distribution over the request sequences, determining the exact distribution of real-life phenomena is a difficult task.

The *Max/Max ratio*, introduced by Borodin and Ben-David [8] compares on-line algorithms based on their amortized worst-case behaviour (here the amortization arises by dividing the cost of the algorithm over the length of the request sequence). The *relative worst order ratio* [11,12,15] combines some of the desirable properties of the Max/Max ratio and the *random order ratio* (introduced in [27] in the context of the on-line bin packing problem). As with the Max/Max ratio, it allows for direct comparison of two on-line algorithms. Informally, for a given request sequence the measure considers the worst-case ordering (permutation) of the sequence, for each of the two algorithms, and compares their behaviour on these orderings. It then finds among all possible sequences the one that maximizes this worst-case performance. Recently, Panagiotou and Souza proposed a model that explains the good performance of LRU in practice [29]. They classify input sequences according to some parameters and prove an upper bound on the competitive ratio of LRU as a function of these parameters. Then they argue that sequences in practice have parameters that lead a to constant competitive ratio for LRU.

There are several models for paging which assume locality of reference Borodin, Raghavan, Irani, and Schieber [10] proposed the *access graph* model in which the universe of possible request sequences is reduced to reflect that the actual sequences that can arise depend heavily on the structure of the program being executed. The space of request sequences can then be modeled by a graph in which paths between vertices correspond to actual sequences. In a generalization of the access graph model, Karlin, Phillips, and Raghavan [26] proposed a model in which the request sequences are distributed according to a Markov chain process. Becchetti [7] refined the diffuse adversary model of Koutsoupias and Papadimitriou by considering only probabilistic distributions in which temporal locality of reference is present. Torng [34] considered the decomposition of input sequences to phases in the same manner as marking algorithms. He then modeled locality of reference by restricting the input to sequences with long average phase length. Using the *full access cost model*, he computed the performance of several paging algorithms on sequences with high locality of reference. Most notably, Albers, Favrholdt, and Giel [2] introduced a model in which input sequences are classified according to a measure of locality of reference. The measure is based on Denning's working set concept [17] which is supported by extensive experimental results. The technique used, which we term *concave analysis*, reflects the fact that efficient algorithms must perform competitively in each class of inputs of similar locality of reference, as opposed to the worst case alone. It should be noted that [2] focuses on the *fault rate* as the measure of the cost of an algorithm, as opposed to the traditional definition of cost as the number of cache misses.

4   Bijective Analysis and Average Analysis

Bijective Analysis and Average Analysis are two models recently proposed by Angelopoulos, Dorrigiv and López-Ortiz [5] for comparing on-line algorithms. In this section, we first provide the formal definitions of Bijective Analysis and Average Analysis and then apply them to the paging algorithms. These models have certain desired characteristics for comparing online algorithms: they allow for direct comparison of two on-line algorithms without appealing to the concept of the off-line "optimal" cost (see [5] for a more detailed discussion). In addition, these measures do not evaluate the performance of the algorithm on a single "worst-case" request, but instead use the cost that the algorithm incurs on each and all request sequences. Informally, Bijective Analysis aims to pair input sequences for two algorithms $\mathcal{A}$ and $\mathcal{B}$ using a bijection in such a way that the cost of $\mathcal{A}$ on input σ is no more than the cost of $\mathcal{B}$ on the image of σ, for all request sequences σ of the same length. In this case, intuitively, $\mathcal{A}$ is no worse than $\mathcal{B}$. On the other hand, Average Analysis compares the average cost of the two algorithms over all request sequences of the same length. For an on-line algorithm $\mathcal{A}$ and an input sequence σ, let $\mathcal{A}(\sigma)$ be the cost incurred by $\mathcal{A}$ on σ. Denote by $\mathcal{I}_n$ the set of all input sequences of length n.

Definition 2. *[5] We say that an on-line algorithm $\mathcal{A}$ is no worse than an on-line algorithm $\mathcal{B}$ according to Bijective Analysis if there exists an integer $n_0 \geq 1$ so that for each $n \geq n_0$, there is a bijection $b : \mathcal{I}_n \leftrightarrow \mathcal{I}_n$ satisfying $\mathcal{A}(\sigma) \leq \mathcal{B}(b(\sigma))$ for each $\sigma \in \mathcal{I}_n$. We denote this by $\mathcal{A} \preceq_b \mathcal{B}$. Otherwise we denote the situation by $\mathcal{A} \npreceq_b \mathcal{B}$. Similarly, we say that $\mathcal{A}$ and $\mathcal{B}$ are the same according to Bijective Analysis if $\mathcal{A} \preceq_b \mathcal{B}$ and $\mathcal{B} \preceq_b \mathcal{A}$. This is denoted by $\mathcal{A} \equiv_b \mathcal{B}$. Lastly we say $\mathcal{A}$ is better than $\mathcal{B}$ according to Bijective Analysis if $\mathcal{A} \preceq_b \mathcal{B}$ and $\mathcal{B} \npreceq_b \mathcal{A}$. We denote this by $\mathcal{A} \prec_b \mathcal{B}$.*

Definition 3. *[5] We say that an on-line algorithm $\mathcal{A}$ is no worse than an on-line algorithm $\mathcal{B}$ according to Average Analysis if there exists an integer $n_0 \geq 1$ so that for each $n \geq n_0$, $\sum_{I \in \mathcal{I}_n} \mathcal{A}(I) \leq \sum_{I \in \mathcal{I}_n} \mathcal{B}(I)$. We denote this by $\mathcal{A} \preceq_a \mathcal{B}$. Otherwise we denote the situation by $\mathcal{A} \npreceq_a \mathcal{B}$. $\mathcal{A} \equiv_a \mathcal{B}$, and $\mathcal{A} \prec_a \mathcal{B}$ are defined as for Bijective Analysis.*

In [5] it is shown that LRU is strictly better than FWF under Bijective Analysis. Additionally, lookahead is beneficial in Bijective Analysis model: more specifically, LRU with lookahead as small as one (namely the sequence is revealed to the algorithm as consecutive pairs of requests) is strictly better than LRU without any lookahead. Both of these results describe natural, "to-be-expected" properties of the corresponding paging strategies which competitive analysis nevertheless fails to yield.

Also it turns out that a very large class of natural paging strategies known as *lazy* algorithms (including LRU and FIFO, but not FWF) are in fact strongly equivalent under this rather strict bijective measure. The strong equivalence of lazy algorithms is evidence of an inherent difficulty to separate these algorithms in any general unrestricted setting. In fact, it implies that to obtain theoretical separation between algorithms we must either induce a partition of the request sequence space (e.g. as in Albers et al. [2]) or assume a distribution (or a set of distributions) on the sequence space (e.g. as in Koutsoupias and Papadimitriou [28], Young [36] and Becchetti [7]). The latter group of approaches use probabilistic assumptions on the sequence space. However, we are interested in measures that separate algorithms under a deterministic model.

Next we briefly describe concave analysis. In this model a request sequence has high locality of reference if the number of distinct pages in a window of size n is small. Consider a function that represents the maximum number of distinct pages in a window of size n within a given request sequence. Extensive experiments with real data show that this function can be bounded by a concave function for most practical request sequences [2]. Let f be an increasing concave function. We say that a request sequence is *consistent* with f if the number of distinct pages in any window of size n is at most $f(n)$, for any $n \in \mathcal{N}$. Now we can model locality by considering only those request sequences that are consistent with f.

Using a combination of Average Analysis and concave analysis, Angelopoulos et al. [5] show that LRU is never outperformed in any possible subpartition on the request sequence space induced by concave analysis, while it always outperforms *any other paging algorithm* in at least one subpartition of the sequence space. This result proves separation between LRU and all other algorithms and provides theoretical backing to the observation that LRU is preferable in practice.

To be more precise we restrict the input sequences to those consistent with a given concave function f. Let $\mathcal{I}^f$ denote the set of such sequences. We can easily modify the definitions of Bijective Analysis and Average Analysis (Definition 2 and Definition 3) by considering $\mathcal{I}^f$ instead of $\mathcal{I}$. We denote the corresponding relations by $\mathcal{A} \preceq_b^f \mathcal{B}$, $\mathcal{A} \preceq_a^f \mathcal{B}$, etc. Note that we can make any sequence consistent with f by repeating every request a sufficient number of times. Therefore even if we restrict the input to sequences with high locality of reference, there is a worst case sequence for LRU that is consistent with f and therefore the competitive ratio of LRU is the same as in the standard model. Observe that the performance of a paging algorithm is now evaluated within the subset of request sequences of a given length whose locality of reference is consistent with f, i.e. $\mathcal{I}_n^f$.

Theorem 1 (Unique optimality of LRU). *[5] For any concave function f and any paging algorithm $\mathcal{A}$, LRU $\preceq_a^f \mathcal{A}$. Furthermore, let $\mathcal{A}$ be a paging algorithm other than LRU. Then there is a concave function f so that $\mathcal{A} \npreceq_a^f$ LRU which implies $\mathcal{A} \npreceq_b^f$ LRU.*

5   List Update and Cooperative Analysis

List update is a fundamental problem in the context of on-line computation. Consider an unsorted list of l items. The input to the algorithm is a sequence of n requests that should be served in an on-line manner. Let $\mathcal{A}$ be an arbitrary on-line list update algorithm. To serve a request to an item x, $\mathcal{A}$ should linearly search the list until it finds x. If x is the ith item in the list, $\mathcal{A}$ incurs cost i to access x. Immediately after accessing x, $\mathcal{A}$ can move x to any position closer to the front of the list at no extra cost. This is called a *free exchange*. Also $\mathcal{A}$ can exchange any two consecutive items at a cost of 1. These are called *paid exchanges*. An efficient algorithm should use free and paid exchanges so as to minimize the overall cost of serving a sequence. This is called the *standard cost model* [3]. Three well-known deterministic on-line algorithms are *Move-To-Front* (MTF), *Transpose*, and *Frequency-Count* (FC). MTF moves the requested item to the front of the list whereas Transpose exchanges the requested item with the item that immediately precedes it. FC maintains a frequency count for each item, updates this count after each access, and makes necessary moves so that the list always contains items in non-increasing order of frequency count. Sleator and Tarjan showed that MTF is 2-competitive, while Transpose and FC do not have constant competitive ratios [33].

The competitive analysis of list update algorithms does not have as many drawbacks as paging and at first it gives promising results: list update algorithms with better competitive ratio tend to have better performance in practice. However, in terms of separation list update algorithms have similar drawbacks to paging: while algorithms can generally be more easily distinguished than in the paging case, the experimental study of list update algorithms by Bachrach and El-Yaniv suggests that the relative performance hierarchy as computed by the competitive ratio does not correspond to the observed relative performance of the algorithms in practice [6].

Like paging, "real-life" input sequences for list update problem usually exhibit *locality of reference*. As stated before, for the paging problem, several models for capturing locality of reference have been proposed [2,7,34]. Likewise, many researchers have pointed out that input sequences of list update algorithms in practice show locality of reference [9,23,32] and actually on-line list update algorithms try to take advantage of this property [23,31]. Hester and Hirschberg [23] posed the question of providing a good definition of locality of accesses for the list update problem as an open problem. In addition, it has been commonly assumed, based on intuition and experimental evidence, that MTF is the best algorithm on sequences with high locality of reference, e.g., Hester and Hirschberg [23] claim: "move-to-front performs best when the list has a high degree of locality". However, to the best of our knowledge, locality of reference for list update algorithms had not been formally studied, until recently [4,19].

In [4], Angelopoulos, Dorrigiv and López-Ortiz extended the concave analysis model [2] to the list problem. The validity of the extended model was supported by experimental results obtained on the Calgary Corpus, which is frequently used as a standard benchmark for evaluating the performance of compression algorithms (and by extension list update algorithms, e.g. [6]). They combined Average Analysis with concave analysis and proved that under this model MTF is never outperformed, while it always outperforms *any other on-line list update algorithm.* Thus, [4] resolved the open problem posed by Hester and Hirschberg [23].

Based on adaptive analysis ideas, Dorrigiv and López-Ortiz [19] proposed cooperative analysis for analyzing on-line algorithms. The idea behind *cooperative analysis* is to give more weight to "well-behaved" input sequences. Informally, an on-line algorithm has good *cooperative* ratio if it performs well on good sequences and not too poorly on bad sequences. For example, as stated before, input sequences for paging and list update have *locality of reference* in practice, therefore one possibility is to relate goodness of sequences to their amount of locality. In [19], we showed that cooperative analysis of paging and list update algorithms gives promising results. Here we just briefly describe the results for list update. We use a measure of badness that is related to locality of reference as follows. For a sequence σ of length n, define $d_\sigma[i]$ for $1 \leq i \leq n$ as either 0 if this is the first request to item $\sigma[i]$, or otherwise, the number of distinct items that are requested since the last request to $\sigma[i]$ (including $\sigma[i]$). Define $\overline{\ell(\sigma)}$, the non-locality of a sequences σ, as $\overline{\ell(\sigma)} = \sum_{1 \leq i \leq n} d_\sigma[i]$.

Definition 4. *[19] We say that an on-line list update algorithm $\mathcal{A}$ has locality-cooperative ratio α if there is a constant β so that for every sequence σ, $\mathcal{A}(\sigma) \leq \alpha \times \overline{\ell(\sigma)} + \beta$. We define locality-cooperative ratio of $\mathcal{A}$, $LCR(\mathcal{A})$, as the smallest number α so that $\mathcal{A}$ has locality-cooperative ratio α.*

The following theorem summarizes the results proved for locality-cooperative ratio of list update algorithms.

Theorem 2. *[19] For any on-line list update algorithm $\mathcal{A}$, $1 \leq LCR(\mathcal{A}) \leq l$; furthermore:*

1. $LCR(MTF) = 1$.
2. $LCR(Transpose) \geq l/2$.
3. $LCR(FC) \geq \frac{l+1}{2} \approx l/2$.
4. $LCR(TS) \geq \frac{2l}{l+1} \approx 2$.
5. $LCR(Bit) \geq \frac{3l+1}{2l+2} \approx 3/2$.

6   Conclusions

In this paper, we highlighted the gap between theoretical and experimental results for some on-line problems and possible ways to close this gap. We observed that standard measure for analysis of on-line algorithms, i.e., competitive analysis, leads to results that are not consistent with practice for paging and list update. Then we described reasons for the shortcomings of competitive analysis and described several new models for analysis of on-line algorithms that do not have these drawbacks. Bijective Analysis and Average Analysis directly compare two on-line algorithms on all sequences of the same length and lead to satisfactory results when applied to paging and list update. The new concept of cooperative ratio applies adaptive analysis ideas to the analysis of on-line algorithms and divides the cost of the algorithm on a sequence to some property of that sequence.

References

1. Aggarwal, A., Vitter, J.S.: The Input/Output complexity of sorting and related problems. Communications of the ACM 31(9), 1116–1127 (1988)
2. Albers, S., Favrholdt, L.M., Giel, O.: On paging with locality of reference. Journal of Computer and System Sciences 70(2), 145–175 (2005)
3. Albers, S., Westbrook, J.: Self-organizing data structures. In: Fiat, A. (ed.) Online Algorithms. LNCS, vol. 1442, pp. 13–51. Springer, Heidelberg (1998)
4. Angelopoulos, S., Dorrigiv, R., López-Ortiz, A.: List update with locality of reference: Mtf outperforms all other algorithms. Technical Report CS-2006-46, University of Waterloo, Cheriton School of Computer science (November 2006)
5. Angelopoulos, S., Dorrigiv, R., López-Ortiz, A.: On the separation and equivalence of paging strategies. In: SODA 2007. Proceedings of the 18th ACM-SIAM Symposium on Discrete Algorithms, pp. 229–237 (2007)
6. Bachrach, R., El-Yaniv, R.: Online list accessing algorithms and their applications: Recent empirical evidence. In: SODA 1997. Proceedings of the 8th Annual ACM-SIAM Symposium on Discrete Algorithms, pp. 53–62 (1997)
7. Becchetti, L.: Modeling locality: A probabilistic analysis of LRU and FWF. In: Albers, S., Radzik, T. (eds.) ESA 2004. LNCS, vol. 3221, pp. 98–109. Springer, Heidelberg (2004)
8. Ben-David, S., Borodin, A.: A new measure for the study of on-line algorithms. Algorithmica 11, 73–91 (1994)
9. Borodin, A., El-Yaniv, R.: Online Computation and Competitive Analysis. Cambridge University Press, Cambridge (1998)
10. Borodin, A., Irani, S., Raghavan, P., Schieber, B.: Competitive paging with locality of reference. Journal of Computer and System Sciences 50, 244–258 (1995)
11. Boyar, J., Favrholdt, L.M.: The relative worst order ratio for on-line algorithms. In: Proceedings of the 5th Italian Conference on Algorithms and Complexity (2003)
12. Boyar, J., Favrholdt, L.M., Larsen, K.S.: The relative worst order ratio applied to paging. In: SODA 2005. Proceedings of the 16th ACM-SIAM Symposium on Discrete Algorithms, pp. 718–727 (2005)
13. Boyar, J., Larsen, K.S.: The Seat Reservation Problem. Algorithmica 25(4), 403–417 (1999)
14. Boyar, J., Larsen, K.S., Nielsen, M.N.: The Accommodating Function: A generalization of the competitive ratio. SIAM Journal on Computing 31(1), 233–258 (2001)
15. Boyar, J., Medvedev, P.: The relative worst order ratio applied to seat reservation. In: Hagerup, T., Katajainen, J. (eds.) SWAT 2004. LNCS, vol. 3111, pp. 90–101. Springer, Heidelberg (2004)
16. Chrobak, M., Noga, J.: LRU is better than FIFO. Algorithmica 23(2), 180–185 (1999)
17. Denning, P.J.: The working set model for program behaviour. Communications of the ACM, 11(5) (May 1968)
18. Dorrigiv, R., López-Ortiz, A.: A survey of performance measures for on-line algorithms. SIGACT News (ACM Special Interest Group on Automata and Computability Theory) 36(3), 67–81 (2005)
19. Dorrigiv, R., López-Ortiz, A.: The cooperative ratio of on-line algorithms. Technical Report CS-2007-39, University of Waterloo, Cheriton School of Computer science (October 2007)

20. Dorrigiv, R., López-Ortiz, A.: On certain new models for paging with locality of reference. In: WALCOM 2008. Proceedings of the 2nd Workshop on Algorithms and Computation (to appear, 2008)
21. Dorrigiv, R., López-Ortiz, A., Munro, J.I.: On the relative dominance of paging algorithms. In: ISAAC 2007. Proceedings of the 18th International Symposium on Algorithms and Computation (to appear, 2007)
22. Fiat, A., Woeginger, G.J.: Competitive odds and ends. In: Fiat, A. (ed.) Online Algorithms. LNCS, vol. 1442, pp. 385–394. Springer, Heidelberg (1998)
23. Hester, J.H., Hirschberg, D.S.: Self-organizing linear search. ACM Computing Surveys 17(3), 295 (1985)
24. Irani, S.: Competitive analysis of paging. In: Fiat, A., Woeginger, G.J. (eds.) Online Algorithms. LNCS, vol. 1442, pp. 52–73. Springer, Heidelberg (1998)
25. Irani, S., Karlin, A.R., Phillips, S.: Strongly competitive algorithms for paging with locality of reference. SIAM Journal on Computing 25, 477–497 (1996)
26. Karlin, A.R., Phillips, S.J., Raghavan, P.: Markov paging. SIAM Journal on Computing 30(3), 906–922 (2000)
27. Kenyon, C.: Best-fit bin-packing with random order. In: SODA 1996. Proceedings of the 7th Annual ACM-SIAM Symposium on Discrete Algorithms, pp. 359–364 (1996)
28. Koutsoupias, E., Papadimitriou, C.: Beyond competitive analysis. SIAM Journal on Computing 30, 300–317 (2000)
29. Panagiotou, K., Souza, A.: On adequate performance measures for paging. In: STOC 2006. Proceedings of the 38th Annual ACM Symposium on Theory of Computing, pp. 487–496 (2006)
30. Prokop, H.: Cache-oblivious algorithms. Master's thesis, Massachusetts Institute of Technology, Dept. of Electrical Engineering and Computer Science (1999)
31. Reingold, N., Westbrook, J., Sleator, D.D.: Randomized competitive algorithms for the list update problem. Algorithmica 11, 15–32 (1994)
32. Schulz, F.: Two new families of list update algorithms. In: Chwa, K.-Y., Ibarra, O.H. (eds.) ISAAC 1998. LNCS, vol. 1533, pp. 99–108. Springer, Heidelberg (1998)
33. Sleator, D.D., Tarjan, R.E.: Amortized efficiency of list update and paging rules. Communications of the ACM 28, 202–208 (1985)
34. Torng, E.: A unified analysis of paging and caching. Algorithmica 20(2), 175–200 (1998)
35. Young, N.E.: The k-server dual and loose competitiveness for paging. Algorithmica 11(6), 525–541 (1994)
36. Young, N.E.: Bounding the diffuse adversary. In: SODA 1998. Proceedings of the 9th Annual ACM-SIAM Symposium on Discrete Algorithms, pp. 420–425 (1998)
37. Young, N.E.: On-line paging against adversarially biased random inputs. Journal of Algorithms 37(1), 218–235 (2000)
38. Young, N.E.: On-line file caching. Algorithmica 33(3), 371–383 (2002)

Simple Geometrical Intersection Graphs

Ryuhei Uehara

Department of Information Processing, School of Information Science,
JAIST, Ishikawa 923-1292, Japan
uehara@jaist.ac.jp

Abstract. A graph $G = (V, E)$ is said to be an intersection graph if and only if there is a set of objects such that each vertex v in V corresponds to an object O_v and $\{u, v\} \in E$ if and only if O_v and O_u have a nonempty intersection. Interval graphs are typical intersection graph class, and widely investigated since they have simple structures and many hard problems become easy on the graphs. In this paper, we survey known results and investigate (unit) grid intersection graphs, which is one of natural generalized interval graphs. We show that the graph class has so rich structure that some typical problems are still hard on the graph class.

Keywords: graph isomorphism, grid intersection graphs, Hamiltonian path problem, interval graphs.

1 Introduction

A graph $G = (V, E)$ is said to be an intersection graph if and only if there is a set of objects such that each vertex v in V corresponds to an object O_v and $\{u, v\} \in E$ if and only if O_v and O_u have a nonempty intersection. Interval graphs are typical intersection graph class, and widely investigated. One reason is that interval graphs have wide applications including scheduling and bioinformatics [6]. Another reason is that an interval graph has a simple structure, and hence we can solve many problems efficiently, whereas the problems are hard in general [6,16].

Some natural generalizations of interval graphs have been investigated (see, e.g., [4,11,16]). Among them, we focus on grid intersection graphs. Grid intersection graphs are natural bipartite analogy and 2D generalization of interval graphs; a bipartite graph $G = (X, Y, E)$ is a grid intersection graph if and only if G is an intersection graph of X and Y, where X corresponds to a set of horizontal line segments, and Y corresponds to a set of vertical line segments. Recently, Otachi, Okamoto, and Yamazaki investigate relationships between the class of grid intersection graphs and other bipartite graph classes [13]. In this paper, we show that grid intersection graphs have a rich structure. More precisely, we show two hardness results. First, the Hamiltonian cycle problem is still $\mathcal{NP}$-complete even if graphs are restricted to unit length grid intersection graphs. The Hamiltonian cycle problem is one of the classic and basic $\mathcal{NP}$-complete problems. Second, the graph isomorphism problem is still GI-complete even if graphs are restricted to grid intersection graphs. (We say the graph isomorphism problem is GI-complete if the problem is as hard as to solve the problem on general graphs.) The results imply that (unit length) grid intersection graphs have so rich structure that many other hard problems may still hard even on (unit length) grid intersection graphs.

S.-i. Nakano and Md. S. Rahman (Eds.): WALCOM 2008, LNCS 4921, pp. 25–33, 2008.

On an interval graph, those problems are easy to solve; in fact, the Hamiltonian cycle and path problems, and the graph isomorphism problem are solvable in linear time [9,10]. Comparing to the results, we can observe that the generalized interval graphs have so rich structure that some problems become hard on the graphs in general. However, the author hopes that the paper involves in investigating efficient algorithms on the graph classes; showing hardness and investigating efficient algorithms lead us to capture the essence of the difficulty of the graph classes and the problems.

2   Preliminaries

The *neighborhood* of a vertex v in a graph $G = (V, E)$ is the set $N_G(v) = \{u \in V \mid \{u, v\} \in E\}$, and the *degree* of a vertex v is $|N_G(v)|$ denoted by $d_G(v)$. If no confusion can arise we will omit the index G. For a subset U of V, the subgraph of G induced by U is denoted by $G[U]$. Given a graph $G = (V, E)$, its *complement* $\bar{G} = (V, \bar{E})$ is defined by $\bar{E} = \{\{u, v\} \mid \{u, v\} \notin E\}$. A vertex set I is an *independent set* if and only if $G[I]$ contains no edges, and then the graph $\bar{G}[I]$ is said to be a *clique*.

For a graph $G = (V, E)$, a sequence of distinct vertices $v_0, v_1, \ldots, v_l$ is a *path*, denoted by $(v_0, v_1, \ldots, v_l)$, if $\{v_j, v_{j+1}\} \in E$ for each $0 \le j < l$. The *length* of a path is the number of edges on the path. For two vertices u and v, the *distance* of the vertices, denoted by $dist(u, v)$, is the minimum length of the paths joining u and v. A *cycle* consists of a path $(v_0, v_1, \ldots, v_l)$ of length at least 2 with an edge $\{v_0, v_l\}$, and denoted by $(v_0, v_1, \ldots, v_l, v_0)$. The *length* of a cycle is the number of edges on the cycle (equal to the number of vertices). A path P in G is said to be *Hamiltonian* if P visits every vertex in G exactly once. The *Hamiltonian path problem* is to determine if a given graph has a Hamiltonian path. The *Hamiltonian cycle problem* is defined similarly for a cycle. The problems are well known $\mathcal{NP}$-complete problem (see, e.g., [5]).

An edge that joins two vertices of a cycle but is not itself an edge of the cycle is a *chord* of that cycle. A graph is *chordal* if every cycle of length at least 4 has a chord. A graph $G = (V, E)$ is *bipartite* if and only if V can be partitioned into two sets X and Y such that every edge joins a vertex in X and the other vertex in Y. A bipartite graph is *chordal bipartite* if every cycle of length at least 6 has a chord.

In this paper, we will discuss about intersection graphs of geometrical objects. Interval graphs are characterized by intersection graphs of intervals, and it is well known that chordal graphs are intersection graphs of subtrees of a tree (see, e.g., [16]).

A natural bipartite analogy of interval graphs are called *interval bigraphs* which are intersection graphs of two-colored intervals so that we do not join two vertices if they have the same color. Based on the definition, Müller showed that the recognition problem for interval bigraphs can be solved in polynomial time [12]. Later, Hell and Huang show an interesting characterization of interval bigraphs, which is based on the idea to characterize the complements of the graphs [7]. Based on their characterization, we can construct an $O(n^2)$ time simple recognition algorithm for interval bigraphs.

A bipartite graph $G = (X, Y, E)$ is a *grid intersection graph* if every vertex $x \in X$ and $y \in Y$ can be assigned line segments I_x and J_y in the plane, parallel to the horizontal and vertical axis so that for all $x \in X$ and $y \in Y$, $\{x, y\} \in E$ if and only if I_x and J_y

cross each other. We call $(\mathcal{I}, \mathcal{J})$ a *grid representation* of G, where $\mathcal{I} = \{I_x \mid x \in X\}$ and $\mathcal{J} = \{J_y \mid y \in Y\}$. A grid representation is *unit* if all line segments in the representation have the same (unit) length. A bipartite graph is a *unit grid intersection graph* if it has a unit grid representation.

Recently, Otachi, Okamoto, and Yamazaki show some relationship between (unit) grid intersection graphs and other graph classes [13]; for example, interval bigraph is included in the intersection of unit grid intersection graphs and chordal bipartite graphs. It is worth mentioning that it is open whether chordal bipartite graphs are included in grid intersection graphs or not.

Two graphs $G = (V, E)$ and $G' = (V', E')$ are *isomorphic* if and only if there is a one-to-one mapping $\phi : V \to V'$ such that $\{u, v\} \in E$ if and only if $\{\phi(u), \phi(v)\} \in E'$ for every pair of vertices $u, v \in V$. We denote by $G \sim G'$ if G and G' are isomorphic. The *graph isomorphism (GI) problem* is to determine if $G \sim G'$ for given graphs G and G'. A graph class C is said to be *GI-complete* if there is a polynomial time reduction from the graph isomorphism problem for general graphs to the graph isomorphism problem for C. Intuitively, the graph isomorphism problem for the class C is as hard as the problem for general graphs if C is GI-complete. The graph isomorphism problem is GI-complete for several graph classes; for example, chordal bipartite graphs, and strongly chordal graphs [19]. On the other hand, the graph isomorphism problem can be solved efficiently for many graph classes; for example, interval graphs [10], probe interval graphs [17], permutation graphs [2], directed path graphs [1], and distance hereditary graphs [8].

3 Hard Problems

We give two hardness results for grid intersection graphs in this section.

Theorem 1. *The Hamiltonian cycle problem is $\mathcal{NP}$-complete for unit grid intersection graphs.*

Proof. It is clear that the problem is in $\mathcal{NP}$. Hence we show $\mathcal{NP}$-hardness. We show a similar reduction in [18]. We start from the Hamiltonian cycle problem in planar directed graph with degree bound two, which is still $\mathcal{NP}$-hard [14]. Let $G_0 = (V_0, A)$ be a planar directed graph with degree bound two. (We deal with directed graphs only in this proof; we will use (u, v) as a directed edge, called *arc*, which is distinguished from $\{u, v\}$.) As shown in [14,18], we can assume that G_0 consists of two types of vertices: (type $\triangle$) with two indegrees and one outdegree, and (type $\triangledown$) with one indegree and two outdegrees. Hence, the set V_0 of vertices can be partitioned into two sets $V_\triangle$ and $V_\triangledown$ that consist of the vertices in type $\triangle$ and $\triangledown$, respectively.

Moreover, we have two more claims; (1) the unique arc from a type $\triangle$ vertex has to be the unique arc to a type $\triangledown$ vertex, and (2) each of two arcs from a type $\triangledown$ vertex has to be one of two arcs to a type $\triangledown$ vertex. If the unique arc from a type $\triangle$ vertex v is into one of a type $\triangle$ vertex u, The vertex u has to be visited from v to make a Hamiltonian cycle. Hence the vertex u can be replaced by an arc from v to the vertex w which is pointed from u. On the other hand, if one of two arcs from a type $\triangledown$ vertex v is reach another type $\triangledown$ vertex u, the vertex u should be visited from v. Hence the other arc a from v can be removed from G_0. Then the vertex w incident to a has degree 2. Hence we have two cases; w can be replaced by an arc, or we can conclude G_0 does not have a Hamiltonian cycle. Repeating these processes, we have the claims (1) and (2), which imply that we have $|V_\triangle| = |V_\triangledown|$, the underlying graph of G_0 is bipartite (with two sets $V_\triangle$ and $V_\triangledown$), and any cycle contains two types of vertices alternately.

By the claims, we can partitioned into arcs into two groups; (1) arcs from a type $\triangle$ vertex to a type $\triangledown$ vertex called *thick arcs*, and (2) arcs from a type $\triangledown$ vertex to a type $\triangle$ vertex called *thin arcs*. By above discussion, we can observe that any Hamiltonian cycle has to contain all thick arcs[1].

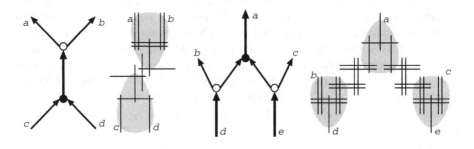

Fig. 1. Reduction of thick arcs   **Fig. 2.** Reduction of thin arcs

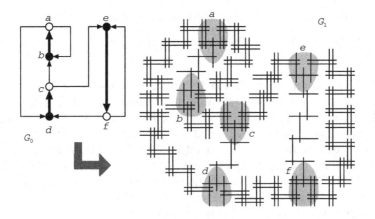

Fig. 3. Reduction of a graph G_0

---

[1] Moreover, contracting thick arcs, we can show $\mathcal{NP}$-completeness of the Hamiltonian cycle problem even if we restrict ourselves to the directed planar graphs that only consist of vertices of two outdegrees and two indegrees.

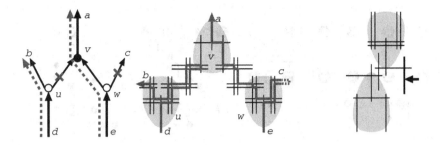

Fig. 4. How to sweep thin arcs   **Fig. 5.** Hamiltonian path problem

Now, we construct a unit grid intersection graph $G_1 = (V_1, E_1)$ from $G_0 = (V_0, A)$ which satisfies the above conditions. One type $\triangledown$ vertex is represented by five vertical lines and two horizontal lines, and one type $\triangle$ vertex is represented by three vertical lines and one horizontal line in Fig. 1 (each corresponding line segments are in gray area). Each thick arc is represented by alternations of one parallel vertical line and one parallel horizontal line in Fig. 1, and each thin arc is represented by alternations of two parallel vertical lines and two parallel horizontal lines in Fig. 2. The vertices are joined by the arcs in a natural way. An example is illustrated in Fig. 3.

For the resultant graph G_1, it is obvious that the reduction can be done in a polynomial time, and G_1 is a unit grid intersection graph. Hence we show G_0 has a Hamiltonian cycle if and only if G_1 has a Hamiltonian cycle. First, we assume that G_0 has a Hamiltonian cycle C_0, and show that G_1 also has a Hamiltonian cycle C_1. C_1 visits the vertices (or line segments) in G_1 along C_0 as follows. For each thick arc in G_0, the corresponding segments in G_1 are visited straightforwardly. We show how to visit the segments corresponding to thin arcs (Fig. 4). For each thin arc not on C_0, they are visited by C_1 as shown in the left side of Fig. 4 (between u and v); a pair of parallel lines are used to sweep the arc twice, and the endpoints are joined by one line segment in the gadget of a type $\triangle$ vertex (v). On the other hand, for each thin arc on C_0, they are visited by C_1 as shown in the right side of Fig. 4 (between w and v); a pair of parallel lines are used to sweep the arc once, and the path goes from e to a. Hence from a given Hamiltonian cycle C_0 on G_0, we can construct a Hamiltonian cycle C_1 on G_1.

Now we assume that G_1 has a Hamiltonian cycle C_1, and show that G_0 also has a Hamiltonian cycle C_0. By observing that there are no ways for C_1 to visit lines corresponding to thick arcs described above, and the unique center horizontal line of a type $\triangle$ vertex can be used exactly once, we can see that C_1 forms a Hamiltonian cycle of G_1 as in the same manner represented above. Hence C_0 can be constructed from C_1 in the same way. □

Corollary 1. *The Hamiltonian path problem is $\mathcal{NP}$-complete for unit grid intersection graphs.*

Proof. We reduce the graph G_1 in the proof of Theorem 1 to G_1' as follows; pick up any line segment in a thick arc, and add one more line segment as in Fig. 5. Then, it is easy to see that G_1 has a Hamiltonian cycle if and only if G_1' has a Hamiltonian path

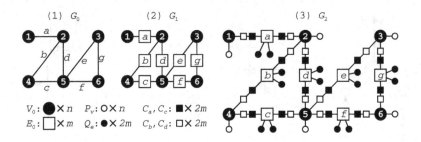

Fig. 6. Reduction for *GI*-completeness

(with an endpoint corresponding to the additional line segment). Hence we have the corollary.   □

Theorem 2. *The graph isomorphism problem is GI-complete for grid intersection graphs.*

Proof. We show a similar reduction in [1,19]. We first start the graph isomorphism problem for general graph $G_0 = (V_0, E_0)$ with $|V_0| = n$ and $|E_0| = m$. (we will refer the graph in Fig. 6(1) as an example). Without loss of generality, we assume that G_0 is connected. From G_0, we define a bipartite graph $G_1 = (V_0, E_0, E_1)$ with two vertex sets V_0 and E_0 by $E_1 := \{\{v, e\} \mid v$ is one endpoint of $e\}$. (Intuitively, each edge is divided into two edges joined by a new vertex; see Fig. 6(2)). Then, $e \in E_0$ have degree 2 by its two endpoints in V_0. It is easy to see that $G_0 \sim G_0'$ if and only if $G_1 \sim G_1'$ for any graphs G_0 and G_0' with resultant graphs G_1 and G_1'.

Now, we construct a grid intersection graph $G_2 = (V_2, E_2)$ from the bipartite graph $G_1 = (V_0, E_0, E_1)$ such that $G_1 \sim G_1'$ if and only if $G_2 \sim G_2'$ in the same manner. The vertex set V_2 consists of the following sets (see Fig. 6(3)):

V_0, E_0; we let $V_0 = \{v_1, v_2, \ldots, v_n\}$, $E_0 = \{e_1, e_2, \ldots, e_m\}$, where $e_i = \{v_i, v_j\}$ for some $1 \le i, j \le n$.

P_v, Q_e; each vertex in $P_v \cup Q_e$ is called *pendant* and $P_v := \{p_1, p_2, \ldots, p_n\}$, $Q_e := \{q_1, q_2, \ldots, q_m, q_1', q_2', \ldots, q_m'\}$. That is, we have $|P_v| = n$ and $|Q_e| = 2m$.

C_a, C_b, C_c, C_d; each vertex in $C_a \cup C_b \cup C_c \cup C_d$ is called *connector*, and $C_a := \{a_1, a_2, \ldots, a_m\}$, $C_b := \{b_1, b_2, \ldots, b_m\}$, $C_c := \{c_1, c_2, \ldots, c_m\}$, and $C_d := \{d_1, d_2, \ldots, d_m\}$.

The edge set E_2 contains the following edges (Fig. 6(3)):

1. For each i with $1 \le i \le n$, each pendant p_i is joined to v_i. That is, $\{p_i, v_i\} \in E_2$ for each i with $1 \le i \le n$.
2. For each j with $1 \le j \le m$, two pendants q_j and q_j' are joined to e_j. That is, $\{q_j, e_j\}, \{q_j', e_j\} \in E_2$ for each j with $1 \le j \le m$.
3. For each e_j with $1 \le j \le m$, we have two vertices v_i and $v_{i'}$ with $\{v_i, e_j\}, \{v_{i'}, e_j\} \in E_1$. For the three vertices $e_j, v_i, v_{i'}$, we add $\{e_j, a_j\}, \{v_i, b_j\}, \{a_j, b_j\}, \{e_j, c_j\}, \{v_{i'}, d_j\}, \{c_j, d_j\}$ into E_2. Intuitively, each edge in G_1 is replaced by a path of length 3 that consists of one vertex in $C_a \cup C_c$ and the other one in $C_b \cup C_d$.

The edge set E_2 also contains the edges $\{v_i, e_j\}$ for each i, j with $1 \leq i \leq n$ and $1 \leq j \leq m$. In other words, every vertex in V_0 is connected to all vertices in E_0 (the edges are omitted in Fig. 6(3) to simplify).

Let G_0 and G_0' be any two graphs. Then, it is easy to see that $G_0 \sim G_0'$ implies $G_2 \sim G_2'$. Hence, we have to show that G_2 is a grid intersection graph, and G_0 can be reconstructed from G_2 uniquely up to isomorphism.

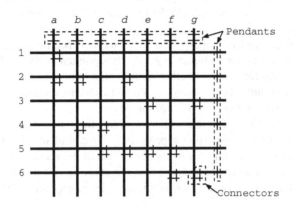

Fig. 7. Grid representation of G_2

We can represent the vertices in $V_0 \cup Q_e \cup C_a \cup C_c$ (white vertices in Fig. 6(3)) as horizontal segments and the vertices in $E_0 \cup P_v \cup C_b \cup C_d$ (black vertices in Fig. 6(3)) as vertical segments as follows (Fig. 7): First, all vertices in V_0 correspond to unit length horizontal segments, and placed in parallel. All vertices in E_0 correspond to unit length vertical segments and placed in parallel, and the segments corresponding to vertices in V_0 and E_0 make a mesh structure (as in Fig. 7). Each pendant vertex in P_v and Q_e corresponds to small segment, and attached to its neighbor in an arbitrary way, for example, as in Fig. 7. Each pair of connectors in C_a and C_b (or C_c and C_d) joins corresponding vertices in V_0 and E_0 as in Fig. 7. Then it is easy to see that the resultant grid representation gives G_2.

Next, we show that G_0 can be reconstructed from G_2 uniquely up to isomorphism. First, any vertex of degree 1 is a pendant in G_2. Hence we can distinguish $P_v \cup Q_e$ from the other vertices. Then, for each vertex $v \in V_2 \setminus (P_v \cup Q_e)$, $|N(v) \cap (P_v \cup Q_e)| = 1$ if and only if $v \in V_0$, and $|N(v) \cap (P_v \cup Q_e)| = 2$ if and only if $v \in E_0$. Hence two sets V_0 and E_0 are distinguished, and then $P_v \cup Q_e$ can be divided into P_v and Q_e. Moreover, we have $C_a \cup C_b \cup C_c \cup C_d = V_2 \setminus (P_v \cup Q_e \cup V_0 \cup E_0)$. Thus, tracing the paths induced by $C_a \cup C_b \cup C_c \cup C_d$, we can reconstruct each edge $e_j = (v_i, v_{i'})$ with $e_j \in E_0$ and $v_i, v_{i'} \in V_0$. Therefore, we can reconstruct G_0 from G_2 uniquely up to isomorphism.

Hence the graph isomorphism problem for grid intersection graphs is as hard as the graph isomorphism problem for general graphs. Thus the graph isomorphism problem is *GI*-complete for grid intersection graphs.   □

4   Concluding Remarks

In this paper, we focus on hardness of two problems on geometrical intersection graphs. From the viewpoint of the parameterized complexity (see Downey and Fellow [3]), it is interesting to investigate efficient algorithms for these graph classes with some constraints. What if the number of vertical lines (or the possible positions on the coordinate of vertical lines) is bounded by a constant? In this case, we can use the dynamic programming technique for the graphs. Do the restrictions make some problems solvable in polynomial time?

From the graph theoretical point of view, there are two open problems. The first one is a geometric model for chordal bipartite graphs. It is pointed by Spinrad in [15], but it is not solved yet. The other one is whether chordal bipartite graphs are included in grid intersection graphs or not. This problem is pointed by Otachi, Okamoto, and Yamazaki in [13], and it seems to be true, but it is not solved.

It is also interesting the graph isomorphism problem for unit grid intersection graphs. By Theorem 2, the graph isomorphism problem is *GI*-complete for grid intersection graphs. We cannot make all line segments unit length in the reduction in the proof of Theorem 2. It is worth mentioning again that the graph isomorphism can be solved efficiently for interval bigraphs, which is another bipartite generalization of interval graphs.

References

1. Babel, L., Ponomarenko, I.N., Tinhofer, G.: The Isomorphism Problem For Directed Path Graphs and For Rooted Directed Path Graphs. Journal of Algorithms 21, 542–564 (1996)
2. Colbourn, C.J.: On Testing Isomorphism of Permutation Graphs. Networks 11, 13–21 (1981)
3. Downey, R.G., Fellows, M.R.: Parameterized Complexity. Springer, Heidelberg (1999)
4. Fishburn, P.C.: Interval Orders and Interval Graphs. Wiley & Sons, Inc., Chichester (1985)
5. Garey, M.R., Johnson, D.S.: Computers and Intractability — A Guide to the Theory of NP-Completeness. Freeman, San Francisco (1979)
6. Golumbic, M.C.: Algorithmic Graph Theory and Perfect Graphs, 2nd edn. Elsevier, Amsterdam (2004)
7. Hell, P., Huang, J.: Interval Bigraphs and Circular Arc Graphs. J. of Graph Theory (to appear), http://www.cs.sfu.ca/~pavol/intBig.ps
8. Nakano, S., Uehara, R., Uno, T.: A New Approach to Graph Recognition and Applications to Distance Hereditary Graphs. In: TAMC 2007. LNCS, vol. 4484, pp. 115–127. Springer, Heidelberg (2007)
9. Keil, J.M.: Finding Hamiltonian Circuits in Interval Graphs. Information Processing Letters 20(4), 201–206 (1985)
10. Lueker, G.S., Booth, K.S.: A Linear Time Algorithm for Deciding Interval Graph Isomorphism. Journal of the ACM 26(2), 183–195 (1979)
11. McKee, T.A., McMorris, F.R.: Topics in Intersection Graph Theory. SIAM (1999)
12. Müller, H.: Recognizing Interval Digraphs and Interval Bigraphs in Polynomial Time. Disc. Appl. Math. 78, 189–205 (1997), http://www.comp.leeds.ac.uk/hm/pub/node1.html
13. Otachi, Y., Okamoto, Y., Yamazaki, K.: Relationships between the class of unit grid intersection graphs and other classes of bipartite graphs. Discrete Applied Mathematics (accepted)

14. Plesník, J.: The NP-completeness of the Hamiltonian Cycle Problem in Planar Digraphs with Degree Bound Two. Information Processing Letters 8(4), 199–201 (1979)
15. Spinrad, J.P.: Open Problem List (1995), http://www.vuse.vanderbilt.edu/~spin/open.html
16. Spinrad, J.P.: Efficient Graph Representations. American Mathematical Society (2003)
17. Uehara, R.: Canonical Data Structure for Interval Probe Graphs. In: Fleischer, R., Trippen, G. (eds.) ISAAC 2004. LNCS, vol. 3341, pp. 859–870. Springer, Heidelberg (2004)
18. Uehara, R., Iwata, S.: Generalized Hi-Q is NP-Complete. The Transactions of the IEICE E73(2), 270–273 (1990), http://www.jaist.ac.jp/~uehara/pdf/phd7.ps.gz
19. Uehara, R., Toda, S., Nagoya, T.: Graph Isomorphism Completeness for Chordal Bipartite Graphs and Strongly Chordal Graphs. Discrete Applied Mathematics 145(3), 479–482 (2004)

On the Approximability of Comparing Genomes with Duplicates

Sébastien Angibaud, Guillaume Fertin, and Irena Rusu

Laboratoire d'Informatique de Nantes-Atlantique (LINA), FRE CNRS 2729
Université de Nantes, 2 rue de la Houssinière, 44322 Nantes Cedex 3 - France
{sebastien.angibaud,guillaume.fertin,irena.rusu}@univ-nantes.fr

Abstract. A central problem in comparative genomics consists in computing a (dis-)similarity measure between two genomes, e.g. in order to construct a phylogenetic tree. A large number of such measures has been proposed in the recent past: *number of reversals, number of breakpoints, number of common* or *conserved intervals, SAD* etc. In their initial definitions, all these measures suppose that genomes contain no duplicates. However, we now know that genes can be duplicated within the same genome. One possible approach to overcome this difficulty is to establish a one-to-one correspondence (i.e. a matching) between genes of both genomes, where the correspondence is chosen in order to optimize the studied measure. Then, after a gene relabeling according to this matching and a deletion of the unmatched signed genes, two genomes without duplicates are obtained and the measure can be computed.

In this paper, we are interested in three measures (*number of breakpoints, number of common intervals* and *number of conserved intervals*) and three models of matching (*exemplar* model, *maximum matching* model and *non maximum matching* model). We prove that, for each model and each measure, computing a matching between two genomes that optimizes the measure is **APX-Hard**. We show that this result remains true even for two genomes G_1 and G_2 such that G_1 contains no duplicates and no gene of G_2 appears more than twice. Therefore, our results extend those of [5,6,8]. Finally, we propose a 4-approximation algorithm for a measure closely related to the *number of breakpoints*, the *number of adjacencies*, under the *maximum matching* model, in the case where genomes contain the same number of duplications of each gene.

Keywords: genome rearrangement, APX-Hardness, duplicates, breakpoints, adjacencies, common intervals, conserved intervals.

1   Introduction and Preliminaries

In comparative genomics, computing a measure of (dis-)similarity between two genomes is a central problem; such a measure can be used for instance to construct phylogenetic trees. The measures defined so far fall into two categories: the first one contains distances, for which we count the number of operations needed to transform a genome into another (see for instance *edit distance* [13] or

S.-i. Nakano and Md. S. Rahman (Eds.): WALCOM 2008, LNCS 4921, pp. 34–45, 2008.
© Springer-Verlag Berlin Heidelberg 2008

number of reversals [3]). The second one contains (dis-)similarity measures based on the genome structure, such as *number of breakpoints* [5], *conserved intervals distance* [4], *number of common intervals* [6], *SAD* and *MAD* [16] etc.

When genomes contain no duplicates, most measures can be computed in polynomial time. However, assuming that genomes contain no duplicates is too limited, as it has been shown that a great number of duplicates exists in some genomes. For example, in [12], authors estimate that fifteen percent of genes are duplicated in the human genome. A possible approach to overcome this difficulty is to specify a one-to-one correspondence (i.e. a matching) between genes of both genomes and to remove the remaining genes, thus obtaining two genomes with identical gene composition and no duplicates. This matching is chosen in order to optimize the studied measure. Three models achieving this correspondence have been proposed : *exemplar* model [15], *maximum matching* model [17] and *non maximum matching* model [2].

Let $\mathcal{F}$ be a set of *genes*, where each gene is represented by an integer. A genome G is a sequence of signed elements (*signed genes*) from $\mathcal{F}$. Let $occ(g, G)$ be the number of occurrences of a gene g in a genome G and let $occ(G) = \max\{occ(g, G)|g$ is present in $G\}$. Two genomes G_1 and G_2 are called *balanced* iff, for each gene g, we have $occ(g, G_1) = occ(g, G_2)$. Denote η_G the size of genome G. Let $G[p], 1 \leqslant p \leqslant \eta_G$, be the signed gene that occurs at position p on genome G. For any signed gene g, let $\bar{g}$ be the signed gene having the opposite sign. Given a genome G without duplicates and two signed genes a, b such that a is located before b, let $G[a, b]$ be the set of genes located between genes a and b in G. We also note $[a, b]_{G_1}$ the substring (i.e. the sequence of consecutive elements) of G_1 starting by a and finishing by b.

Breakpoints, adjacencies, common and conserved intervals. Let us now define the four measures we will study in this paper. Let G_1, G_2 be two genomes without duplicates and with the same gene composition.

Breakpoint and Adjacency. Let (a, b) be a pair of consecutive signed genes in G_1. We say that the pair (a, b) induces a *breakpoint* of (G_1, G_2) if neither (a, b) nor $(\bar{b}, \bar{a})$ is a pair of consecutive signed genes in G_2. Otherwise, we say that (a, b) induces an *adjacency* of (G_1, G_2). For example, when $G_1 = +1 + 2 + 3 + 4 + 5$ and $G_2 = +5 - 4 - 3 + 2 + 1$, the pair $(2, 3)$ in G_1 induces a breakpoint of (G_1, G_2) while $(3, 4)$ in G_1 induces an adjacency of (G_1, G_2). We note $B(G_1, G_2)$ the number of breakpoints that exist between G_1 and G_2.

Common interval. A *common interval* of (G_1, G_2) is a substring of G_1 such that G_2 contains a permutation of this substring (not taking signs into account). For example, consider $G_1 = +1 + 2 + 3 + 4 + 5$ and $G_2 = +2 - 4 + 3 + 5 + 1$. The substring $[+3, +5]_{G_1}$ is a common interval of (G_1, G_2). We notice that the notion of common interval does not consider the sign of genes.

Conserved interval. Consider two signed genes a and b of G_1 such that a precedes b, where the precedence relation is large in the sense that, possibly, $a = b$. The substring $[a, b]_{G_1}$ is called a *conserved interval* of (G_1, G_2) if it satisfies the two following properties: first, either a precedes b or $\bar{b}$ precedes $\bar{a}$ in G_2; second, the set of genes located between genes a and b in G_2 is equal to

$G_1[a, b]$. For example, if $G_1 = +1 + 2 + 3 + 4 + 5$ and $G_2 = -5 - 4 + 3 - 2 + 1$, the substring $[+2, +5]_{G_1}$ is a conserved interval of (G_1, G_2).

Note that a conserved interval is actually a common interval, but with additional restrictions on its extremities. An interval of a genome G which is either of length one (i.e. a singleton) or the whole genome G is called a *trivial interval*.

Dealing with duplicates in genomes. When genomes contain duplicates, we cannot directly compute the measures defined previously. A solution consists in finding a one-to-one correspondence (i.e. a matching) between duplicated genes of G_1 and G_2, and use this correspondence to rename genes of G_1 and G_2 and to delete the unmatched signed genes in order to obtain two genomes G'_1 and G'_2 such that G'_2 is a permutation of G'_1; thus, the measure computation becomes possible. In this paper, we will focus on three models of matching : the *exemplar*, *maximum matching* and *non maximum matching* models.

- The *exemplar model* [15]: for each gene g, we keep in the matching only one occurrence of g in G_1 and in G_2, and we remove all the other occurrences. Hence, we obtain two genomes G_1^E and G_2^E without duplicates. The pair (G_1^E, G_2^E) is called an *exemplarization* of (G_1, G_2).
- The *maximum matching model* [17]: in this case, we keep in the matching the maximum number of genes in both genomes. More precisely, we look for a one-to-one correspondence between genes of G_1 and G_2 that, for each gene g, matches exactly $min(occ(g, G_1), occ(g, G_2))$ occurrences. After this operation, we delete each unmatched signed genes. The pair (G_1^E, G_2^E) obtained by this operation is called a *maximum matching* of (G_1, G_2).
- The *non maximum matching model* [2]: this model is an intermediate between the *exemplar* and the *maximum matching* models. In this new model, for each gene family g, we keep an arbitrary number k_g, such that $1 \leqslant k_g \leqslant min(occ(g, G_1), occ(g, G_2))$, of genes in G_1^E and in G_2^E. We call the pair (G_1^E, G_2^E) a *non maximum matching* of (G_1, G_2).

Problems studied in this paper. Let G_1 and G_2 be two genomes with duplicates. Let *EBD* (resp. *MBD*, *NMBD*) be the problem which consists in finding an exemplarization (G_1^E, G_2^E) of (G_1, G_2) (resp. maximum matching, non maximum matching) that minimizes the number of breakpoints between G_1^E and G_2^E. *EBD* is proved to be **NP-Complete** even if $occ(G_1) = 1$ and $occ(G_2) = 2$ [5]. Some inapproximability results are given: it has been proved in [8] that, in the general case, *EBD* cannot be approximated within a factor $c \log n$, where $c > 0$ is a constant, and cannot be approximated within a factor 1.36 when $occ(G_1) = occ(G_2) = 2$. Likewise, the problem consisting in deciding if there exists an exemplarization (G_1^E, G_2^E) of (G_1, G_2) such that there is no breakpoint between G_1^E and G_2^E is **NP-Complete** even when $occ(G_1) = occ(G_2) = 3$. Moreover, for two balanced genomes G_1 and G_2 such that $k = occ(G_1) = occ(G_2)$, several approximation algorithms for *MBD* are given. Those approximation algorithms admit respectively a ratio of 1.1037 when $k = 2$ [10], 4 when $k = 3$ [10] and $4k$ in the general case [11].

Let $EComI$ (resp. $MComI$, $NMComI$) be the problem which consists in finding an exemplarization (G_1^E, G_2^E) of (G_1, G_2) (resp. maximum matching, non maximum matching) such that the number of common intervals of (G_1^E, G_2^E) is maximized. $EComI$ and $MComI$ are proved to be **NP-Complete** even if $occ(G_1) = 1$ and $occ(G_2) = 2$ in [6].

Let $EConsI$ (resp. $MConsI$, $NMConsI$) be the problem which consists in finding an exemplarization (G_1^E, G_2^E) of (G_1, G_2) (resp. maximum matching, non maximum matching) such that the number of conserved intervals of (G_1^E, G_2^E) is maximized. In [4], Blin and Rizzi have studied the problem of computing a *distance* built on the number of conserved intervals. This distance differs from the *number of conserved intervals* we study in this paper, mainly in the sense that (i) it can be applied to two *sets* of genomes (as opposed to two genomes in our case), and (ii) the distance between two identical genomes of length n is equal to 0 (as opposed to $\frac{n(n+1)}{2}$ in our case). Blin and Rizzi [4] proved that finding the minimum distance is **NP-Complete**, under both the *exemplar* and *maximum matching* models. A closer analysis of their proof shows that it can be easily adapted to prove that $EConsI$ and $MConsI$ are NP-complete, even in the case $occ(G_1) = 1$.

We can conclude from these results that the MBD, $NMBD$, $NMComI$ and $NMConsI$ problems are also **NP-Complete**, since when one genome contains no duplicates, *exemplar*, *maximum matching* and *non maximum matching* models are equivalent.

In this paper, we study the approximation complexity of three measure computations: *number of breakpoints*, *number of conserved intervals* and *number of common intervals*. In Sections 2 and 3, we prove the **APX**-Hardness of $EComI$, $EConsI$ and EBD even when applied on genomes G_1 and G_2 such that $occ(G_1) = 1$ and $occ(G_2) = 2$, which induce the **APX**-Hardness under the other models. These results extend those of papers [5,6,8]. In Section 4, we consider the *maximum matching* model and a fourth measure, the *number of adjacencies* for which we give a 4-approximation algorithm when genomes are balanced. Hence, we are able to provide an approximation algorithm with *constant* ratio, even when the number of occurrences of genes is unbounded.

2 $EComI$ and $EConsI$ Are APX-Hard

Theorem 1. *The $EComI$ and $EConsI$ problems are **APX-Hard** even when applied to genomes G_1, G_2 such that $occ(G_1) = 1$ and $occ(G_2) = 2$.*

In this section, we prove Theorem 1 by using an *L-reduction* [14] from the *Minimum Vertex Cover* problem on cubic graphs, denoted by VC_3 and proved **APX-Complete** in [1]. Let $G = (V, E)$ be a cubic graph, i.e. for all $v \in V, degree(v) = 3$. A set of vertices $V' \subseteq V$ is called a *vertex cover* of G if for each edge $e \in E$, there exists a vertex $v \in V'$ such that e is incident to v. The problem VC_3 is defined as follows:

> **Problem:** VC_3
> **Input:** A cubic graph $G = (V, E)$, an integer k.
> **Question:** Does there exist a vertex cover V' of G such that $|V'| \leqslant k$?

Reduction. Let (G, k) be an instance of VC_3, where $G = (V, E)$ is a cubic graph with $V = \{v_1 \ldots v_n\}$ and $E = \{e_1 \ldots e_m\}$. Consider the transformation R which associates to the graph G two genomes G_1 and G_2 in the following way, where each gene has a positive sign.

$$G_1 = b_1, b_2 \ldots b_m, x, a_1, C_1, f_1, a_2, C_2, f_2 \ldots a_n, C_n, f_n, y, b_{m+n}, b_{m+n-1} \ldots b_{m+1}$$
$$G_2 = y, a_1, D_1, f_1, b_{m+1}, a_2, D_2, f_2, b_{m+2} \ldots b_{m+n-1}, a_n, D_n, f_n, b_{m+n}, x$$

with :

- for each i, $1 \leqslant i \leqslant n$, $a_i = 6i - 5$, $f_i = 6i$ and $C_i = (a_i + 1), (a_i + 2), (a_i + 3), (a_i + 4)$
- for each i, $1 \leqslant i \leqslant n + m$, $b_i = 6n + i$
- $x = 7n + m + 1$ and $y = 7n + m + 2$
- for each i, $1 \leqslant i \leqslant n$, $D_i = a_i + 3, b_{j_i}, a_i + 1, b_{k_i}, a_i + 4, b_{l_i}, a_i + 2$ where e_{j_i}, e_{k_i} and e_{l_i} are the edges which are incident to v_i in G, with $j_i < k_i < l_i$.

In the following, genes b_i, $1 \leqslant i \leqslant m$, are called *markers*. There is no duplicated gene in G_1 and the markers are the only duplicated genes in G_2; these genes occur twice in G_2. Hence, we have $occ(G_1) = 1$ and $occ(G_2) = 2$.

Preliminary results. In order to prove Theorem 1, we first give four intermediate lemmas. In the following, a common interval for the *EComI* problem or a conserved interval for *EConsI* is called a *robust interval*.

Lemma 1. *For any exemplarization (G_1, G_2^E) of (G_1, G_2), the non trivial robust intervals of (G_1, G_2^E) are necessarily contained in some sequence $a_i C_i f_i$ of G_1 $(1 \leqslant i \leqslant n)$.*

Lemma 2. *Let (G_1, G_2^E) be an exemplarization of (G_1, G_2) and $i \in [1 \ldots n]$. Let Δ_i be a substring of $[a_i + 3, a_i + 2]_{G_2^E}$ that does not contain any marker. If $|\Delta_i| \in \{2, 3\}$, then there is no robust interval I of (G_1, G_2^E) such that Δ_i is a permutation of I.*

For more clarity, let us now introduce some notations. Given a graph $G = (V, E)$, let $VC = \{v_{i_1}, v_{i_2} \ldots v_{i_k}\}$ be a vertex cover of G. Let $R(G) = (G_1, G_2)$ be the pair of genomes defined by the construction described in (1) and (2). Now, let F be the function which associates to VC, G_1 and G_2 an exemplarization $F(VC)$ of (G_1, G_2) as follows. In G_2, all the markers are removed from the sequences D_i for all $i \neq i_1, i_2 \ldots i_k$. Next, for each marker which is still present twice, one of its occurrences is arbitrarily removed. Since in G_2 only markers are duplicated, we conclude that $F(VC)$ is an exemplarization of (G_1, G_2).

Given a cubic graph G and genomes G_1 and G_2 obtained by the transformation $R(G)$, let us define the function S which associates to an exemplarization (G_1, G_2^E) of (G_1, G_2) the vertex cover VC of G defined as follows: $VC = \{v_i | 1 \leqslant i \leqslant n \wedge \exists j \in \{1 \ldots m\}, b_j \in G_2^E[a_i, f_i]\}$. In other words, we keep in VC the vertices v_i of G for which there exists some gene b_j such that b_j is in $G_2^E[a_i, f_i]$. We now prove that VC is a vertex cover. Consider an edge e_p of G. By construction of G_1 and G_2, there exists some i, $1 \leqslant i \leqslant n$, such that gene b_p is located between a_i and f_i in G_2^E. The presence of gene b_p between a_i and f_i implies that vertex v_i belongs to VC. We conclude that each edge is incident to at least one vertex of VC.

Let W be the function defined on $\{EConsI, EComI\}$ by $W(pb) = 1$ if $pb = EConsI$ and $W(pb) = 4$ if $pb = EComI$. Let $OPT_P(A)$ be the optimum result of an instance A for an optimization problem P, $P \in \{EComI, EConsI, VC_3\}$.

We define the function T which associates to a problem $pb \in \{EConsI, EComI\}$ and a cubic graph G, the number of robust trivial intervals of an exemplarization of both genomes G_1 and G_2 obtained by $R(G)$ for the problem pb. Let n and m be respectively the number of vertices and the number of edges of G. We have $T(EConsI, G) = 7n + m + 2$ and $T(EComI, G) = 7n + m + 3$. Indeed, for $EComI$, there are $7n + m + 2$ singletons and we also need to consider the whole genome.

Lemma 3. *Let $pb \in \{EcomI, EConsI\}$. Let G be a cubic graph and $R(G) = (G_1, G_2)$. Let (G_1, G_2^E) be an exemplarization of (G_1, G_2) and let i, $1 \leqslant i \leqslant n$. Then only two cases can occur:*

1. *Either in G_2^E, all the markers from D_i were removed, and in this case, there are exactly $W(pb)$ non trivial robust intervals involving D_i.*
2. *Or in G_2^E, at least one marker was kept in D_i, and in this case, there is no non trivial robust interval involving D_i.*

Lemma 4. *Let $pb \in \{EcomI, EConsI\}$. Let $G = (V, E)$ be a cubic graph with $V = \{v_1 \ldots v_n\}$ and $E = \{e_1 \ldots e_m\}$ and let G_1, G_2 be the two genomes obtained by $R(G)$.*

1. *Let VC be a vertex cover of G and denote $k = |VC|$. Then the exemplarization $F(VC)$ of (G_1, G_2) has at least $N = W(pb) \cdot n + T(pb, G) - W(pb) \cdot k$ robust intervals.*
2. *Let (G_1, G_2^E) be an exemplarization of (G_1, G_2) and let VC' be the vertex cover of G obtained by $S(G_1, G_2^E)$. Then $|VC'| = \frac{W(pb) \cdot n + T(pb, G) - N}{W(pb)}$, where N is the number of robust intervals of (G_1, G_2^E).*

Main result. Let us first define the notion of *L-reduction* [14]: let A and B be two optimization problems and c_A, c_B be respectively their cost functions. An *L-reduction* from problem A to problem B is a pair of polynomial functions R and S with the following properties:

(a) If x is an instance of A, then $R(x)$ is an instance of B ;
(b) If x is an instance of A and y is a solution of $R(x)$, then $S(y)$ is a solution of x;
(c) If x is an instance of A whose optimum is $OPT(x)$, then $R(x)$ is an instance of B such that $OPT(R(x)) \leqslant \alpha.OPT(x)$, where α is a positive constant ;
(d) If s is a solution of $R(x)$, then:
$|OPT(x) - c_A(S(s))| \leqslant \beta |OPT(R(x)) - c_B(s)|$ where β is a positive constant.

We prove Theorem 1 by showing that the pair (R, S) defined previously is an *L-reduction* from VC_3 to $EConsI$ and from VC_3 to $EComI$. First note that properties (a) and (b) are obviously satisfied by R and S.

Consider $pb \in \{EComI, EConsI\}$. Let $G = (V, E)$ be a cubic graph with n vertices and m edges. We now prove properties (c) and (d). Consider the genomes G_1 and G_2 obtained by $R(G)$. First, we need to prove that there exists $\alpha \geqslant 0$ such that $OPT_{pb}(G_1, G_2) \leqslant \alpha.OPT_{VC_3}(G)$.

Since G is a cubic graph, we have the three following properties: $n \geqslant 4$ (**I1**), $m = \frac{1}{2}\sum_{i=1}^{n} degree(v_i) = \frac{3n}{2}$ (**I2**) and $OPT_{VC_3}(G) \geqslant \frac{m}{3} = \frac{n}{2}$ (**I3**).

To explain property (**I3**), remark that, in a cubic graph G with n vertices and m edges, each vertex covers three edges. Thus, a set of k vertices covers at most $3k$ edges. Hence, any vertex cover of G must contain at least $\frac{m}{3}$ vertices.

By Lemma 3, we know that sequences of the form $a_i C_i f_i$, $1 \leqslant i \leqslant n$ contain either zero or $W(pb)$ non trivial robust intervals. By Lemma 1, there are no other non trivial robust intervals. So, we have the following inequality:

$$OPT_{pb}(G_1, G_2) \leqslant \underbrace{T(pb, G)}_{trivial\ robust\ intervals} + W(pb) \cdot n.$$

If $pb = EComI$, we have $OPT_{EComI}(G_1, G_2) \leqslant 7n + m + 3 + 4n$ and by (**I1**) and (**I2**), we obtain $OPT_{EComI}(G_1, G_2) \leqslant \frac{27n}{2}$ (**I4**). If $pb = EConsI$, we have $OPT_{EConsI}(G_1, G_2) \leqslant 7n + m + 2 + n$ and by (**I1**) and (**I2**), we obtain $OPT_{EConsI}(G_1, G_2) \leqslant \frac{21n}{2}$ (**I5**). Altogether, by (**I3**), (**I4**) and (**I5**), we prove property (c) with $\alpha = 27$.

Now, let us prove property (d). Let $VC = \{v_{i_1}, v_{i_2} \ldots v_{i_P}\}$ be a minimum vertex cover of G. Denote $P = OPT_{VC_3}(G) = |VC|$ and let G_1 and G_2 be the genomes obtained by $R(G)$. Let (G_1, G_2^E) be an exemplarization of (G_1, G_2) and let k' be the number of robust intervals of (G_1, G_2^E). Finally, let VC' be the vertex cover of G such that $VC' = S(G_1, G_2^E)$. We need to find a positive constant β such that $|P - |VC'|| \leqslant \beta |OPT_{pb}(G_1, G_2) - k'|$.

For $pb \in \{EComI, EConsI\}$, let N_{pb} be the number of robust intervals between the two genomes obtained by $F(VC)$. By the first property of Lemma 4, we have $OPT_{pb}(G_1, G_2) \geqslant N_{pb} \geqslant W(pb) \cdot n + T(pb, G) - W(pb) \cdot P$

By the second property of Lemma 4, we have $|VC'| = \frac{W(pb) \cdot n + T(pb, G) - k'}{W(pb)}$.

Recall that $OPT_{pb}(G_1, G_2) \geqslant W(pb) \cdot n + T(pb, G) - W(pb) \cdot P$. So, it is sufficient to prove $\exists \beta \geqslant 0, |P - |VC'|| \leqslant \beta |W(pb) \cdot n + T(pb, G) - W(pb) \cdot P - k'|$. Since $P \leqslant |VC'|$, we have $|P - |VC'|| = |VC'| - P = \frac{W(pb) \cdot n + T(pb, G) - k'}{W(pb)} - P$ and then

$$|P - |VC'|| = \frac{1}{W(pb)}(W(pb) \cdot n + T(pb, G) - W(pb) \cdot P - k')$$

So $\beta = 1$ is sufficient in both cases, since $W(EComI) = 4$ and $W(EConsI) = 1$, which implies $\frac{1}{W(pb)} \leqslant 1$. Altogether, we then have:

$$\left| OPT_{VC_3}(G) - |VC'| \right| \leqslant 1 \cdot \left| OPT_{pb}(G_1, G_2) - k' \right|$$

We proved that the reduction (R, S) is an *L-reduction*. This implies that for two genomes G_1 and G_2, both problems *EConsI* and *EComI* are **APX-Hard** even if $occ(G_1) = 1$ and $occ(G_2) = 2$. Theorem 1 is proved. □

We extend in Corollary 1 our results for the *maximum matching* and *non maximum matching* models.

Corollary 1. *MComI, NMComI, MConsI and NMConsI are* **APX-Hard** *even when applied to genomes G_1, G_2 such that $occ(G_1) = 1$ and $occ(G_2) = 2$.*

3   *EBD* is APX-Hard

In this section, we prove the following theorem:

Theorem 2. *EBD is* **APX-Hard** *even when applied to genomes G_1, G_2 such that $occ(G_1) = 1$ and $occ(G_2) = 2$.*

To prove Theorem 2, we use an *L-Reduction* from the VC_3 problem to the *EBD* problem. Let $G = (V, E)$ be a cubic graph with $V = \{v_1 \ldots v_n\}$ and $E = \{e_1 \ldots e_m\}$. For each i, $1 \leqslant i \leqslant n$, let e_{f_i}, e_{g_i} and e_{h_i} be the three edges which are incident to v_i in G with $f_i < g_i < h_i$. Let R' be the polynomial transformation which associates to G the following genomes G_1 and G_2, where each gene has a positive sign:

$G_1 = a_0 \ a_1 \ b_1 \ a_2 \ b_2 \ldots a_n \ b_n \ c_1 \ d_1 \ c_2 \ d_2 \ldots c_m \ d_m \ c_{m+1}$
$G_2 = a_0 \ a_n \ d_{f_n} \ d_{g_n} \ d_{h_n} \ b_n \ldots a_2 \ d_{f_2} \ d_{g_2} \ d_{h_2} \ b_2 \ a_1 \ d_{f_1} \ d_{g_1} \ d_{h_1} \ b_1 \ c_1 \ c_2 \ldots c_m \ c_{m+1}$
with :

- $a_0 = 0$, and for each i, $0 \leqslant i \leqslant n$, $a_i = i$ and $b_i = n + i$
- $c_{m+1} = 2n+m$, and for each i, $1 \leqslant i \leqslant m+1$, $c_i = 2n+i$ and $d_i = 2n+m+1+i$

We remark that there is no duplication in G_1, so $occ(G_1) = 1$. In G_2, only the genes d_i, $1 \leqslant i \leqslant m$, are duplicated and occur twice. Thus $occ(G_2) = 2$.

Let G be a cubic graph and VC be a vertex cover of G. Let G_1 and G_2 be the genomes obtained by $R'(G)$. We define F' to be the polynomial transformation which associates to VC, G_1 and G_2 the exemplarization (G_1, G_2^E) of (G_1, G_2) as follows. For each i such that $v_i \notin VC$, we remove from G_2 the genes d_{f_i}, d_{g_i} and d_{h_i}. Then, for each $1 \leqslant j \leqslant m$ such that d_j still has two occurrences in G_2, we arbitrarily remove one of these occurrences in order to obtain the genome G_2^E. Hence, (G_1, G_2^E) is an exemplarization of (G_1, G_2).

Given a cubic graph G, we construct G_1 and G_2 by the transformation $R'(G)$. Given an exemplarization (G_1, G_2^E) of (G_1, G_2), let S' be the polynomial transformation which associates to (G_1, G_2^E) the set $VC = \{v_i | 1 \leqslant i \leqslant n,$

a_i and b_i are not consecutive in G_2^E}. We claim that VC is a vertex cover of G. Indeed, let e_p, $1 \leqslant p \leqslant m$, be an edge of G. Genome G_2^E contains one occurrence of gene d_p since G_2^E is an exemplarization of G_2. By construction, there exists i, $1 \leqslant i \leqslant n$, such that d_p is in $G_2^E[a_i, b_i]$ and such that e_p is incident to v_i. The presence of d_p in $G_2^E[a_i, b_i]$ implies that vertex v_i belongs to VC. We can conclude that each edge of G is incident to at least one vertex of VC.

Lemmas 5 and 6 below are used to prove that (R', S') is an *L-Reduction* from the VC_3 problem to the EBD problem. Let $G = (V, E)$ be a cubic graph with $V = \{v_1, v_2 \ldots v_n\}$ and $E = \{e_1, e_2 \ldots e_m\}$ and let us construct (G_1, G_2) by the transformation $R'(G)$.

Lemma 5. *Let VC be a vertex cover of G and (G_1, G_2^E) the exemplarization given by $F'(VC)$. Then $|VC| = k \Rightarrow B(G_1, G_2^E) \leqslant n + 2m + k + 1$, where $B(G_1, G_2^E)$ is the number of breakpoints between G_1 and G_2^E.*

Lemma 6. *Let (G_1, G_2^E) be an exemplarization of (G_1, G_2) and VC' be the vertex cover of G obtained by $S'(G_1, G_2^E)$. We have $B(G_1, G_2^E) = k' \Rightarrow |VC'| = k' - n - 2m - 1$.*

Lemma 7. *The inequality $OPT_{EBD}(G_1, G_2) \leqslant 12 \cdot OPT_{VC_3}(G)$ holds.*

Lemma 8. *Let (G_1, G_2^E) be an exemplarization of (G_1, G_2) and let VC' be the vertex cover of G obtained by $S'(G_1, G_2^E)$. Then, we have $|OPT_{VC_3}(G) - |VC'|| \leqslant |OPT_{EBD}(G_1, G_2) - B(G_1, G_2^E)|$.*

Lemmas 7 and 8 prove that the pair (R', S') is an *L-reduction* from VC_3 to EBD. Hence, EBD is **APX-Hard** even if $occ(G_1) = 1$ and $occ(G_2) = 2$, and Theorem 2 is proved. We extend in Corollary 2 our results for the *maximum matching* and *non maximum matching* models.

Corollary 2. *The $MEBD$ and $NMEBD$ problems are **APX-Hard** even when applied to genomes G_1, G_2 such that $occ(G_1) = 1$ and $occ(G_2) = 2$.*

4   Approximating the Number of Adjacencies

For two balanced genomes G_1 and G_2, several approximation algorithms for computing the number of breakpoints between G_1 and G_2 are given for the *maximum matching* model [10,11]. We propose in this section a 4-approximation algorithm to compute a maximum matching of two balanced genomes that maximizes a new measure of similarity: the number of adjacencies (as opposed to minimizing the number of breakpoints). Remark that, as opposed to the results in [10,11], our approximation ratio is independent of the maximum number of duplicates. Note also that in [7], some recent results have been proposed for this measure of similarity. We first define the problem $AdjD$ we are interested in as follows:

Problem: $AdjD$
Input: Two balanced genomes G_1 and G_2.
Question: Find a maximum matching (G_1', G_2') of (G_1, G_2) which maximizes the number of adjacencies between G_1' and G_2'.

In [9], a 4-approximation algorithm for the weighted 2-*interval Pattern* problem $(W2IP)$ is given. In the following, we first define $W2IP$, and then we present how we can relate any instance of $AdjD$ to an instance of $W2IP$.

The weighted 2-interval Pattern problem. A 2-*interval* is the union of two disjoint intervals defined over a single sequence. For a 2-interval $D = (I, J)$, we suppose that the interval I does not overlap J and that I precedes J. We will denote this relation by $I < J$. We say that two 2-intervals $D_1 = (I_1, J_1)$ and $D_2 = (I_2, J_2)$ are *disjoint* if D_1 and D_2 have no common point (i.e. $(I_1 \cup J_1) \cap (I_2 \cup J_2) = \emptyset$). Three possible relations exist between two disjoint 2-intervals: (1) $D_1 \prec D_2$, if $I_1 < J_1 < I_2 < J_2$; (2) $D_1 \sqsubset D_2$, if $I_2 < I_1 < J_1 < J_2$; (3) $D_1 \between D_2$, if $I_1 < I_2 < J_1 < J_2$. We say that a pair of 2-intervals (D_1, D_2) is R-*comparable* for some $R \in \{\prec, \sqsubset, \between\}$, if either $(D_1, D_2) \in R$ or $(D_2, D_1) \in R$. A set of 2-intervals $\mathcal{D}$ is $\mathcal{R}$-comparable for some $\mathcal{R} \subseteq \{\prec, \sqsubset, \between\}$, $\mathcal{R} \neq \emptyset$, if any pair of distinct 2-intervals in $\mathcal{D}$ is R-comparable for some $R \in \mathcal{R}$. The non-empty set $\mathcal{R}$ is called a $\mathcal{R}$-*model*. We can define $W2IP$ as follows:

Problem: Weighted 2-interval Pattern $(W2IP)$
Input: A set $\mathcal{D}$ of 2-intervals, a $\mathcal{R}$-model $\mathcal{R} \subseteq \{\prec, \sqsubset, \between\}$ with $\mathcal{R} \neq \emptyset$, a weighted function $w : \mathcal{D} \mapsto \mathbb{R}$.
Question: Find a maximum weight $\mathcal{R}$-comparable subset of $\mathcal{D}$.

Transformation. We now describe how to transform any instance of $AdjD$ into an instance of $W2IP$. Let G_1 and G_2 be two balanced genomes. Two intervals I_1 of G_1 and I_2 of G_2 are said to be *identical* if they correspond to the same string (up to a complete reversal, where a reversal also changes all the signs). We denote by $Make2I$ the construction of the 2-intervals set obtained from the concatenation of G_1 and G_2. $Make2I$ is defined as follows: for any pair (I_1, I_2) of identical intervals of G_1, G_2, we construct a 2-interval $D = (I_1, I_2)$ of weight $|I_1| - 1$. We note $\mathcal{D} = Make2I(G_1, G_2)$ the set of all 2-*intervals* obtained in this way. We now define how to transform any solution of $W2IP$ into a solution of $AdjD$. Let G_1 and G_2 be two balanced genomes and let $\mathcal{D} = Make2I(G_1, G_2)$. Let S be a solution of $W2IP$ over the $\{\prec, \sqsubset, \between\}$-model for $\mathcal{D}$. We denote by $W2IP\_to\_AdjD$ the transformation of S into a maximum matching (G_1', G_2') of $G_1, G_2)$ defined as follows. First, for each 2-interval $D = (I_1, I_2)$ of S, we match the genes of I_1 and I_2 in the natural way; then, in order to achieve a maximum matching (since each gene is not necessarily covered by a 2-interval of S), we apply the following greedy algorithm: iteratively, we match arbitrarily two unmatched genes present in both G_1 and G_2, until no such gene exist. After a relabeling of signed genes, we obtain a maximum matching (G_1', G_2') of (G_1, G_2).

Lemma 9. *Let G_1, G_2 be two balanced genomes and let $\mathcal{D} = Make2I(G_1, G_2)$. Let S be a solution of $W2IP$ over the $\{\prec, \sqsubset, \lozenge\}$-model and let W_S be the weight of S. Then the maximum matching (G_1', G_2') of (G_1, G_2) obtained by the transformation $W2IP\_to\_AdjD(S)$ induces at least W_S adjacencies.*

Lemma 10. *Let G_1 and G_2 be two balanced genomes and let (G_1', G_2') be a maximum matching of (G_1, G_2). Let $\mathcal{D} = Make2I(G_1, G_2)$. Let W be the number of adjacencies induced by (G_1', G_2') between G_1 and G_2. Then there exists a solution S of $W2IP$ over the $\{\prec, \sqsubset, \lozenge\}$-model for $\mathcal{D}$ with weight equal to W.*

We now describe the algorithm $ApproxAdjD$ and then prove that it is a 4-approximation of the problem $AdjD$ by Theorem 3.

Algorithm 1. *ApproxAdjD*

Require: Two balanced genomes G_1 and G_2.
Ensure: A maximum matching (G_1', G_2') of (G_1, G_2).
   – Construct the set of weighted 2-intervals $\mathcal{D} = Make2I(G_1, G_2)$
   – Invoke the 4-approximation algorithm of Crochemore et al. [9] to obtain a solution S of $W2IP$ over the $\{\prec, \sqsubset, \lozenge\}$-model for $\mathcal{D}$
   – Construct the maximal matching $(G_1', G_2') = W2IP\_to\_AdjD(S)$

Theorem 3. *Algorithm $ApproxAdjD$ is a 4-approximation algorithm for $AdjD$.*

Proof. Let G_1 and G_2 be two balanced genomes and let $\mathcal{D} = Make2I(G_1, G_2)$. We first prove that the optimum of $AdjD$ for (G_1, G_2) is equal to the optimum of $W2IP$. Let OPT_{AdjD} be the optimum of $AdjD$ for (G_1, G_2). By Lemma 10, we know that there exists a solution S for $W2IP$ with weight $W_S = OPT_{AdjD}$. Now, suppose that there exists a solution S' for $W2IP$ with weight $W_{S'} > W_S$. Then, by Lemma 9, there exists a solution for $AdjD$ with weight $W \geqslant W_{S'}$. However, $W_{S'} > W_S$ by hypothesis, a contradiction to the fact that $W_S = OPT_{AdjD}$. Therefore, the two problems have the same optimum and, as a result, any approximation ratio for $W2IP$ implies the same approximation ratio for $AdjD$. In [9], a 4-approximation algorithm is proposed for $W2IP$; this directly implies that$ApproxAdjD$ is a 4-approximation algorithm for $AdjD$. $\qquad\square$

5   Conclusions and Future Work

The problems studied in this paper are shown to be **APX-Hard**, but some approximation algorithms exist when genomes are balanced [10,11]. However, it remains open whether approximation algorithms exist when genomes are not balanced. It has been shown in [8] that deciding if two genomes G_1 and G_2 have zero breakpoint under the *exemplar* model is **NP-Complete** even when $occ(G_1) = occ(G_2) = 3$ (problem $ZEBD$). This result implies that the EBD problem cannot be approximated in that case. Another open question is the complexity of $ZEBD$ when no gene appears more that twice in the genome.

References

1. Alimonti, P., Kann, V.: Some APX-completeness results for cubic graphs. Theoretical Computer Science 237(1-2), 123–134 (2000)
2. Angibaud, S., Fertin, G., Rusu, I., Vialette, S.: A general framework for computing rearrangement distances between genomes with duplicates. Journal of Computational Biology 14(4), 379–393 (2007)
3. Bafna, V., Pevzner, P.: Sorting by reversals: genome rearrangements in plant organelles and evolutionary history of X chromosome. Molecular Biology and Evolution, 239–246 (1995)
4. Blin, G., Rizzi, R.: Conserved interval distance computation between non-trivial genomes. In: Wang, L. (ed.) COCOON 2005. LNCS, vol. 3595, pp. 22–31. Springer, Heidelberg (2005)
5. Bryant, D.: The complexity of calculating exemplar distances. In: Comparative Genomics: Empirical and Analytical Approaches to Gene Order Dynamics, Map Alignement, and the Evolution of Gene Families, pp. 207–212. Kluwer Academic Publishers, Dordrecht (2000)
6. Chauve, C., Fertin, G., Rizzi, R., Vialette, S.: Genomes containing duplicates are hard to compare. In: Alexandrov, V.N., van Albada, G.D., Sloot, P.M.A., Dongarra, J.J. (eds.) ICCS 2006. LNCS, vol. 3992, pp. 783–790. Springer, Heidelberg (2006)
7. Chen, Z., Fu, B., Xu, J., Yang, B., Zhao, Z., Zhu, B.: Non-breaking similarity of genomes with gene repetitions. In: CPM 2007. LNCS, vol. 4580, pp. 119–130. Springer, Heidelberg (2007)
8. Chen, Z., Fu, B., Zhu, B.: The approximability of the exemplar breakpoint distance problem. In: Cheng, S.-W., Poon, C.K. (eds.) AAIM 2006. LNCS, vol. 4041, pp. 291–302. Springer, Heidelberg (2006)
9. Crochemore, M., Hermelin, D., Landau, G.M., Vialette, S.: Approximating the 2-interval pattern problem. In: Brodal, G.S., Leonardi, S. (eds.) ESA 2005. LNCS, vol. 3669, pp. 426–437. Springer, Heidelberg (2005)
10. Goldstein, A., Kolman, P., Zheng, Z.: Minimum common string partition problem: Hardness and approximations. In: Fleischer, R., Trippen, G. (eds.) ISAAC 2004. LNCS, vol. 3341, pp. 473–484. Springer, Heidelberg (2004)
11. Kolman, P., Waleń, T.: Reversal distance for strings with duplicates: Linear time approximation using hitting set. In: Erlebach, T., Kaklamanis, C. (eds.) WAOA 2006. LNCS, vol. 4368, pp. 279–289. Springer, Heidelberg (2007)
12. Li, W., Gu, Z., Wang, H., Nekrutenko, A.: Evolutionary analysis of the human genome. Nature (409), 847–849 (2001)
13. Marron, M., Swenson, K.M., Moret, B.M.E.: Genomic distances under deletions and insertions. Theoretical Computer Science 325(3), 347–360 (2004)
14. Papadimitriou, C., Yannakakis, M.: Optimization, approximation, and complexity classes. Journal of Computer and System Sciences 43(3), 425–440 (1991)
15. Sankoff, D.: Genome rearrangement with gene families. Bioinformatics 15(11), 909–917 (1999)
16. Sankoff, D., Haque, L.: Power boosts for cluster tests. In: McLysaght, A., Huson, D.H. (eds.) RECOMB 2005. LNCS (LNBI), vol. 3678, pp. 121–130. Springer, Heidelberg (2005)
17. Tang, J., Moret, B.M.E.: Phylogenetic reconstruction from gene-rearrangement data with unequal gene content. In: Dehne, F., Sack, J.-R., Smid, M. (eds.) WADS 2003. LNCS, vol. 2748, pp. 37–46. Springer, Heidelberg (2003)

Indexing Circular Patterns

Costas S. Iliopoulos[1,*] and M. Sohel Rahman[1,2,**]

[1] Algorithm Design Group
Department of Computer Science, King's College London,
Strand, London WC2R 2LS, England
{csi,sohel}@dcs.kcl.ac.uk
http://www.dcs.kcl.ac.uk/adg
[2] Department of Computer Science and Engineering
Bangladesh University of Engineering and Technology
Dhaka-1000, Bangladesh

Abstract. This paper deals with the Circular Pattern Matching Problem (CPM). In CPM, we are interested in pattern matching between the text $\mathcal{T}$ and the circular pattern $\mathcal{C}(\mathcal{P})$ of a given pattern $\mathcal{P} = \mathcal{P}_1 \ldots \mathcal{P}_m$. The circular pattern $\mathcal{C}(\mathcal{P})$ is formed by concatenating $\mathcal{P}_1$ to the right of $\mathcal{P}_m$. We can view $\mathcal{C}(\mathcal{P})$ as a set of m patterns starting at positions $j \in [1..m]$ and wrapping around the end and if any of these patterns matches $\mathcal{T}$, we find a match for $\mathcal{C}(\mathcal{P})$. In this paper, we present two efficient data structures to index circular patterns. This problem has applications in pattern matching in geometric and astronomical data as well as in computer graphics and bioinformatics.

1 Introduction

The classical pattern matching problem is to find all the occurrences of a given pattern $\mathcal{P}$ of length m in a text $\mathcal{T}$ of length n, both being sequences of characters drawn from a finite character set Σ. This problem is interesting as a fundamental computer science problem and is a basic need of many practical applications. In this paper, we study the circular pattern matching (CPM) problem. The circular pattern, denoted $\mathcal{C}(\mathcal{P})$, corresponding to a given pattern $\mathcal{P} = \mathcal{P}_1 \ldots \mathcal{P}_m$, is the string formed by concatenating $\mathcal{P}_1$ to the right of $\mathcal{P}_m$. In CPM, we are interested in pattern matching between the text $\mathcal{T}$ and the circular pattern $\mathcal{C}(\mathcal{P})$ of a given pattern $\mathcal{P}$. We can view $\mathcal{C}(\mathcal{P})$ as a set of m patterns starting at positions $j \in [1..m]$ and wrapping around the end. In other words, in CPM, we search for all 'conjugates'[1] of a given pattern in a given text.

The study of the conjugates of words have received fair amount of attention in the literature, although, mostly with different objectives than ours. Probably, the first such study was due to Booth in [6], who devised a linear algorithm

---

* Supported by EPSRC and Royal Society grants.
** Supported by the Commonwealth Scholarship Commission in the UK under the Commonwealth Scholarship and Fellowship Plan (CSFP).
[1] Two words x, y are conjugate if there exist words u, v such that $x = uv$ and $y = vu$.

S.-i. Nakano and Md. S. Rahman (Eds.): WALCOM 2008, LNCS 4921, pp. 46–57, 2008.
© Springer-Verlag Berlin Heidelberg 2008

to compute the 'least', i.e. lexicographically smallest conjugate of a word[2]. Several refinements were proposed in [21,9,5]. There exists many other results on related fields, but mostly from combinatorics point of view. Finally, a number of solutions to Problem CPM have been reported in [17,7,10]. However, these solutions doesn't address the indexing version of the problem, which is the focus of this paper. In this version of the problem, the goal is to construct an efficient index data structure to facilitate subsequent batch of queries as efficiently as possible. In this paper, we address the indexing version of Problem CPM and present two efficient data structures, namely, CPI-I and CPI-II. Our first data structure, CPI-I, can be constructed in $O(n \log^{1+\epsilon} n)$ time and space, for any constant $0 < \epsilon < 1$, where n is the length of text. Using CPI-I, the subsequent queries for a circular pattern $\mathcal{C}(\mathcal{P})$, corresponding to a given pattern $\mathcal{P}$, can be answered in (near optimal) $O(|\mathcal{P}| \log \log n + K)$ time, where K is the output size. Our second data structure, CPI-II, exhibits better time and space complexity in terms of construction ($O(n)$ construction time requiring $O(n \log n)$ bits of space) but suffers a little in the query time ($O(|\mathcal{P}| \log n + K)$).

Apart from being interesting from the pure combinatorial point view, Problem CPM has applications in pattern matching in geometric and astronomical data. For example, in geometry, a polygon may be encoded spelling its co-ordinates. Now, given the data stream of a number of polygons, we may need to find out whether a desired polygon exists in the data stream. The difficulty in this situation lies in the fact that the same polygon may be encoded differently depending on its "starting" coordinate and hence, there exists k possible encodings where k is the number of vertices of the polygon. Therefore, instead of traditional pattern matching, we need to resort to Problem CPM. This problem seems to be useful in Computer Graphics as well and hence may be used as a built in function in Graphics Cards handling polygon rendering. Another potential use of Problem CPM can be found in Computational Biology as follows. The genome of the harpes virus circularizes upon infection [23]. This makes Problem CPM better suited for motif matching in this case. Furthermore, this problem seems to be related to the much studied swap matching problem[3] [4] and also to the very recent studies of pattern matching with address error[4] [3]. The goal of this paper is to present efficient indexing schemes to solve the circular pattern matching problem.

2   Preliminaries

A *string* is a sequence of zero or more symbols from an alphabet Σ. A string X of length n is denoted by $X[1..n] = X_1 X_2 \ldots X_n$, where $X_i \in \Sigma$ for $1 \leq i \leq n$. The *length* of X is denoted by $|X| = n$. The string $\overleftarrow{X}$ denotes the reverse of the string X, i.e., $\overleftarrow{X} = X_n X_{n-1} \ldots X_1$. A string w is called a *factor* of X if

---

[2] Also known as the 'Lyndon' word.

[3] In CPM, the patterns can be thought of having a swap of two parts of it.

[4] The circular pattern can be thought of as having a special type of address error.

$X = uwv$ for $u, v \in \Sigma^*$; in this case, the string w occurs at position $|u| + 1$ in X. The factor w is denoted by $X[|u| + 1..|u| + |w|]$. A k-factor is a factor of length k. A prefix (or suffix) of X is a factor $X[x..y]$ such that $x = 1$ ($y = n$), $1 \le y \le n$ ($1 \le x \le n$). We define i-th prefix to be the prefix ending at position i i.e. $X[1..i], 1 \le i \le n$. On the other hand, i-th suffix is the suffix starting at position i i.e. $X[i..n], 1 \le i \le n$. A string ρ is called a period of X if X can be written as $X = \rho^k \rho'$ where $k \ge 1$ and ρ' is a prefix of ρ. The shortest period of X is called the period of X. If X has a period ρ such that $|\rho| \le n/2$, then X is said to be periodic. Further, if setting $X = \rho^k$ implies $k = 1$, X is said to be primitive; if $k > 1$, then $X = \rho^k$ is called a repetition.

The circular string, denoted $\mathcal{C}(X)$, corresponding to a string X, is the string formed by concatenating X_1 to the right of X_n. For a circular string, the notion of a factor is extended in the sense that, now, factors can wrap around the end of the string. In particular, a k-factor of $\mathcal{C}(X)$, now, can start at position $j \in [k + 1..n]$ and can wrap around the end of X. To accommodate this notion, we use the notation $X[j..(j + k - 1) \bmod n]$ (instead of $X[j..(j + k - 1)])$ to denote k-factor at position j. Now we are ready to formally define the Circular Pattern Matching problem.

Problem "CPM" (Circular Pattern Matching). *Given a text $\mathcal{T}$ of length n and a pattern $\mathcal{P}$ of length m, find the positions $i \in [1..n - m + 1]$ where $\mathcal{C}(\mathcal{P})$ matches $\mathcal{T}$. $\mathcal{C}(\mathcal{P})$ is said to match $\mathcal{T}$ at position $i \in [1..n - m + 1]$, if, and only if, there exist some $j \in [1..m]$, such that $\mathcal{P}[j..(j + m - 1) \bmod m] = \mathcal{T}[i..i + m - 1]$.*

We can view a circular pattern $\mathcal{C}(\mathcal{P})$ to be a set such that $\mathcal{C}(\mathcal{P}) = \{\mathcal{P}[i..(i + m - 1) \bmod m \mid 1 \le i \le m\}$[5]. So, alternatively, we can say that, $\mathcal{C}(\mathcal{P})$ matches $\mathcal{T}$ at position $i \in [1..n - m + 1]$, if and only if, there is a $Q \in \mathcal{C}(\mathcal{P})$ that matches $\mathcal{T}$ at position i. We define $Q_j = P[j..(j + m - 1) \bmod m], 1 \le j \le m$. It would be useful to define $Q_j = Q_j^f Q_j^\ell$, where $Q_j^f = P[j..m]$ and $Q_j^\ell = P[1..(j + m - 1) \bmod m]$ for $1 < j \le m$ (See Fig. 1). Note that, the text $\mathcal{T}$ is not circular in our problem.

Example 1. Suppose we have a text $\mathcal{T} = dbcadacadab$ and a pattern $\mathcal{P} = abcad$. Then $\mathcal{C}(\mathcal{P}) = \{abcad, bcada, cadab, adabc, dabca\}$. It is easy to verify that, according to our definition, $\mathcal{C}(\mathcal{P})$ matches $\mathcal{T}$ at positions 2 and 7. To elaborate, at position 2, we have a match for $Q_2 = bcada$ and at position 7, the match is due to $Q_3 = cadab$.

To the best of our knowledge, the best solution in the literature for Problem CPM requires $O(n \log |\Sigma|)$ time and $O(m)$ space [17]. In traditional pattern matching problem, indexing has always received particular attention. This is because in many practical problems, we need to handle batch of queries and, hence, it is computationally advantageous to preprocess the text in such a way that allows efficient query processing afterwards. The indexing version of Problem CPM is formally defined below.

---

[5] Note that $\mathcal{C}(\mathcal{P})$ can be a multi-set, i.e., where more than one member can have same value.

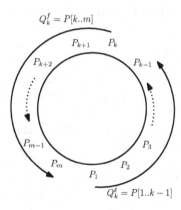

Fig. 1. $Q_k(= Q_k^f Q_k^l) \in \mathcal{C}(P)$

Problem "ICPM" (Indexing for Circular Pattern Matching). *Given a text T of length n, preprocess T to answer the following form of queries:*
Query: *Given a pattern P of length m, find the indices $i \in [1..n-m+1]$ at which $\mathcal{C}(P)$ matches T.*

This paper deals with Problem ICPM, i.e. the indexing version of Problem CPM. In traditional indexing problem one of the basic data structures used is the suffix tree. In our indexing problem we make use of this suffix tree data structure. A complete description of a suffix tree is beyond the scope of this paper, and can be found in [19,22] or in any textbook on stringology (e.g. [8,10]). However, for the sake of completeness, we define the suffix tree data structure as follows. Given a string T of length n over an alphabet Σ, the suffix tree ST_T of T is the compacted trie of all suffixes of $T\$$, where $\$ \notin \Sigma$. Each leaf in ST_T represents a suffix $T[i..n]$ of T and is labeled with the index i. We refer to the list (in left-to-right order) of indices of the leaves of the subtree rooted at node v as the leaf-list of v; it is denoted by $LL(v)$. Each edge in ST_T is labeled with a nonempty substring of T such that the path from the root to the leaf labeled with index i spells the suffix $T[i..n]$. For any node v, we let ℓ_v denote the string obtained by concatenating the substrings labeling the edges on the path from the root to v in the order they appear. Several algorithms exist that can construct the suffix tree ST_T in $O(n \log \sigma)$ time, where $\sigma = \min(n, |\Sigma|)$ [19,22]. Given the suffix tree ST_T of a text T we define the "locus" μ^P of a pattern P as the node in ST_T such that ℓ_{μ^P} has the prefix P and $|\ell_{\mu^P}|$ is the smallest of all such nodes. Note that the locus of P does not exist if P is not a substring of T. Therefore, given P, finding μ^P suffices to determine if P occurs in T. Given a suffix tree of a text T, a pattern P, one can find its locus and hence the fact whether T has an occurrence of P in optimal $O(|P|)$ time. In addition to that all such occurrences can be reported in constant time per occurrence.

3   CPI-I: An Index for Circular Pattern

In this section, we present a data structure to index the circular patterns. Using the traditional indexing scheme, we can easily solve Problem ICPM as follows. We use the suffix tree data structure as the index. Using the suffix tree, we search each distinct $Q \in \mathcal{C}(\mathcal{P})$ in $\mathcal{T}$. In total, the query processing then requires $O(m^2 + |Occ_{\mathcal{T}}^{\mathcal{C}(\mathcal{P})}|)$ time in the worst case where $Occ_{\mathcal{T}}^{\mathcal{C}(\mathcal{P})}$ is the set of occurrences of $\mathcal{C}(\mathcal{P})$ in $\mathcal{T}$. The goal of this paper is to construct a data structure that would facilitate a better query time in the worst case. In the rest of this section, we present the construction of our new data structure CPI-I to index circular patterns.

Our basic idea is to build an index that would solve the problem in two steps. At first, it will give us (implicitly) the occurrences of the two parts of each $Q_i \in \mathcal{C}(\mathcal{P}), 1 < i \leq m$, i.e. Q_i^f and Q_i^{ℓ}. Then, the index would give us the intersection of those occurrences, so as to provide us with the desired result. In order to do that, we maintain two suffix tree data structures $ST_{\mathcal{T}}$ and $ST_{\overleftarrow{\mathcal{T}}}$. We use $ST_{\mathcal{T}}$ to find the occurrences of Q_i^{ℓ}. We use the suffix tree of the reverse string of $\mathcal{T}$, i.e. $ST_{\overleftarrow{\mathcal{T}}}$, to find the occurrences of $\overleftarrow{Q_i^f}$. By doing so, in effect, we get the end positions of the occurrences of Q_i^f in $\mathcal{T}$. Now, we have to do a bit of "*shifting*", because, we want to find the intersection of the two set of occurrences. This is handled as follows. According to the definition of suffix tree, each leaf in $ST_{\mathcal{T}}$ is labeled by the starting location of its suffix. However, to achieve the desired alignment, each leaf in $ST_{\overleftarrow{\mathcal{T}}}$ is labeled by $(n + 1) - i + 1$, where i is the starting position of the leaf's suffix in $ST_{\overleftarrow{\mathcal{T}}}$. It is easy to see that, now, getting the occurrences of $\overleftarrow{Q_i^f}$ in $ST_{\overleftarrow{\mathcal{T}}}$ is equivalent to getting the occurrences of Q_i^f in $ST_{\mathcal{T}}$ according to our desired alignment. So it remains to show how we could perform the intersection efficiently in the context of indexing. In order to do that, we first do some preprocessing on $ST_{\mathcal{T}}$ and $ST_{\overleftarrow{\mathcal{T}}}$ as follows. For each of the two suffix trees, we maintain a linked list of all leaves in a left-to-right order. In other words, we realize the list $LL(\mathcal{R})$ in the form of a linked list where $\mathcal{R}$ is the root of the suffix tree. In addition to that, for each of the two suffix trees, we set pointers $v.left$ and $v.right$ from each tree node v to its leftmost leaf v_{ℓ} and rightmost leaf v_r (considering the subtree rooted at v) in the linked list. It is easy to realize that, with these set of pointers at our disposal, we can indicate the set of occurrences of a pattern $\mathcal{P}$ by the two leaves $\mu_{\ell}^{\mathcal{P}}$ and $\mu_r^{\mathcal{P}}$, because, all the leaves between and including $\mu_{\ell}^{\mathcal{P}}$ and $\mu_r^{\mathcal{P}}$ in $LL(\mathcal{R})$ correspond to the occurrences of $\mathcal{P}$ in $\mathcal{T}$.

In what follows, we define the term ℓ_T and r_T such that $LL(\mathcal{R})[\ell_T] = \mu_{\ell}^{Q_i^f}$ and $LL(\mathcal{R})[r_T] = \mu_r^{Q_i^f}$, where $\mathcal{R}$ is the root of $ST_{\mathcal{T}}$. Similarly we define $\ell_{\overleftarrow{T}}$ and $r_{\overleftarrow{T}}$ such that $LL(\overleftarrow{\mathcal{R}})[\ell_{\overleftarrow{T}}] = \mu_{\ell}^{Q_i^{\ell}}$ and $LL(\overleftarrow{\mathcal{R}})[r_{\overleftarrow{T}}] = \mu_r^{Q_i^{\ell}}$, where $\overleftarrow{\mathcal{R}}$ is the root of $ST_{\overleftarrow{\mathcal{T}}}$. Now, we have two lists $LL(\mathcal{R})$ and $LL(\overleftarrow{\mathcal{R}})$ and two intervals $[\ell_T..r_T]$ and $[\ell_{\overleftarrow{T}}..r_{\overleftarrow{T}}]$ respectively. Now, our problem is to find the intersection of the indices within these two intervals. We call this problem *Range Set Intersection* Problem. We first define the problem formally below.

Problem "RSI" (Range Set Intersection Problem). *Let* $V[1..n]$ *and* $W[1..n]$ *be two permutations of* $[1..n]$. *Preprocess* V *and* W *to answer the following form of queries:*
Find the intersection of the elements of $V[i..j]$ *and* $W[k..\ell]$, $1 \leq i \leq j \leq n, 1 \leq k \leq \ell \leq n$.

In order to solve the above problem, we reduce it to the well-studied *Range Search Problem* on a Grid.

Problem "RSG" (Range Search Problem on Grid). *Let* $A[1..n]$ *be a set of* n *points on the grid* $[0..U] \times [0..U]$. *Preprocess* A *to answer the following form of queries:*
Given a query rectangle $q \equiv (a,b) \times (c,d)$ *find the set of points contained in* q.

We can see that Problem RSI is just a different formulation of the Problem RSG as follows. We set $U = n$. Since V and W in Problem RSI are permutations of $[1..n]$, every number in $[1..n]$ appears precisely once in each of them. We define the coordinates of every number $i \in [1..n]$ to be (x, y), where $V[x] = W[y] = i$. Thus we get the n points on the grid $[0..n] \times [0..n]$, i.e. the array A of Problem RSG. The query rectangle q is deduced from the two intervals $[i..j]$ and $[k..\ell]$ as follows: $q \equiv (i, k) \times (j, \ell)$. It is easy to verify that the above reduction is correct and hence we can solve Problem RSI using the solution of Problem RSG. There has been significant research work on Problem RSG. We are going to use the data structure of Alstrup et al. [2]. We can use this data structure to answer the query of Problem RSG in $O(\log \log n + k)$ time where k is the number of points contained in the query rectangle q. The data structure requires $O(n \log^{1+\epsilon} n)$ time and space, for any constant $0 < \epsilon < 1$. Algorithm 1, formally states the steps to build our data structure, CPI-I, to index the circular patterns.

Let us now analyze the the running time of Algorithm 1. The algorithm can be divided into 3 main parts. Part 1 deals with the suffix tree of the text $\mathcal{T}$ and comprises of Steps 1 to 6. Part 2 consists of Steps 7 to 12 and deals with the suffix tree of the reverse text $\overleftarrow{\mathcal{T}}$. Part 3 deals with the reduction to Problem RSG from Problem RSI and the subsequent preprocessing step. The computational effort spent for Parts 1 and 2 are $O(n \log \sigma)$ as follows. Step 1 (resp. Step 7) builds the traditional suffix tree and hence can be done in $O(n \log \sigma)$ time. Step 2 (resp. Step 8) can be done easily while building the suffix tree. Step 3 and Step 4 (resp. Step 9 and Step 10) can be done together in $O(n)$ by traversing $ST_{\mathcal{T}}$ (resp. $ST_{\overleftarrow{\mathcal{T}}}$) using a breadth first or in order traversal. So, in total, Part 1 and Part 2, i.e. Step 1 to 12 requires $O(n \log \sigma)$ time.

In Part 3, we first construct the set A of points in the grid $[0..n] \times [0..n]$, on which we will apply the range search. This step can also be done in $O(n)$ as follows. Assume that, $\mathcal{L}$ (resp. $\overleftarrow{\mathcal{L}}$) is the linked list realizing $LL(\mathcal{R})$ (resp. $LL(\overleftarrow{\mathcal{R}})$). Each element in $\mathcal{L}$ (resp. $\overleftarrow{\mathcal{L}}$) is the label of the corresponding leaf in $LL(\mathcal{R})$ (resp. $LL(\overleftarrow{\mathcal{R}})$). We construct $\mathcal{L}^{-1}$ such that $\mathcal{L}^{-1}[\mathcal{L}[i]] = i$. Similarly, we construct $\overleftarrow{\mathcal{L}}^{-1}$. It is easy to see that, with $\mathcal{L}^{-1}$ and $\overleftarrow{\mathcal{L}}^{-1}$ in our hand, we can easily construct A in $O(n)$. A detail is that, in our case, there may exist

Algorithm 1. Algorithm to build the index, CPI-I, for the circular patterns

1: Build a suffix tree $ST_\mathcal{T}$ of $\mathcal{T}$. Let the root of $ST_\mathcal{T}$ is $\mathcal{R}$.
2: Label each leaf of $ST_\mathcal{T}$ by the starting location of its suffix.
3: Construct a linked list $\mathcal{L}$ realizing $LL(\mathcal{R})$. Each element in $\mathcal{L}$ is the label of the corresponding leaf in $LL(\mathcal{R})$.
4: **for** each node v in $ST_\mathcal{T}$ **do**
5:     Store $v.left = i$ and $v.right = j$ such that $\mathcal{L}[i]$ and $\mathcal{L}[j]$ corresponds to, respectively, (leftmost leaf) v_l and (rightmost leaf) v_r of v.
6: **end for**
7: Build a suffix tree $ST_{\overleftarrow{\mathcal{T}}}$ of $\overleftarrow{\mathcal{T}}$. Let the root of $ST_{\overleftarrow{\mathcal{T}}}$ is $\overleftarrow{\mathcal{R}}$.
8: Label each leaf of $ST_{\overleftarrow{\mathcal{T}}}$ by $(n+1) - i + 1$ where i is the starting location of its suffix.
9: Construct a linked list $\overleftarrow{\mathcal{L}}$ realizing $LL(\overleftarrow{\mathcal{R}})$. Each element in $\overleftarrow{\mathcal{L}}$ is the label of the corresponding leaf in $LL(\overleftarrow{\mathcal{R}})$.
10: **for** each node v in $ST_{\overleftarrow{\mathcal{T}}}$ **do**
11:     Store $v.left = k$ and $v.right = \ell$ such that $\overleftarrow{\mathcal{L}}[k]$ and $\overleftarrow{\mathcal{L}}[\ell]$ corresponds to, respectively, (leftmost leaf) v_l and (rightmost leaf) v_r of v.
12: **end for**
13: **for** $i = 1$ to n **do**
14:     Set $A[i] = \epsilon$
15: **end for**
16: **for** $i = 1$ to n **do**
17:     **if** there exists (x, y) such that $\mathcal{L}[x] = \overleftarrow{\mathcal{L}}[y] = i$ **then**
18:         $A[i] = (x, y)$
19:     **end if**
20: **end for**
21: Preprocess A for Range Search on a Grid $[0..n] \times [0..n]$.

$i, 1 \leq i \leq n$ such that $\overleftarrow{\mathcal{L}}[j] \neq i$ for all $1 \leq j \leq n$. This is because $\overleftarrow{\mathcal{L}}$ is a permutation of $[2..n+1]$ instead of $[1..n]$. So, we need to assume $U = n + 1$. After A is constructed, we perform Step 21, which requires $O(n \log^{1+\epsilon} n)$ time and space, for any constant $0 < \epsilon < 1$. Therefore, the data structure CPI-I is built in $O(n \log^{1+\epsilon} n)$ time and space.

Now, we discuss the query processing. Suppose we are given the CPI-I of a text $\mathcal{T}$ and a query for a circular pattern $\mathcal{C}(\mathcal{P})$ corresponding to the pattern $\mathcal{P}$. In what follows, we consider only a particular $Q_i \in \mathcal{C}(\mathcal{P}), 1 < i \leq m$.[6] We first find the locus $\mu^{Q_i^\ell}$ in $ST_\mathcal{T}$. Let $i = \mu^{Q_i^\ell}.left$ and $j = \mu^{Q_i^\ell}.right$. Now we find the locus $\mu^{\overleftarrow{Q_i^f}}$ in $ST_{\overleftarrow{\mathcal{T}}}$. Let $k = \mu^{\overleftarrow{Q_i^f}}.left$ and $\ell = \mu^{\overleftarrow{Q_i^f}}.right$. Then, we find all the points in A that are inside the rectangle $q \equiv (i, k) \times (j, \ell)$. Let B is the set of those points. Then, it is easy to verify that $Occ_\mathcal{T}^{Q_i} = \{(\mathcal{L}[x] - |Q_i^f|) \mid (x, y) \in B\}$. So, in this way we find the occurrences of Q_i for $1 < i \leq m$, which, along with $Occ_\mathcal{T}^\mathcal{P}$, gives us the desired set of occurrences i.e. $Occ_\mathcal{T}^{\mathcal{C}(\mathcal{P})}$. The steps are formally presented in the form of Algorithm 3.

---

[6] The occurrences of $Q_1 = \mathcal{P}$ can be easily found using traditional techniques.

Algorithm 2. Query Processing for CPI-I

1: Set $Occ_T^{\mathcal{C}(\mathcal{P})} = Occ_\mathcal{P}^T.\{$We find out the occurrences of $Q_1 = \mathcal{P}$ and initialize the $Occ_T^{\mathcal{C}(\mathcal{P})}$ with that$\}$

2: **for** $d = 1$ *to* $m - 1$ **do**

3:     Find $\mu^{\mathcal{P}[1..d]}$ in ST_T.

4:     Set $i[d] = \mu^{Q_{d+1}^\ell}.left,\ j[d] = \mu^{Q_{d+1}^\ell}.right$.

5:     Find $\mu^{\overleftarrow{\mathcal{P}}[1..d]}$ in $ST_{\overleftarrow{T}}$.

6:     Set $k[d] = \mu^{\overleftarrow{Q_d^f}}.left$ and $\ell[d] = \mu^{\overleftarrow{Q_d^f}}.right$.

7: **end for**

8: **for** $d_1 = 1$ *to* $m - 1, d_2 = m - 1$ *downto* 1 **do**

9:     Set $B = \{(x,y) \mid (x,y) \in A$ and (x,y) *is contained in* $q \equiv (i[d_1],k[d_2]) \times (j[d_1],\ell[d_2])\}$.

10:     Set $Occ_T^{\mathcal{C}(\mathcal{P})} = Occ_T^{\mathcal{C}(\mathcal{P})} \bigcup \{(\mathcal{L}[x] - |Q_{d_1}^f|) \mid (x,y) \in B\}$.

11: **end for**

12: **return** $Occ_T^{\mathcal{C}(\mathcal{P})}$.

The running time of the query processing (Algorithm 3) is deduced as follows. Apart from Step 1[7], which takes $O(m)$, Algorithm 3 is divided into two main parts. The first part comprises of the for loop of Step 2 and the second part consists of the loop of Step 8. In Step 2, we find and store, for each prefix of $\mathcal{P}$ (resp. $\overleftarrow{\mathcal{P}}$), the leftmost and rightmost entry in $\mathcal{L}$ (resp. $\overleftarrow{\mathcal{L}}$), so that, we can use them in Step 8 with appropriate combination. This can be done in $O(m)$ as follows. We first traverse ST_T matching $\mathcal{P}$. Suppose at an (explicit) node u, of ST_T, we are at $\mathcal{P}_i$ and when we reach the next (explicit) node v, we are at $\mathcal{P}_j$. It is easy to realize that, for all k such that $i < k \leq j$, the leftmost and rightmost entry for $\mathcal{P}[1..k]$ in $\mathcal{L}$ are, respectively, $v.left$ and $v.right$. So, we can find (and store) them in $O(m)$ time as we traverse ST_T matching $\mathcal{P}$. The same technique is used for $\overleftarrow{\mathcal{P}}$ with $ST_{\overleftarrow{T}}$. The goal of Step 8 is to get the intersection of the ranges of the appropriate prefix-suffix combinations of $\mathcal{P}$ derived in Step 2. The construction of the set B in Step 9 is done by performing the range query and hence requires $O(\log\log n + |B|)$ time. Note that $|B| = |Occ_{Q_i}^T|$ and hence, in total, the time required by Step 8, per Q_i, is $O(\log\log n + |Occ_{Q_i}^T|)$. So, in total, Step 8 requires, $O(m\log\log n + \sum_{1<i\leq m}|Occ_{Q_i}^T|)$. So, the total worst case running time of Algorithm 3 is:

$$O(m + |Occ_\mathcal{P}^T| + m\log\log n + \sum_{1<i\leq m}|Occ_{Q_i}^T|) = O(m\log\log n + \sum_{1\leq i\leq m}|Occ_{Q_i}^T|).$$

One subtle but important point is that, if there exists $Q_i = Q_j$ such that $Q_i, Q_j \in \mathcal{C}(\mathcal{P}), i \neq j$, then we will have:

$$\sum_{1\leq i\leq m}|Occ_{Q_i}^T| > |Occ_T^{\mathcal{C}(\mathcal{P})}|.$$

---

[7] This step can be easily incorporated in the two main loops of Algorithm 3; this, however, is kept for ease of exposition.

Hence, in such a case, our algorithm would require more than constant time per occurrence to report them, which is not desirable. However, we note that, this can only happen when $\mathcal{P}$ is a repetition. We handle this issue as follows. We need to realize that, if ρ is the period of $\mathcal{P}$, then there exists only $|\rho|$ distinct members in $\mathcal{C}(\mathcal{P})$. So, to solve the issue, the query processing (Algorithm 3) is preceded by a further pre-processing to find out ρ. This can be done easily in $O(m)$ by using, for example, the preprocessing of KMP algorithm [15]. Then, the two loops in Algorithm 3 (Step 2 and Step 8) are executed $|\rho|$ times instead of $m - 1$, which solves the problem.

4   CPI-II: Another Index for Circular Pattern

In Section 3, we have presented the data structure CPI-I requiring $O(n \log^{1+\epsilon} n)$ space. Since we need to use two suffix trees, along with the range search data structure, from practical point of view, the space usage of CPI-I is quite high. Note that, we can use the space efficient alternative suffix arrays, instead of suffix trees, to build CPI-I with some standard modifications in Algorithm 1; but, still, we can't get rid of the range search data structure. In this section, however, we use the suffix array to build another index for circular pattern, employing different techniques. The resulting data structure, CPI-II, will have better time and space complexity. The query time, however, will suffer by a $\log n / \log \log n$ factor.

We start with a very concise definition of suffix array. The suffix array $SA[1..n]$ of a text $\mathcal{T}$ is an array of integers $j \in [1..n]$ such that $SA[i] = j$ if, and only if, $\mathcal{T}[j..n]$ is the i-th suffix of $\mathcal{T}$ in (ascending) lexicographic order. Suffix arrays were first introduced in [18], where an $O(n \log n)$ construction algorithm and $O(m + \log n + |Occ_{\mathcal{P}}^{\mathcal{T}}|)$ query time were presented. Recently, linear time construction algorithms for space efficient suffix arrays have been presented [16,14,13]. The query time is also improved to optimal $O(m + |Occ_{\mathcal{P}}^{\mathcal{T}}|)$ in [1]. We recall that, the result of a query for a pattern $\mathcal{P}$ on a suffix array SA of $\mathcal{T}$, is given in the form of an interval $[s..e]$ such that $Occ_{\mathcal{P}}^{\mathcal{T}} = \{SA[s], SA[s+1], \ldots, SA[e]\}$. In this case, the interval $[s..e]$ is denoted by $\mathcal{I}nt_{\mathcal{P}}^{\mathcal{T}}$.

Now, the data structure CPI-II consists of the suffix array SA and the inverse suffix array SA^{-1} of $\mathcal{T}$. The inverse suffix array of $\mathcal{T}$ is denoted by $SA^{-1}[1..n]$, where $SA^{-1}[i]$ equals the number of suffixes that are lexicographically smaller than $\mathcal{T}[i..n]$. Both SA and SA^{-1} require $O(n \log n)$ bits of space and can be constructed in linear time [16,14,13]. We next discuss how we perform the query on CPI-II and analyze the running time. We use the following well-known results from [10] and [12].

Lemma 1. ([10]) *Given a text $\mathcal{T}$ and the suffix array of $\mathcal{T}$, assume $[s..e] = \mathcal{I}nt_{\mathcal{P}}^{\mathcal{T}}$ is already computed. Then for any character c, the interval $[s'..e'] = \mathcal{I}nt_{\mathcal{P}c}^{\mathcal{T}}$ can be computed in $O(\log n)$ time.*

Lemma 2. [12] *Given a text T along with the suffix array and inverse suffix array of T, assume that, the interval $[s'..e'] = Int_{P'}^T$, and the interval $[s''..e''] = Int_{P''}^T$ are already computed. Then, the interval $[s..e] = Int_{P'P''}^T$ can be computed in $O(\log n)$ time.*

Lemma 3. [12] *Given a text T and the suffix array and inverse suffix array of T, assume $[s..e] = Int_P^T$ is already computed. Also, assume that an array C is given such that $C[c]$ stores the total number of occurrences of all $c' \leq c$.[8] Then for any character c, the interval $[s'..e'] = Int_{cP}^T$ can be computed in $O(\log n)$ time.*

Algorithm 3. Query Processing for CPI-II

1: Compute the interval $[s..e]$ for $\mathcal{P}_1$ in T
2: $Pref[1].start = s, Pref[1].end = e$
3: **for** $i = 2$ to m **do**
4:     Using $Pref[i-1]$, compute the interval $[s..e]$ of $\mathcal{P}[1..i]$ in T
5:     $Pref[i].start = s, Pref[i].end = e$
6: **end for**
7: Compute the interval $[s..e]$ for $\mathcal{P}_m$ in T
8: $Suf[m].start = s, Suf[m].end = e$
9: **for** $i = m - 1$ downto 1 **do**
10:     Using $Suf[i+1]$, compute the interval $[s..e]$ of $\mathcal{P}[i..m]$ in T
11:     $Suf[i].start = s, Suf[i].end = e$
12: **end for**
13: $\mathcal{I} = \epsilon$
14: **for** $j = 2$ to m **do**
15:     Using $Suf[j]$ and $Pref[j + m - 1 \mod m]$, compute the interval $[s_j..e_j]$ for $Q_j = Q_j^f Q_j^\ell$
16:     $\mathcal{I} = \mathcal{I} \bigcup [s_j..e_j]$
17: **end for**
18: $\mathcal{I} = \mathcal{I} \bigcup Pref[m]$ {Include the interval of $\mathcal{P}$}
19: Sort $\mathcal{I}$ according to $s_j, 1 \leq j \leq m$.
20: **for** $j = 2$ to m **do**
21:     **if** $s_j < e_{j-1}$ **then**
22:         $s_j = \min\{s_j, s_{j-1}\}$
23:         $e_j = \max\{s_j, s_{j-1}\}$
24:         $\mathcal{I} = \mathcal{I} - [s_{j-1}..e_{j-1}]$
25:     **end if**
26: **end for**
27: $Occ_T^{\mathcal{C}(\mathcal{P})} = \epsilon$
28: **for** $[s..e] \in \mathcal{I}$ **do**
29:     $Occ_T^{\mathcal{C}(\mathcal{P})} = Occ_T^{\mathcal{C}(\mathcal{P})} \bigcup \{SA[s]..SA[e]\}$
30: **end for**
31: **return** $Occ_T^{\mathcal{C}(\mathcal{P})}$

---

[8] Here '$\leq$' implies a lexicographical relation.

Now the query is done as follows. We first compute the intervals for all the prefixes and suffixes of $\mathcal{P}$ and store them in the arrays $Pref[1..m]$ and $Suf[1..m]$ respectively. To compute the intervals of prefixes, we can use Lemma 1 as follows. We first compute the interval for the character $\mathcal{P}_1$ and store the interval in $Pref[1]$. This can be done in $O(\log n)$. Next, we compute the interval for $\mathcal{P}[1..2]$ using $Pref[1]$ in $O(\log n)$ time (Lemma 1). Then, we compute the interval for $\mathcal{P}[1..3]$ using $Pref[2]$ and so on. So, in this way, we can compute the array $Pref[1..m]$ in $O(m \log n)$ time. Similarly, we compute the interval of suffixes in $Suf[1..m]$. Courtesy to Lemma 3, we can do this in $O(m \log n)$ time as well. With $Pref$ and Suf at our disposal, we can find the intervals for the circular pattern $\mathcal{C}(\mathcal{P})$ as follows. We use the result of Lemma 2. Note that, for $Q_j \in \mathcal{C}(\mathcal{P}), 1 < j \leq n$, we have the interval of Q_j^f in $Suf[j]$ and the interval for Q_j^ℓ in $Pref[j + m - 1 \bmod m]$. Now according to Lemma 2, using $Suf[j]$ and $Pref[j + m - 1 \bmod m]$, we can find the interval for Q_j in $\log n$ time. Finally, interval for $\mathcal{P}(= Q_1)$ can also be found in $Pref[m]$ or $Suf[1]$. Therefore, we can find intervals for all the $Q_j, 1 \leq j \leq n$ in $O(m \log n)$ time. Now we simply report all the occurrences in this set of intervals. To avoid reporting a particular occurrence more than once, we sort the m intervals according to the start of the interval requiring a further $O(m \log m)$ time. And then do a linear traversal on the two end of the sorted intervals to get a equivalent set of 'disjoint' intervals. Then, we do the reporting. Therefore, in total, the query processing requires $O(m \log n + |Occ_T^{\mathcal{C}(\mathcal{P})}|)$ time per query.

To reduce the space usage further, we can implement the above algorithm using the compressed suffix array (CSA) [11], which requires $O(n)$ bits of space and can be constructed in $O(n)$ time. Also, using a supporting data structure, requiring $O(n)$ bits of space, we can evaluate $SA[i]$ and $SA^{-1}[i]$ from CSA in $O(\log n)$ time [20]. Therefore, the time complexity for query processing becomes $O(m \log^2 n + |Occ_T^{\mathcal{C}(\mathcal{P})}|)$ per query.

5   Conclusion

In this paper, we have studied the Circular Pattern Matching (CPM) Problem and have presented two efficient data structures to index circular patterns. This problem seems to be interesting and is motivated by practical applications. Given a text $\mathcal{T}$ of length n, our first data structure, CPI-I, can be constructed in $O(n \log^{1+\epsilon} n)$ time and space, for any constant $0 < \epsilon < 1$ and the subsequent queries for a circular pattern $\mathcal{C}(\mathcal{P})$, corresponding to a given pattern $\mathcal{P}$, can be answered in (near optimal) $O(m \log \log n + |Occ_T^{\mathcal{C}(\mathcal{P})}|)$ time. Our second data structure, CPI-II, exhibits better time and space complexity in terms of construction but suffers a little in query time. In particular, CPI-II can be constructed in $O(n)$ time requiring $O(n \log n)$ bits of space. The query time of CPI-II is $O(m \log n + |Occ_T^{\mathcal{C}(\mathcal{P})}|)$. Finally, using the compressed suffix array, we can reduce the space usage of CPI-II to linear at the cost of a $\log n$ factor rise in the query time.

References

1. Abouelhoda, M.I., Ohlebusch, E., Kurtz, S.: Optimal exact string matching based on suffix arrays. In: Laender, A.H.F., Oliveira, A.L. (eds.) SPIRE 2002. LNCS, vol. 2476, pp. 31–43. Springer, Heidelberg (2002)
2. Alstrup, S., Brodal, G.S., Rauhe, T.: New data structures for orthogonal range searching. In: FOCS, pp. 198–207 (2000)
3. Amir, A., Aumann, Y., Benson, G., Levy, A., Lipsky, O., Porat, E., Skiena, S., Vishne, U.: Pattern matching with address errors: rearrangement distances. In: SODA, pp. 1221–1229. ACM Press, New York (2006)
4. Amir, A., Aumann, Y., Landau, G.M., Lewenstein, M., Lewenstein, N.: Pattern matching with swaps. J. Algorithms 37(2), 247–266 (2000)
5. Apostolico, A., Crochemore, M.: Optimal canonization of all substrings of a string. Inf. Comput. 95(1), 76–95 (1991)
6. Booth, K.S.: Lexicographically least circular substrings. Inf. Process. Lett. 10(4/5), 240–242 (1980)
7. Crochemore, M., Hancart, C., Lecroq, T.: Algorithmique du texte. Vuibert Informatique (2001)
8. Crochemore, M., Rytter, W.: Jewels of Stringology. World Scientific, Singapore (2002)
9. Duval, J.-P.: Factorizing words over an ordered alphabet. J. Algorithms 4(4), 363–381 (1983)
10. Gusfield, D.: Algorithms on Strings, Trees, and Sequences - Computer Science and Computational Biology. Cambridge University Press, Cambridge (1997)
11. Hon, W.-K., Sadakane, K., Sung, W.-K.: Breaking a time-and-space barrier in constructing full-text indices. In: FOCS, pp. 251–260. IEEE Computer Society Press, Los Alamitos (2003)
12. Huynh, T.N.D., Hon, W.-K., Lam, T.W., Sung, W.-K.: Approximate string matching using compressed suffix arrays. Theor. Comput. Sci. 352(1-3), 240–249 (2006)
13. Kärkkäinen, J., Sanders, P., Burkhardt, S.: Simple linear work suffix array construction. J. ACM 53(6), 918–936 (2006)
14. Kim, D.K., Sim, J.S., Park, H., Park, K.: Constructing suffix arrays in linear time. J. Discrete Algorithms 3(2-4), 126–142 (2005)
15. Knuth, D.E., Morris Jr., J.H., Pratt, V.R.: Fast pattern matching in strings. SIAM J. Comput. 6(2), 323–350 (1977)
16. Ko, P., Aluru, S.: Space efficient linear time construction of suffix arrays. J. Discrete Algorithms 3(2-4), 143–156 (2005)
17. Lothaire, M. (ed.): Applied Combinatorics on Words. In: Encyclopedia of Mathematics and its Applications. Cambridge University Press, Cambridge (2005)
18. Manber, U., Myers, E.W.: Suffix arrays: A new method for on-line string searches. SIAM J. Comput. 22(5), 935–948 (1993)
19. McCreight, E.M.: A space-economical suffix tree construction algorithm. J. ACM 23(2), 262–272 (1976)
20. Sadakane, K., Shibuya, T.: Indexing huge genome sequences for solving various problems. Genome Informatics 12, 175–183 (2001)
21. Shiloach, Y.: Fast canonization of circular strings. J. Algorithms 2(2), 107–121 (1981)
22. Ukkonen, E.: On-line construction of suffix trees. Algorithmica 14(3), 249–260 (1995)
23. Wagner, E.K., Hewlett, M.J.: Basic Virology, 2nd edn. Blackwell Publishing, Malden (2003)

A Fast Algorithm to Calculate Powers of a Boolean Matrix for Diameter Computation of Random Graphs

Md. Abdur Razzaque, Choong Seon Hong, M. Abdullah-Al-Wadud,
and Oksam Chae

Department of Computer Engineering, Kyung Hee University
1 Seocheon-ri, Kiheung-eup, Yongin-si, Gyonggi-do, South Korea, 449-701
m_a_razzaque@yahoo.com, cshong@khu.ac.kr,
awsujon@yahoo.com, oschae@khu.ac.kr

Abstract. In this paper, a fast algorithm is proposed to calculate k^{th} power of an $n \times n$ Boolean matrix that requires $O(kn^3p)$ addition operations, where p is the probability that an entry of the matrix is 1. The algorithm generates a single set of inference rules at the beginning. It then selects entries (specified by the same inference rule) from any matrix A^{k-1} and adds them up for calculating corresponding entries of A^k. No multiplication operation is required. A modification of the proposed algorithm[1] can compute the diameter of any graph and for a massive random graph, it requires only $O(n^2(1-p)E[q])$ operations, where q is the number of attempts required to find the first occurrence of 1 in a column in a linear search. The performance comparisons say that the proposed algorithms outperform the existing ones.

Keywords: Boolean Matrix, Random Graphs, Adjacency Matrix, Graph Diameter, Computational Complexity.

1 Introduction

Boolean matrix and its powers play a major role in mathematical research, electrical engineering, computer programming, networking, biometrics, economics, marketing and communications - the list can go on and on [1, 2, 3]. In these applications, Boolean matrices interpret the relationship among the nodes of a network, genes of different organisms, points of a circuit etc. For drawing and comparison of RNA secondary structure, [1] builds a path matrix by calculating higher powers of the input distance matrix. [2] presents an algorithm to calculate the network capacity in terms of the maximum number of k-hop paths based on the k-hop adjacency matrix of the network. [3] produces an adjacency matrix from a large database containing compiled gene ontology information for the

---

[1] This research was supported by the MIC, Korea, under the ITRC support program supervised by the IITA, Grant no - (IITA-2006 - (C1090-0602-0002)).

S.-i. Nakano and Md. S. Rahman (Eds.): WALCOM 2008, LNCS 4921, pp. 58–69, 2008.

genes of several model organisms. Higher powers of this matrix are then calculated to determine how closely the genes are related on biological processes and molecular functions.

Adjacency matrix is a Boolean matrix that represents the dependency among vertices of a graph. If A is the adjacency matrix of a random graph $G(n,p)$, the entries in its kth power gives the number of walks of length k between each pair of vertices [4]. To find out A^k from A, $\forall k \geq 2$, is one of the most fundamental problems in graph theory. Strassen [5] made the astounding discovery that one can multiply two $n \times n$ matrices recursively in only $O(n^{2.81})$ multiplication operations, compared with $O(n^3)$ for the standard algorithm. The constant factor implied in the big O notation of this algorithm is about 4.695. A sequence of improvements have been done to Strassen's original algorithm [6]. The best one is achieved by Coppersmith and Winograd [7], which requires at most $O(n^{2.376})$ multiplication and addition operations [8]. It also improves on the constant factor in Strassen's algorithm, reducing it to 4.537. These approaches require increasingly sophisticated mathematics and are intimidating reading. Yet, viewed at a high level, the methods all rely on the same framework: for some k, they provide a way to multiply $k \times k$ matrices using $m \lll k^3$ multiplications, and apply the technique recursively to show that the exponent, $\omega < \log_k m$. The main challenge in devising these methods is in the design of the recursion for very large values of k. Moreover, the accuracy of these recursive algorithms are often worse than those of the standard algorithm [9].

Usually, to find out A^k from A, k-1 number of matrix multiplications is required, but the use of doubling trick reduces the number approximately to $\lceil log_2 k \rceil$. For instance, using doubling trick, A^{16} can be computed by 4 matrix multiplications only, $A \to A^2 \to A^4 \to A^8 \to A^{16}$ and A^{40} can be computed by 6 matrix multiplications only, $A \to A^2 \to A^4 \to A^8 \to A^{16} \to A^{32} \to A^{40}(A^{32} \times A^8)$. Hence, Coppersmith and Winograd's algorithm along with doubling trick necessitates almost $O(\lceil \log_2 k \rceil n^{2.376})$ multiplications.

In this paper, we propose a fast algorithm for finding out the higher powers, k, of a Boolean matrix A that runs in $O(kn^3 p)$ addition operations, faster than that of [7] even with doubling trick. We observe that 0 entries in the matrix have no contribution in this computation and therefore we exclude this overhead and develop inference rules using 1 entries. The proposed algorithm selects entries (specified by the inference rule) from the matrix A^{k-1} and adds them up only for updating corresponding entries of A^k. Consequently, our algorithm completely avoids multiplication operations and executes faster. Our second contribution is graph diameter computation algorithm. Let us consider, $G(n,p)$ is a random graph of n vertices in which a pair of vertices appears as an edge with probability p. Diameter of $G(n,p)$ is the longest of the shortest walks in between any two vertices of the graph [10]. In other words, the graph diameter is the maximum number of nodes traversed along an optimal path connecting two arbitrary nodes. Floyd-Warshall's shortest path algorithm [11, 12] gives the shortest paths in between each pair of vertices both for directed and undirected graphs. By finding out the longest one of them, we can solve diameter computation problem.

The time complexity of this algorithm is $O(n^3)$. To the best of our knowledge, there has been no further improvement in computation time required for computing exact diameter of a graph, even though there is rich literature for estimating diameter of random graphs [10, 13, 14, 15, 16]. The proposed graph diameter computation algorithm is much faster, for a massive random graph [14] having diameter d, our algorithm requires $O(n^2(1\text{-}p)E[q])$ operations, where q is the number of attempts required for finding out the first occurrence of 1 in a column.

The rest of the paper is organized as follows. Section 2 introduces the algorithm for computing higher powers of a Boolean matrix and section 3 describes the diameter computation algorithm. Correctness proof and complexity analysis of the algorithms are presented in section 4 and 5 respectively. Section 6 carries out the performance comparisons and the paper is concluded in section 7.

2   Higher Powers of Boolean Matrix

In this section, we describe an algorithm that takes a Boolean matrix A as input and produces the powers of it. The proposed algorithm works as follows. At first, it produces rules (we call it inference rule) for all rows of A. We define an $n \times n$ integer array $Irule$, which stores the column indices having 1 entries in the respective rows of A. The inference rule generated for a particular row is used to update all entries of that row to calculate any matrix A^k from A^{k-1}. Therefore, the procedure **GenerateIrule()** in Algorithm 1 is called only once for a matrix.

Algorithm 1. Inference rule generation

Procedure **GenerateIrule(A, n, $Irule$)**
/* r_m is the number of 1's in r^{th} row */
1. **for** $r := 0$ to $n - 1$ **do** /* for each row, r */
2. $r_m := 0$;
3.   **for** $c := 0$ to $n - 1$ **do** /* for each column, c */
4.   **if** $A(r, c) = 1$ **then**
5. $Irule(r, r_m) = c$;
6. $r_m = r_m + 1$;
7.   **endif**
8.   **end for**
9. **end for**

Once the inference rules for a matrix A is generated, $Irule$ contains the column indices having 1 entries in each row. Now, we calculate each entry of higher power matrix $A^2(r,c)$ from A following (1).

$$A^2(r, c) = \sum_{i=0}^{r_m-1} A(Irule(r, i), c) \tag{1}$$

The interesting fact is that the inference rules, we have already developed, to get A^2 from A, can also be applied to get A^3 from A^2, A^4 from A^3 and so on. Hence, each entry of A^k is updated as follows

$$A^k(r,c) = \sum_{i=0}^{r_m-1} A^{k-1}(Irule(r,i),c), \forall k \geq 2 \qquad (2)$$

The procedure **BoolMatMul()** presented in Algorithm 2 calculates A^k from A^{k-1}.

Algorithm 2. Boolean matrix multiplication using addition operations only

---

Procedure **BoolMatMul(A, n, $Irule$, k)**
/* r_m is the number of 1's in r^{th} row */
1. **for** $r := 0$ to $n-1$ **do** /* for each row, r */
2.   **for** $c := 0$ to $n-1$ **do** /* for each column, c */
3.   **for** $m := 0$ to $r_m - 1$ **do**
4. $x := Irule(r,m)$;
5. $A^k(r,c) = A^k(r,c) + A^{k-1}(x,c)$;
6.   **end for**
7.   **end for**
8. **end for**

At every step, we need to store only the immediate lower powered matrix in memory. The procedure **HigherPower()** in Algorithm 3 shows how it iterates the above steps for each power k of the matrix A to compute A^P, where P is any integer greater than or equal to 2.

Algorithm 3. Calculating P^{th} power of Boolean matrix A

Procedure **HigherPower(A, n, P)**
Irule: $n \times n$ integer array, initialized to 0
1. GenerateIrule(A, n, $Irule$);
2. **for** $k := 2$ to P **do** /* Calculates A^P */
3.   BoolMatMul(A, n, $Irule$, k);
4. **end for**

The mathematical reasoning behind the proposed algorithm is not so difficult to understand. Its correctness proof is given in section 4.

3   Diameter Computation Algorithm

The adjacency matrix A of $G(n,p)$ is a Boolean matrix with n rows and n columns labeled by graph vertices having entries as follows

$$A(i,j) = \begin{cases} 1, \text{ if there is an edge from } v_i \text{ to } v_j \\ 0, \text{ otherwise} \end{cases} \tag{3}$$

For a simple graph, the adjacency matrix must have 0's on the diagonal. In this section, we propose an algorithm that takes such an adjacency matrix A of a random graph and computes its diameter. For simplicity we consider connected graph here.

The entries of the matrix A^k contain the number of walks of length k between corresponding vertices. For pair of vertices having walk length other than k, or having no path connecting them, entries in A^k become zero. Therefore, if cycles are avoided in computing the walk lengths, only shortest paths are calculated here. Eventually, all entries of A^{d+1}, where $k \geq 1$, become zero if diameter of the graph is d. For solving diameter computation problem, we are concerned only with the shortest paths between pair of vertices. Therefore, we define a matrix A_d whose entries are determined using (4).

$$A_d(i,j) = \begin{cases} 1, \text{ if } dist(i,j)=d \\ 0, \text{ otherwise} \end{cases} \tag{4}$$

where $dist(i,j)$ denotes the minimum path length between vertices v_i and v_j. We take an auxiliary Boolean matrix $PathFound_d$ of size $n \times n$, which keeps track of the paths that are already found. $PathFound_1$ is initialized to the matrix A and diagonal entries are set to 1. Accordingly, we modify the procedure **Bool-MatMul(k)** and rename as **FindNewPaths(d)** in Algorithm 4. At each step, it finds a path of length d for each pair of vertices v_i and v_j, unless it is already computed or $i{=}j$ (diagonal entries). Therefore, the entries of $PathFound_d$ are observed as follows

$$PathFound_d(i,j) = \begin{cases} 1, \text{ if } dist(i,j)\leq d \\ 0, \text{ otherwise} \end{cases} \tag{5}$$

Algorithm 4. Searching for existence of paths of length d

Function **FindNewPaths(d)**
Input: $PathFound_{d-1}(r,c), A_{d-1}$
Return type: Boolean
1. $flag = False$;
2. **for** $r := 0$ to $n-1$ **do** /* for each row, r */
3. $Found = False$;
4.   **for** $c := 0$ to $n-1$ **do** /* for each column, c */
5.   **if** $r = c$ **OR** $PathFound_{d-1}(r,c) = 1$ **then**

6.   Skip operations below and continue with next value of c;
7.   **end if**
8.   **for** $m := 0$ to $r_m - 1$ **do** /* each x in inference rule of r */
9. $x := Irule(r, m)$;
10.   **if** $A_{d-1}(x, c) = 1$ **then**
11. $A_d(r, c) := 1$;
12. $PathFound_d(r, c) := 1$;
13. $flag := True$;
14. $Found := True$;
15.   Break loop for m;
16.   **end if**
17.   **end for**
18.   **if** $Found = True$ **then**
19.   Break loop for c;
20.   **end if**
21.   **end for**
22. **end for**
23. return $flag$;

Eventually, for a connected graph, all entries of matrix $PathFound_d$ will become 1 for a certain value of d and **FindNewPaths($d+1$)** returns false. For a disconnected graph $G(n, p)$, we have used the convention that the diameter of G is the maximum diameter of its connected components. In that case, some 0 entries will still be remaining when **FindNewPaths($d+1$)** returns false. The function **ComputeDiameter()** in Algorithm 5 repeatedly calls **FindNewPaths(k)** until it returns false, which means the diameter is k-1.

Algorithm 5. Computation of diameter of a graph, G

Function **ComputeDiameter()**
Input: Adjacency matrix A of G
Return value: Diameter of G as integer
1. GenerateIrule(A, n, $Irule$);
2. $k := 1$;
3. **repeat**
4. $k := k + 1$;
5. **untill** FindNewPaths(k) = $false$;
6. **return** $k - 1$;

---

4   Proof of Correctness

Theorem 1. *For an* $n \times n$ *Boolean matrix,* A, *inference rules correctly pick up necessary and sufficient entries to calculate* A^k *from* A^{k-1}, $\forall k \geq 2$.

Proof: Traditional multiplication algorithm uses (6) to calculate A^k

$$A^k(r,c) = \sum_{x=0}^{x=n-1} A(r,x) \times A^{k-1}(x,c) \tag{6}$$

Here for the entries $A(r,c)=0$, the multiplication as well as the addition operations cannot contribute anything to the sum. Hence our inference rules take only 1 entries of the respective rows. Again since $A(r,x) = 1, \forall x \in Irule(r)$, each multiplication simply yields $A^{k-1}(x,c)$. Therefore, avoiding multiplication operations, the (6) correctly reduces to

$$A^k(r,c) = \sum_{\forall x \in Irule(r)} A^{k-1}(x,c) \tag{7}$$

Lemma 1. *Generating a single set of inference rules is sufficient for calculating any A^k, where $k \geq 2$.*

Proof: Theorem 1 follows that inference rules produced from A picks up the entries from the other matrix, with which A is being multiplied. Any A^k can be generated by adding entries of A^{k-1}. Hence, the set of inference rules created from A can serve all necessary multiplications.

Theorem 2. *FindNewPaths(d)* updates the entries so that $A_d(i,j)=1$ denotes that the length of the shortest path in between vertices v_i and v_j is d.

Proof: We prove it by induction. $A(i,j)=1$ certainly denotes that the shortest path length between i and j is 1. Now, let us suppose that $A_{d-1}(i,j)$ denotes that the shortest path length between i and j is d-1. The inner loop of **FindNewPaths(d)**, statement 9, looks for an x where $A_{d-1}(x,c)=1$. $Irule(r)$ guarantees that there is a path of length 1 from r to x. Hence, there exists a path of length d from r to c via x. If there exists a path between i and j whose length is h, where $h < d$, it would have been found earlier when **FindNewPaths(h)** was called and the matrix *PathFound* would have kept track of it.

5   Complexity Analysis

Let n be the number of rows or columns of the matrix and p be the probability that an arbitrary entry of the matrix is 1. Then the number of operations required for finding out the inference rules for n rows of the matrix is n^2. If X represents the number of 1's that occur in one row (or column), then X is said to be a binomial random variable with parameters (n, p) [17]. Therefore, the expected number of 1's in a row of the matrix A is given by

$$m = E[X] = \sum_{i=0}^{n} i \binom{n}{i} p^i (1-p)^{n-i} = np \tag{8}$$

5.1 Complexity for Calculating Higher Powers of Boolean Matrix

The proposed algorithm requires m addition operations, given by (8), for calculating a single entry of the next higher-powered matrix A^2. Consequently, the complexity for computing all entries of A^2 is given by $O(n^2 + n \times n \times m)= O(n^2(1+m)) = O(n^2 m)$ (constant part is omitted). By replacing m by np, we get the complexity $O(n^3 p)$ and hence for getting A^k from A requires $O(kn^3 p)$ addition operations. No multiplication is required here.

5.2 Complexity for Diameter Computation

In this case, we update only the 0 entries of the target matrix. For each of the $(n\text{-}m)$ 0 entries of a row, the function **FindNewPaths(d)** searches for the first occurrence of 1 in m positions. The probabilistic distribution of getting first success (occurrence of 1) in repeated finite number of trials (m) is truncated geometric random [18]. Let q represent the number of attempts required for getting the first 1 in a column. Using [18], we get the conditional probability mass function (pmf) of q, with the condition that the procedure gets a 1, as

$$P(q) = \frac{(1-p)^{q-1}p}{1-(1-p)^m} \tag{9}$$

Replacing the value of m from (8) and using geometric series equations [19], we find the expected number of attempts required, $E[q]$, as in (10).

$$E[q] = \frac{1-(1-p)^{np}(1+np^2)}{p(1-(1-p)^{np})} \tag{10}$$

Consequently, the number of search operations required for updating a single row is $(n\text{-}m)E[q]$, which follows that getting A_2 involves $n(n\text{-}m)E[q]$ computations. The conditional probability that a single 0 entry of the matrix is updated is given by [18] as follows

$$C_p = \sum_{k=1}^{m} P(q) = \frac{1-(1-p)^{np-1}}{1-(1-p)^{np}} \tag{11}$$

As our diameter computation algorithm finds new paths only, after each step of calculating A_d from A_{d-1}, the number of 0's need to update will be reduced. After the first step, i.e., when A_2 is calculated from A, in each row $(n\text{-}m)C_p$ number of 0's is updated and the estimated number of 0's remaining is $(n - m) - (n - m)C_p = (n - m)(1 - C_p)$. Similarly, after the second step, i.e., A_3 is calculated from A_2, $n - m)(1 - C_p)C_p$ number of 0's of each row is updated and the number of 0's left is given by $(n - m)(1 - C_p) - (n - m)(1 - C_p)C_p = (n - m)(1 - C_p)^2$.

Therefore, for a graph with diameter d, the expected number of search operations required by our diameter computation algorithm is

$$n^2 + n(n - m)E[q](1 + (1 - C_p) + (1 - C_p)^2 + ... + (1 - C_p)^d) \tag{12}$$

$$= n^2 + n(n-m)E[q]\frac{1-(1-C_p)^d}{1-(1-C_p)}, \ using[18] \tag{13}$$

$$= n^2(1-p)E[q]\acute{C}_p, \ where, \acute{C}_p = \frac{1-(1-C_p)^d}{C_p} \tag{14}$$

(14) gives the complexity for diameter computation for any random graph. For massive random graphs, we have seen that the value of $\acute{C}_p$ is computed as even less than $\sqrt{d}$, which is insignificant as compared to n^2. Hence, we can write the complexity of our graph diameter computation algorithm as $O(n^2(1-p)E[q])$.

This complexity is valid for any $p < 1$. For $p = 1$, we need to have a different look. In such a case the matrix will contain 1 entries only. Hence the loop at step 8 of Algorithm 4 will always terminate at the very first iteration, and $FindNewPath(k)$ will be called only once and demonstrates a complexity of $O(n)$. Here, the overall complexity of the proposed diameter computation algorithm will be $O(n^2)$.

The space/time trade-off of the algorithm is stated as follows. The proposed algorithm needs an $n \times n$ array to store the inference rules. However, it may be cut down by using n lists. Then it will need n^2 values to be stored in the worst case. However, this extreme case will occur when $p = 1$, which gives a time complexity of only $O(n^2)$, i.e., a great reduction in computation time complexity is gained with the compensation of space overhead.

6   Performance Comparisons

For performance comparison, we have carried out experiments on Boolean matrices with various densities of 0's and 1's. The experiments are carried out in a Pentium IV PC (2.8GHz, 1GB RAM). The values plotted in all the graphs are mean of 5 individual experiments.

Fig. 1(a) shows the ratio of computation time required by Coppersmith and Winograd's algorithm [7] (T_c) to that of our proposed algorithm (T_p) for computing A^2 from A. As the number of rows (or columns), n, of a matrix increases, the value of $R_t(= \frac{T_c}{T_p})$ slowly increases for all values of p. R_t decreases as p increases. This is because, in our proposed algorithm, the number of addition operations required for updating a single entry of a row is directly proportional to the number of 1's in that row. However, Fig. 1(a) also depicts that even for matrices with all 1 entries ($p = 1.0$), is greater than 1.2, which shows the superiority of the proposed method.

In Fig. 1(b), the computation time ratio R_t is shown for calculating higher powers of A. R_t decreases with higher values of k. This is because of using doubling trick with [7], which reduces the computation time significantly for large values of k. However, even for $k=100$, R_t is found to be at least 2. Both Fig. 1(b) and Fig. 1(c) show that R_t decreases as p increases, the cause is explained above. Fig. 1(c) also points out that n has very less effect on R_t.

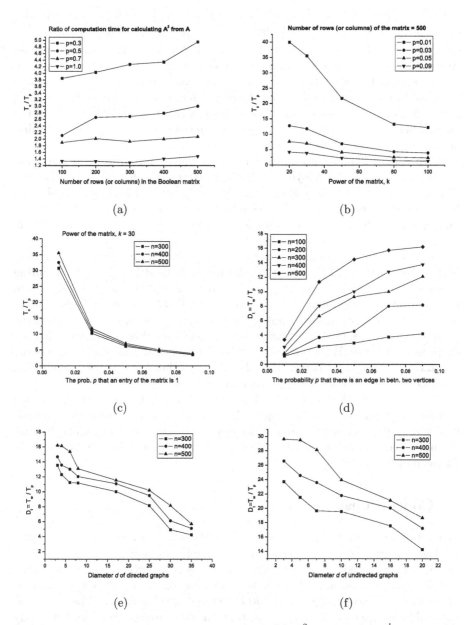

Fig. 1. Ratio of computation time for calculating (a) A^2 from A, (b) A^k from A, (c) A^{30} from A, (d) diameter of directed random graphs (p varies), (e) diameter of directed random graphs (d varies) and (f) diameter of undirected random graphs

The performance of the proposed algorithm may degrade in comparison to [7] for further large values of p and k. However, in natural applications we may seldom require to compute very high power of A. For example, let A represent a random graph on n vertices, where $n=500$ and $p=0.5$. The diameter of this graph is approximately $\frac{\log(n)}{\log(np)}$ [20], which yields only 2. Hence, finding A^2 is sufficient to find relationships in between any pair of vertices of that graph. Again, if the diameter of a graph with 500 vertices is 100, then p should be set to approximately 0.002. This causes each vertex to have degree 1, which is not practical in most applications. Hence we may make comment that finding very high power of matrices representing graphs is not necessary in most of the cases. Therefore, we claim that our proposed method is more applicable to find practically useful powers of matrices.

For practical implementation of our graph diameter computation algorithm and its comparison with Floyd-Warshall's shortest path algorithm [11] that can compute the diameter of a graph exactly, we have randomly placed edges in between vertices ensuring at least one edge to each vertex of the graph. We have taken inputs such that there is no self loop in the graph i.e., the adjacency matrix contains 0 entries in its main diagonal.

Fig. 1(d) and Fig. 1(e) show the ratio of computation time required by Floyd-Warshall's algorithm (T_w) to that of our proposed algorithm (T_p) for computing diameter of directed random graphs. Fig. 4 depicts that the proposed algorithm provides better performance than Floyd-Warshal's algorithm for higher values of p. This is because our algorithm updates 0 entries only and as p increases, the number of 0 entries in the adjacency matrix decreases. Fig. 1(e) shows that D_t drops as d increases. However, it shows that for practical values of diameters the proposed method is still better. It is also observed that the ratio of computation time, $D_t = T_w/T_p$, is higher for larger values of n.

Fig. 1(f) shows that our algorithm provides much better performance for undirected graphs. For all the graphs, the computation time ratio, D_t, is almost double of that for directed graphs as shown in Fig. 1(e). This is because, in case of undirected graphs, we need to deal with only either half part of the main diagonal of the matrix.

7   Conclusions

To the best of the author's knowledge, this is the ever first algorithm for computing higher powers of a matrix that does not require any multiplication operation at all. Our proposed graph diameter computation algorithm neither uses BFS nor dominating sets, needs it only to search the existence of new paths. Our algorithms are easy to implement and have the desired property of being combinatorial in nature and the hidden constants in the running time bound are fairly small.

The applicability of our algorithm, or similar approach, to the problem of diameter computation of weighted graphs, transitive closure and all-pair of shortest paths needs further investigation and analysis. We conjecture that our approach, or simple modification of it, can help solve these problems within $O(n^2(1-p)E[q])$ operations. We leave this as our future work.

References

1. Auber, D., Delest, M., Domenger, J., Dulucq, S.: Efficient drawing of RNA secondary structure. Journal of Graph Algorithms and Applications 10(2) (2006)
2. Li, N., Guo, Y., Zheng, S., Tian, C., Zheng, J.: A matrix based fast calculation algorithm for estimating network capacity of MANETS. In: Proceedings of the 2005 Systems Communications (ICW), IEEE Computer Society, Los Alamitos (2005)
3. Zhou, M., Cui, Y.: Constructing and visualizing gene relation networks, An International Journal on Computational Molecular Biology (2004) ISBN In Silico Biology, 4, 0026
4. Lipschutz, S.: Data Structures and Algorithm. In: Schaum's outline series, McGraw Hill, New York (2000)
5. Strassen, V.: Gaussian elimination is not optimal. Numerical Mathematics 13, 354–356 (1969)
6. Burgisser, P., Clausen, M., Shokrollahi, M.A.: Algebric complexity theory, Grundlehren der Mathematischen Wissenschaften, vol. 315. Springer, Heidelberg (1997)
7. Coppersmith, D., Winograd, S.: Matrix multiplication via arithmetic progressions. J. of Symbolic Computation 9, 251–280 (1990)
8. Bilardi, G., D'Alberto, P., Nicolau, A.: Fractal matrix multiplication: A case study on probability of cache performance. In: Brodal, G.S., Frigioni, D., Marchetti-Spaccamela, A. (eds.) WAE 2001. LNCS, vol. 2141, pp. 26–38. Springer, Heidelberg (2001)
9. Robinson, S.: Toward an optimal algorithm for matrix multiplication. SIAM news 38(9) (2005)
10. Bollobas, B.: The diameter of random graphs. IEEE Trans. Inform. Theory 36(2), 285–288 (1990)
11. Floyd, Robert, W.: Algorithm 97: shortest path. Communications of the ACM 5(6), 345 (1962)
12. Cormen, T.H., Leiserson, C.E., Ronald, L.R.: Introduction to Algorithms, 2nd edn. The MIT press and McGraw-Hill book company, Cambridge (2001)
13. Bollobas, B.: The Evolution of Sparse Graphs. In: Graph Theory and Combinatorics, pp. 35–57. Academic Press, London-New York (1984)
14. Aiello, W., Chung, F., Lu, L.: Random Evolution of Massive Graphs. In: Handbook of Massive Data Sets, pp. 97–122. Kluwer, Dordrecht (2002)
15. Lu, L.: The diameter of random massive graphs. In: Proceedings of the Twelfth Annual ACM-SIAM Symposium on Discrete Algorithms, pp.912–921. ACM/SIAM Press (2000)
16. Marinari, E., Semerjian, G.: On the number of circuits in random graphs, J. of Statistical Mechanics: Theory and Experiment, P06019 (2006) DOI:10.1088/1742 5468/2006/06/P06019
17. Ross, S.M.: Introduction to Probability Models, 8th edn. Academic Press, London (2002)
18. Bertsekas, D., Gallager, R.: Data Netwroks, ex. 2.13, 2nd edn. Prentice Hall, New Jersey (1992)
19. Yates, R.D., Goodman, D.J.: Probability and Stochastic Processes, B.6 (2005)
20. Chung, F., Lu, L.: The diameter of sparse random graph. J. of Advances in Applied Mathematics 26(4), 257–279 (2001)

Cover Ratio of Absolute Neighbor
(Towards an Index Structure for Efficient Retrieval)

Kensuke Onishi[1] and Mamoru Hoshi[2]

[1] Department of Mathematical Sciences, Tokai University, 1117, Kitakaname,
Hiratsuka, Kanagawa, 259-1292, Japan
KensukeOnishi@acm.org
[2] Graduate School of Information Systems, The University of
Electro-Communications, 1-5-1, Chofugaoka, Chofu, Tokyo, 182-8585, Japan
hoshi@is.uec.ac.jp

Abstract. Voronoi diagrams for a fix set of generators are considered
with varying L_p norm. For a generator q in the set, the *absolute neighbor*
of q is defined to be the intersection of all Voronoi regions of q by L_p
norm($p = 1, 2, \ldots, \infty$). Since the shape of Voronoi region is dependent
on the norm used, the collection of absolute neighbors for the set does
not always cover the whole space.

In this paper, we construct absolute neighbors and computed the ratio, called *cover ratio*, of the volume covered by all absolute neighbors
to that of the whole space for some sets of generators by computational
experiments. Computational experiments show that the cover ratio is
higher when a configuration of grid points is used as a set of generators
than when a set of random generators is used. Moreover, we theoretically show that the absolute neighbors for square configuration and for
face-centered configuration cover the whole space. We also discuss an
application of absolute neighbors to constructing an index structure of
the whole space for efficient retrieval.

1 Introduction

Voronoi diagrams have been used in many fields, for example, biology, physics,
vision, archaeology and so on. Theoretical aspects of Voronoi diagrams have
been studied in computational geometry from construction to properties. Many
results for Voronoi diagram are found in [6].

An application of Voronoi diagrams is the construction of an index structure
for databases, when data space is a vector (or at least metric) space([1,3]). In the
application, we select a set Q of generators from the vector space, and compute
the Voronoi diagram for Q with respect to a norm. The Voronoi diagram is
regarded as a partition of the index structure. It is ensured that all points in a
Voronoi region are near to the generator of the region by the norm.

Ordinary index structures support retrieval using *a single norm*. For efficient
ε-similarity search, the partition of the space is constructed through the Voronoi
diagram by the norm used. Recently, new index structures which support retrieval using *various norms* were proposed (Yi *et al.*[9], QIC-m-tree[4], Kimura

S.-i. Nakano and Md. S. Rahman (Eds.): WALCOM 2008, LNCS 4921, pp. 70–80, 2008.

et al.[5] and mm-GNAT[8]). A retrieval using various norms is realized by a scaling of ε in [9,4,5]. On the other hand, we realized such a retrieval by an extension of the existence range of GNAT([2]), called *mm-range*, in [8]. The mm-range is an interval whose lower and upper bounds are distances between a generator and the points in a region measured by L_∞ norm and by L_1 norm, respectively. The smaller the mm-range is, the more efficient the retrieval is. The mm-range depends on the partition by a Voronoi diagram, i.e., on the set of generators and the norm used.

In [8], we experimented three types of partition on mm-GNAT : L_1 norm based, L_2 norm (Euclid distance) based and L_∞ norm based. The results showed that L_2 norm based partition is most efficient among these partitions. The efficiency (the number of distance calculations to get all correct answers) on the mm-GNAT was about 4.0 times as good as on the ordinary index structure (GNAT, [2]) for 4 dimensional artificial data and about 21.0 times for 20 dimensional music data. We investigate efficient partition (Voronoi diagram) such that

- most points in a region are distributed around the generator of the region, and
- distances between a point in a region and the generator of the region change little by various norms.

For constructing such a partition we focus attention on *absolute neighbor* [7]. The absolute neighbor of a generator q is defined by the intersection of $\text{Vor}_p(q; Q)$'s $(p = 1, 2, \ldots)$, where $\text{Vor}_p(q; Q)$ is Voronoi region of a generator q for Q by L_p norm. The absolute neighbor satisfies the first condition of partition, however there is a problem that absolute neighbors do not necessarily cover the whole space. Since the absolute neighbor of q is the intersection, the absolute neighbor is an intersection of each $\text{Vor}_p(q; Q)$. The Voronoi diagrams are different, depending on the L_p norm used. Thus absolute neighbors for all generators of the set Q do not always cover the whole space.

We consider the *cover ratio* of the volume covered by all absolute neighbors to the volume of the whole space. We calculated the cover ratios for sets of random generators and for configurations of grid points: *triangle configuration, square configuration, face-centered configuration*. Computational experiments show that the cover ratios for these configurations are higher than that for a set of random generators. For square configuration and face-centered configuration, the ratios were almost 1.0. We theoretically show that the collections of the absolute neighbors for square configuration and for face-centered configuration cover the whole space.

The content of this paper is as follows. In Section 2, we define the absolute neighbor and explain one of its properties. In Section 3, we describe experimental results of the cover ratio when random points are used as generators. In Section 4, we consider the absolute neighbors for the three grid configurations above. In Section 5, we explain an application of the absolute neighbor as an index structure for databases.

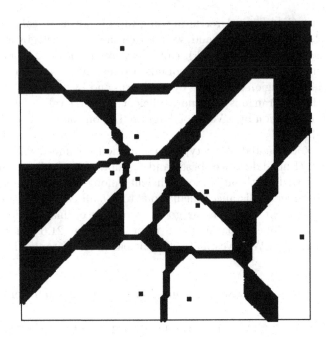

Fig. 1. Absolute neighbors for 10 generators ■

2   Absolute Neighbor

Definition 1 (Absolute Neighbor, [7]). *Let $Q = \{q_1, q_2, \ldots, q_n\}$ be a set of n points, called set of generators, and each q_i is called generator. Absolute neighbor $\mathrm{AN}(q_i; Q)$ of q_i for Q is defined as follows:*

$$\mathrm{AN}(q_i; Q) := \bigcap_{p=1}^{\infty} \mathrm{Vor}_p(q_i; Q),$$

where $\mathrm{Vor}_p(q_i; Q)$ is the Voronoi region of q_i for Q by L_p norm.

Figure 1 shows absolute neighbors for 10 generators in 2 dimensional space. Each of the white regions is the absolute neighbor of a generator (■) in the region. For any L_p norm, every point in an absolute neighbor is nearer to the generator in the absolute neighbor than to all other generators. Colored regions are the sets of points for which the nearest generators are different by L_p norm used.

To construct absolute neighbor from the definition, we need to compute the intersection of infinite sets. This computation is impossible by computer.

Onishi [7] showed the following theorem, which allows us to construct absolute neighbor.

Theorem 1. *Let $Q = \{q_1, q_2, \ldots, q_n\}$ be a set of n generators. Absolute neighbor $\mathrm{AN}(q_i; Q)$ is given by*

$$\mathrm{AN}(q_i; Q) = \mathrm{Vor}_1(q_i; Q) \cap \mathrm{Vor}_\infty(q_i; Q).$$

Table 1. Cover ratio (average)

dim	max	spn: the number of generators					
		10	50	100	500	1000	5000
2	10^2	0.8109	0.7923	0.7883	0.7721	0.7558	0.7355
	10^3	0.7999	0.7807	0.7956	0.7889	0.7863	0.7839
	10^4	0.8230	0.7938	0.7990	0.7860	0.7862	
4	10^2	0.6465	0.6063	0.6025	0.5824	0.5791	
8	10^1	0.4401	0.3778	0.3604	0.3164	0.2962	

3 Computational Experiments

In this section, we describe computational experiments and the results. For measuring the cover ratio, we compute the volume of absolute neighbors and that of the whole space. In general, it is difficult to compute exactly the volume of a region, we approximate the volume by discretization.

Fix the dimension dim, the number of generators spn and the upper bound $max(> 1)$ of each coordinate of points (i.e., each coordinate of point is included in $[1, max]$). We select spn generators from hypercube $[1, max]^{dim}$, which is regarded as a set Q of generators. Suppose a point x in the hypercube. Compute the generators nearest to x by L_1 norm and by L_∞ norm. If these nearest generators are the same, then x is included in the absolute neighbor of the generator. By applying this procedure to all grid points in the hypercube, we decide whether each point is contained in absolute neighbor or not.

For quantitative measurement, we define the *cover ratio* as

$$\frac{\sum_i (\# \text{ points in AN}(q_i; Q))}{\# \text{ grid points in the hypercube}}.$$

In the experiments, we selected 10 generator sets and computed the cover ratio for each set, then calculated the average of the cover ratios. The results are shown in Table 1 and in Figure 2 as semi-logarithmic graph whose vertical axis is the cover ratio and horizontal axis is the ratio of the number of generators to the number of all grid points in the hypercube.

For $dim = 2$, the cover ratios were about 0.8. And, ratios change little by spn and max. For $dim = 4$ and $dim = 8$, the cover ratios were about 0.6 and 0.4, respectively. As the dimension becomes higher, the cover ratio decreases rapidly. This is a kind of *curse of dimension*[1,3].

4 Configuration of Points

In this section, we consider generator sets for which cover ratios are high.

We consider absolute neighbors for the following three types of grid configurations of points:

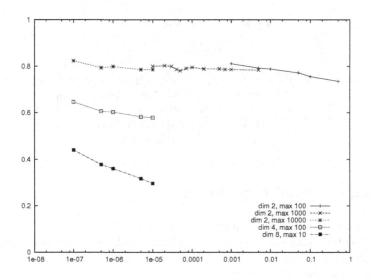

Fig. 2. Cover ratio vs ratio of the number of generators to the number of grid points in the hypercube

- triangle configuration;
- square configuration;
- face-centered configuration.

Triangle Configuration

In a triangle configuration, its grid points form regular triangles. Figure 3 is an example of absolute neighbors for a triangle configuration ($spn = 22$, $max = 100$) in 2 dimensional space. The cover ratio for this case was 0.8852.

Square Configuration

Square configuration is a configuration of grid points whose points form squares and is mathematically defined as follows:

$$\{(\overbrace{2\mathbb{Z}a, 2\mathbb{Z}a, \ldots, 2\mathbb{Z}a}^{d}) \mid a > 0, \mathbb{Z} \text{ is the set of all integers}\}$$

in d dimensional space. Figure 4 is an example of absolute neighbors for a square configuration with $spn = 16$ in 2 dimensional space. In our computational experiment, the cover ratio for $spn = 16$, $max = 100$ was 0.9994. There is a case such that its cover ratio was 1.0000 for $spn = 16$, $max = 100$ in 4 dimensional space. Since the average of the cover ratio was 0.9996 among our computational experiments, the absolute neighbors mostly covered the whole space.

Thereafter, we theoretically show that the whole space is covered by absolute neighbors for square configuration.

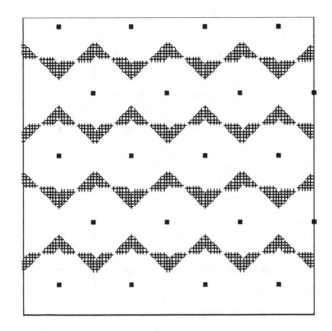

Fig. 3. Absolute neighbors for a triangle configuration ($spn = 22$, $max = 100$, $dim = 2$), cover ratio was 0.8852

Lemma 1. *For a square configuration the Voronoi region of a generator by L_1 norm is the same as that by L_∞ norm.*

Proof. It is sufficient to show that the perpendicular bisector between generators in square configuration by L_1 norm is the same as that by L_∞ norm.

Consider a square configuration in d dimensional space. Without loss of generality, we only show that the perpendicular bisector between two neighbor grid points for L_1 norm are the same as that for L_∞ norm.

[Case 1: L_1 norm] Suppose 2 points

$$o = (0, 0, \ldots, 0), q = (0, \ldots, 0, \overset{j}{2a}, 0, \ldots, 0).$$

The perpendicular bisector is expressed by

$$|x_i - 0| = |x_i - 2a|.$$

This equation means

$$x_i = a \quad (0 < x_i < 2a).$$

[Case 2: L_∞ norm] We show that a point $x = (x_1, x_2, \ldots, x_d)$ ($\forall i$, $0 < x_i < a$) is nearer to o than q by L_∞ norm.

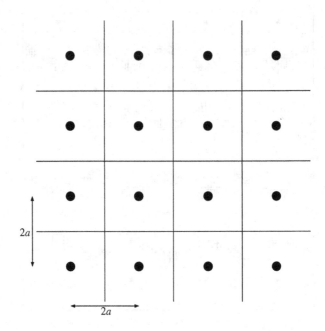

Fig. 4. Absolute neighbors for a square configuration ($spn = 16$, $dim = 2$)

Compute L_∞ norm dist_∞ for these points:

$$\text{dist}_\infty(\boldsymbol{o}, \boldsymbol{x}) = \max_{i=1,\ldots,d} \{|x_i|\} < a,$$

$$\text{dist}_\infty(\boldsymbol{q}, \boldsymbol{x}) = \max_{i(\neq j)} \{|x_i|, |2a - x_j|\}$$

$$= |2a - x_j| > a,$$

where j is the index whose coordinate value is $2a$ in $\boldsymbol{q}$. Since $0 < x_i < a$, the first equation and $|2a - x_j| > a$ are shown. Moreover, $\text{dist}_\infty(\boldsymbol{o}, \boldsymbol{x})(< a)$ is smaller than $\text{dist}_\infty(\boldsymbol{q}, \boldsymbol{x})(> a)$. Therefore $\boldsymbol{x}$ is nearer to $\boldsymbol{o}$ than $\boldsymbol{q}$. So, a region $\{\boldsymbol{x} \mid 0 < x_i < a, i = 1, \ldots, d\}$ is a part of the Voronoi region of $\boldsymbol{o}$.

If $x_j = a$, then

$$\text{dist}_\infty(\boldsymbol{o}, \boldsymbol{x}) = \text{dist}_\infty(\boldsymbol{q}, \boldsymbol{x}) = a$$

and $\boldsymbol{x}$ is on the boundary of the Voronoi regions.

From symmetry, if there exists coordinate such that $-a < x_i < 0$, we can similarly show that

$$\{\boldsymbol{x} \mid |x_i| < a, \ i = 1, \ldots, d\}$$

is included in the Voronoi region of $\boldsymbol{o}$ and its boundary is expressed as $|x_i| = a$.
□

The following theorem is shown from Theorem 1 and Lemma 1.

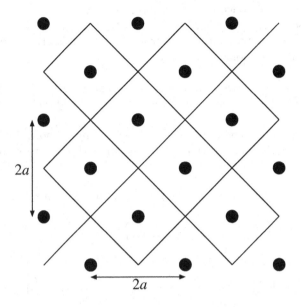

Fig. 5. Absolute neighbors for a face-centered configuration($spn = 18$, $dim = 2$)

Theorem 2. *Consider absolute neighbors for a square configuration. The collection of all absolute neighbors covers the whole space and becomes a partition of the space.*

Proof. From Lemma 1, Voronoi regions by L_1 norm and by L_∞ norm for square configuration are the same. From Theorem 1, absolute neighbor is the intersection of these two Voronoi regions. Thus, the absolute neighbor of a generator is the same to the Voronoi region of the generator by L_1 norm and L_∞ norm. Since the collection of all absolute neighbors becomes Voronoi diagram by L_1 norm and L_∞ norm, the collection of absolute neighbors covers the space.   □

Face-Centered Configuration

Face-centered configuration is defined as follows:

$$\left\{ (n_1, n_2, \ldots, n_d) \ \middle| \ \sum_i n_i = (2\mathbb{Z} + 1)a, n_i = \mathbb{Z}a \right\}.$$

Figure 5 is an example of absolute neighbors for face-centered configuration with $spn = 18$ in 2 dimensional space. In our experiment, the cover ratio for $spn = 18$, $max = 100$ was 1.0000. The cover ratios for a face-centered configurations in 2 dimensional space were almost 1.0. The cover ratios were 0.9991 and 0.9981 for face-centered configuration ($spn = 32$, $max = 100$) in 3 dimensional space and for that ($spn = 40$, $max = 100$) in 4 dimensional space, respectively.

We show that the collection of all absolute neighbors for face-centered configuration also covers the whole space. It should be noted that face-centered configuration cannot be transformed from square configuration by rotation in the 3 or higher dimensional space.

Theorem 3. *The collection of all absolute neighbors for face-centered configuration covers the whole space and becomes a partition of the space.*

Proof. It is sufficient that perpendicular bisectors by L_1 norm for face-centered configuration are the same as that by L_∞ norm. If the bisectors are the same in the region

$$\left\{ (x_1, x_2, \ldots, x_d) \,\middle|\, \sum_i x_i < a, \; 0 < x_i < a \right\}, \tag{1}$$

the bisectors are also the same in the region

$$\left\{ (x_1, x_2, \ldots, x_d) \,\middle|\, \sum_i |x_i| < a, \; 0 < |x_i| < a \right\} \tag{2}$$

from symmetry. Since copies of the region (2) are a tiling of the whole space, the perpendicular bisectors are the same in the whole space.

Hereafter, we show that perpendicular bisectors by L_1 norm for face-centered configuration are the same as that by L_∞ norm in the region (1). It is sufficient to show that perpendicular bisectors between two points

$$q_1 = (a, 0, \ldots, 0), q_j = (0, \ldots, 0, \overset{j}{a}, 0, \ldots, 0)$$

by L_1 norm is the same as that by L_∞ norm. Let j be the index whose coordinate value is a in the point q_j. We compute the perpendicular bisectors in each norm.
[Case 1: L_1 norm]

$$|a - x_1| + |x_2| + \cdots + |x_d| = |x_1| + \cdots + |a - x_j| + \cdots + |x_d|$$
$$(a - x_1) + x_j = x_1 + (a - x_j)$$
$$x_1 = x_j.$$

We have the second equation from the first one by the constraint $0 < x_i < a$.
[Case 2: L_∞ norm]

$$\max\{|a - x_1|, |x_2|, \ldots, |x_d|\} = \max\{|x_1|, \ldots, |a - x_j|, \ldots, |x_d|\}$$
$$(a - x_1) = (a - x_j)$$
$$x_1 = x_j.$$

We have the second equation from the first one by the constraint $\sum_i x_i < a$.
Thus, the perpendicular bisectors are the same.   □

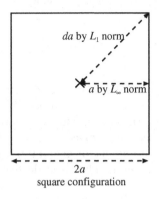

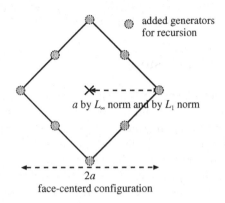

Fig. 6. distances between a generator q_i and a furthest point in $\mathrm{AN}(q_i; Q)$

5   Application

In this section we explain an application of absolute neighbors.

There is an important query to database called ε-*similarity search*: given a query point, report all data point in ε-ball whose center is the query point. Consider the ε-similarity search. Since *all* data point in the ε-ball should be reported in the search, it is better that the collection of the regions covers the whole space. Since the cover ratios for square configuration and for face-centered configuration are 1, we can use these configurations as a set of generators for building an index structure for ε-similarity search.

There are some differences among these configurations. In square configuration, the distance between a generator and a furthest point in the absolute neighbor of the generator is da for L_1 norm and a for L_∞ norm, respectively, where d is the dimension of the space. On the other hand, in face-centered configuration, the distance between a generator and a furthest point in the absolute neighbor is a for both L_1 norm and L_∞ norm (see Figure 6).

We proposed mm-GNAT [8] which is an index structure for fast ε-similarity search by *arbitrary* L_p norm. mm-GNAT realizes the ε-similarity search by arbitrary L_p norm at the cost of regions to be searched being expanded. For reducing the size of the expansion, it is important that distances between two points do not largely change by L_p norm used (second condition of partition, Section 1). Thus, the collection of all absolute neighbors of face-centered configuration is efficient for mm-GNAT.

Here, we consider how to adjust the square or face-centered configurations to underlying data. Compute absolute neighbors for these configurations. If an absolute neighbor contains data points more than a threshold, we recursively consider the square or face-centered configurations in the absolute neighbor. For example, add generators at center of all facet and all vertex of the absolute neighbor in face-centered configuration, then the set of the added generators and the generator of the absolute neighbor become face-centered configuration in the absolute region (see Figure 6).

6 Conclusion

In this paper, we investigated the cover ratios of absolute neighbors for sets of random points and for configurations of grid points as a set of generators. When a set of random points is used as a set of generators, the cover ratios were about 0.8 and 0.6 in 2 dimensional space and in 4 dimensional space, respectively. These cover ratios change little with respect to the number of generators *spn* and the upper bound of coordinate value of points *max*. The ratio decreases rapidly when the dimension becomes higher.

We also investigated configurations of grid points. For square configuration and face-centered configuration the cover ratios are 1, therefore these configurations are best from the viewpoint of cover ratio. We shown that the collection of all absolute neighbors of these configurations is a partition of the space of *any* dimension. Therefore, the collection of all absolute neighbors based on these configurations can be used effectively as a partition (index structure) of high dimensional space.

References

1. Böhm, C., Berchtold, S., Keim, A.D.: Searching in High-Dimensional Spaces: Index Structures for Improving the Performance of Multimedia Databases. ACM Computing Surveys 33(3), 322–373 (2001)
2. Brin, S.: Near Neighbor Search in Large Metric Spaces. In: Proc. of the 21st International Conference on VLDB, pp. 574–584 (1995)
3. Chávez, E., Navarro, G., Baeza-Yates, R., Marroquin, J.L.: Searching in Metric Spaces. ACM Computing Surveys 33(3), 273–321 (2001)
4. Ciaccia, P., Patella, P.: Searching in Metric Spaces with User-Defined and Approximate Distances. ACM Transactions on Database Systems 27(4), 398–437 (2002)
5. Kimura, A., Onishi, K., Kobayakawa, M., Hoshi, M., Ohmori, T.: Distance Conversion Rule for Arbitrary L_p distance. IPSJ Transactions on Databases 46, 93–105 (2005)
6. Okabe, A., Boots, B., Sugihara, K., Chiu, S.N.: Spatial Tessellations: Concepts and Applications of Voronoi Diagrams, 2nd edn. John Wiley & Sons Ltd, Chichester (2000)
7. Onishi, K.: Intersection of Voronoi Regions by L_p distance. In: Proc. of Japan Conference on Discrete and Computational Geometry 1999, pp. 26–28 (1999)
8. Onishi, K., Kobayakawa, M., Hoshi, M.: mm-GNAT: index structure for arbitrary L_p norm. In: MDDM 2007. Proc. of The Second IEEE International Workshop on Multimedia Databases and Data Management, pp. 117–126 (2007)
9. Yi, B.-K., Faloutsos, C.: Fast Time Sequence Indexing for Arbitrary L_p Norms. In: Proc. of the 26th International Conference on VLDB, pp. 385–394 (2000)

Computing β-Drawings of
2-Outerplane Graphs in Linear Time
(Extended Abstract)

Md. Abul Hassan Samee, Mohammad Tanvir Irfan, and Md. Saidur Rahman

Department of Computer Science and Engineering
Bangladesh University of Engineering and Technology (BUET)
Dhaka - 1000, Bangladesh
{samee,mtirfan,saidurrahman}@cse.buet.ac.bd

Abstract. A straight-line drawing of a plane graph G is a drawing of G where
each vertex is drawn as a point and each edge is drawn as a straight-line segment
without edge crossings. A proximity drawing Γ of a plane graph G is a straight-
line drawing of G with the additional geometric constraint that two vertices of G
are adjacent if and only if no other vertex of G is drawn in Γ within a "proximity
region" of these two vertices in Γ. Depending upon how the proximity region
is defined, a given plane graph G may or may not admit a proximity drawing.
In one class of proximity drawings, known as β-drawings, the proximity region
is defined in terms of a parameter β, where $\beta \in [0, \infty)$. A plane graph G is
β-drawable if G admits a β-drawing. A sufficient condition for a biconnected
2-outerplane graph G to have a β-drawing is known. However, the known algorithm for testing the sufficient condition takes time $O(n^2)$. In this paper, we give
a linear-time algorithm to test whether a biconnected 2-outerplane graph G satisfies the known sufficient condition or not. This consequently leads to a linear
algorithm for β-drawing of a wide subclass of biconnected 2-outerplane graphs.

Keywords: Graph Drawing, Proximity Drawing, β-Drawing, Proximity Graph,
2-Outerplane graph, Slicing Path, Good Slicing Path.

1 Introduction

Let Γ be a straight-line drawing of a plane graph G. Let $\Gamma(u)$ be the point on the plane
to which the vertex u of G is mapped in Γ. Then Γ is a *proximity drawing* of the plane
graph G if Γ satisfies the following proximity constraint: two vertices u and v of G are
adjacent if and only if a well-defined "proximity region" corresponding to the points
$\Gamma(u)$ and $\Gamma(v)$ is empty, i.e. the region does not contain $\Gamma(w)$ for any other vertex
w of G. The exact definition of proximity region is problem-specific. As a matter of
fact, there is an infinite number of different types of proximity regions. For example,
an infinite family of parameterized proximity regions has been introduced in [5]. This
family of parameterized proximity regions gives rise to an important class of proximity
drawings, known as β-*drawings*, where β stands for a parameter that can take any real
number value in $[0, \infty)$.

A plane graph G is β-drawable if G admits a β-drawing. Not all graphs are β-
drawable for all values of β. For example, the graph G_1 illustrated in Fig. 1(a) is not

S.-i. Nakano and Md. S. Rahman (Eds.): WALCOM 2008, LNCS 4921, pp. 81–87, 2008.

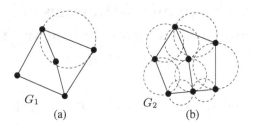

Fig. 1. (a) A graph G_1 which is not β-drawable for $\beta \in (1, 2)$, and (b) a graph G_2 which is β-drawable under the same constraints

β-drawable for $\beta = 1$. This fact can be explained as follows. Suppose we want to achieve a β-drawing of G_1 with $\beta = 1$. The β-region of two vertices u and v for $\beta = 1$ is a circle with $\Gamma(u)$ and $\Gamma(v)$ as its two antipodal points. For the graph G_1, wherever we place the four external vertices, the internal vertex will be inside the β-region of at least one of the four pairs of neighboring external vertices (as shown with the dotted circle in Fig. 1(a)). Therefore the graph G_1 is not β-drawable for $\beta = 1$. Following this same line of reasoning, one can easily work out that the graph G_2 shown in Fig. 1(b) is β-drawable for $\beta = 1$.

The *proximity drawability problem*, i.e. the problem whether a given graph admits a particular proximity drawing or not, has originated from the well-known "proximity graphs." Proximity graphs have wide applications in computer graphics, computational geometry, pattern recognition, computational morphology, numerical analysis, computational biology, GIS, instance-based learning and data-mining [3].

Several research outcomes regarding the proximity drawability of trees and outerplanar graphs are known [1, 2, 6]. One of the problems left open in [6] is to extend the problem of β-drawability of graphs to other nontrivial classes of graphs apart from trees and outerplanar graphs. In [4], the authors gave a sufficient condition for β-drawability of biconnected 2-outerplane graphs, where $\beta \in (1, 2)$. Although their sufficient condition induces a large and non-trivial class of biconnected 2-outerplane graphs, their algorithm for testing whether a given biconnected 2-outerplane graph satisfies those conditions or not, takes time $O(n^2)$.

In this paper, we give a linear-time algorithm for testing whether a biconnected 2-outerplane graph G satisfies the sufficient condition presented in [4]. Our algorithm essentially relies on the sufficient condition presented in [4], but works on a new set of conditions devised by us on "slicing paths" of G.

The rest of this paper is organized as follows. In Section 2, we present some definitions and preliminary results. In Section 3, we give a linear-time algorithm to test whether G satisfies the sufficient condition presented in [4] or not, and in the positive case, to compute a β-drawing of G. Finally, Section 4 is a conclusion.

2   Preliminaries

In this section we give some definitions and present our preliminary results.

A graph G is *connected* if there is a path between every pair of vertices of G, otherwise G is *disconnected*. A connected graph G is *biconnected* if at least two vertices of G are required to be removed to make the resulting graph disconnected. A *component* of G is a maximal connected subgraph of G. A graph G is *planar* if G has an embedding in the plane where no two edges cross each other, except at vertices on which two or more edges are incident. A *plane graph* is a planar graph with a fixed planar embedding. A plane graph partitions the plane into topologically connected regions called *faces*. The unbounded face is called the *external face*, while the remaining faces are called *internal faces*.

An *outerplanar graph* is a graph that has a planar embedding in which all the vertices lie on the external face. An outerplanar graph is also known as a 1-*outerplanar graph*. If all the vertices of a 1-outerplanar graph G appear on the external face of a given embedding of G, then we say that the embedded graph is a 1-*outerplane graph*, otherwise the embedded graph is not 1-outerplane, even though G is 1-outerplanar. These definitions can be generalized as follows. For an integer $k > 1$, an embedded graph is k-*outerplane* if the embedded graph obtained by removing all the vertices of the external face is a $(k - 1)$-outerplane graph. On the other hand, we call a graph k-*outerplanar* if it has an embedding that is k-outerplane.

Let G be a biconnected 2-outerplane graph. The vertices on the external face of G are called the *external vertices* of G. The remaining vertices of G are called the *internal vertices* of G. For any external vertex u of G, the *fan of u*, denoted by F_u, is the subgraph of G induced by the vertices of G that share an internal face with u. The vertex u is called the *apex of F_u*.

For any two distinct points in the plane there is an associated region parameterized by β, where $\beta \in [0, \infty)$, which is called the β-*region* of the two points. There exist two variants of β-regions, namely lune-based and circle-based β-regions [5]. These regions can be further subdivided into two types: *open β-regions* and *closed β-regions*. In an open β-region, the boundary of the region is excluded from the region. However, in a closed β-region, the boundary of the region is included in the region. In [4], the authors studied the lune-based closed β-regions and gave the following sufficient condition for a biconnected 2-outerplane graph G to admit a β-drawing where $\beta \in (1, 2)$.

Theorem 1. *A biconnected 2-outerplane graph G is β-drawable for $\beta \in (1, 2)$ if G satisfies the following conditions 1 and 2.*

1. *There are at least five external vertices; and*
2. *There is an external vertex u such that the fan F_u has all of the following properties: (a) F_u is biconnected 1-outerplane; (b) F_u contains all the internal vertices of G and the internal vertices of G induces a single connected component of G; and (c) every vertex in F_u has at most one neighbor outside F_u and every vertex outside F_u has at most one neighbor in F_u.*

$\square$

However, the sufficient conditions mentioned in Theorem 1 apply for lune-based open β-regions as well [4].

A constructive proof of Theorem 1 has been provided in [4]. In that proof, the authors have first tested whether a biconnected 2-outerplane graph G satisfies the conditions stated in Theorem 1 or not and in the positive case, they have computed a set of external vertices of G such that for each vertex u in that set, the corresponding fan F_u satisfies Conditions $2(a)$–$2(c)$ of Theorem 1. Such a set of external vertices is called the *set of candidate apices*. For the purpose of computing this set, the authors in [4] have adopted the following approach. For each external vertex u of G they have tested whether the fan F_u of u satisfies the Conditions $2(a)$–$2(c)$ of Theorem 1. For a specific fan F_u, this checking will take $O(n)$ time. Since there are $O(n)$ number of external vertices of G, this would require $O(n^2)$ time to compute the set of candidate apices. After computing the set of candidate apices, the authors in [4] have computed a β-drawing of G in time $O(n)$ as follows. They have first drawn the fan F_u of vertex u where u is a candidate apex. They have next drawn the remaining graph $G - (V(F_u))$ and have added edges between the vertices of F_u and the vertices of $G - (V(F_u))$. Thus the constructive proof of Theorem 1 presented in [4] yields an $O(n^2)$ time algorithm which finds candidate apices in $O(n^2)$ time and computes a β-drawing in $O(n)$ time.

In the remainder of this section, we present some fundamental observations which we use to compute the set of candidate apices in linear-time. We first have the following lemma whose proof is omitted in this extended abstract.

Lemma 1. *Let G be a biconnected 2-outerplane graph. Let u be an external vertex of G and F_u be the fan of vertex u. If F_u is 1-outerplane and F_u contains all the internal vertices of G then F_u is biconnected.*   □

It is important here for us to mention the significance of Lemma 1 in our work. Throughout the remainder of this paper, given a biconnected 2-outerplane graph G, we will concentrate on an external vertex u of G and its fan F_u such that: *(a)* the condition given in Condition $2(a)$ of Theorem 1 stating that F_u should be 1-outerplane holds; and *(b)* Conditions $2(b)$ and $2(c)$ of Theorem 1 hold. Lemma 1 ensures that F_u will be biconnected in every such scenario, and hence, all the three conditions $2(a)$–$2(c)$ hold.

Let G be a biconnected 2-outerplane graph which satisfies the sufficient condition given in Theorem 1. We now have the following lemma regarding the subgraph G_{in} of G induced by the internal vertices of G. We have omitted the proof of Lemma 2 in this extended abstract.

Lemma 2. *Let $G = (V, E)$ be a biconnected 2-outerplane graph which satisfies the sufficient condition of Theorem 1. Let G_{in} be the subgraph of G induced by the internal vertices of G. Then G_{in} is a simple path.*   □

3   Linear-Time Algorithm for β-Drawings of G

In this section, we first introduce the concept of a "slicing path" and a "good slicing path" of G. We next use the notion of good slicing paths of G to devise a linear-algorithm to test whether G satisfies the sufficient condition of Theorem 1 or not, and in the positive case, to compute a set of candidate apices of G. By using the linear-time algorithm presented in [4] for computing a β-drawing of such a graph G, we thus give a linear-time algorithm for computing a β-drawing of G.

Let G_{in} be the subgraph of G induced by the internal vertices of G. If G satisfies Theorem 1, then Lemma 2 implies that G_{in} is a simple path. Let $P_{st} = u_s, u_{s+1}, \ldots, u_t$, denote the path induced by the internal vertices of G. Clearly, every neighbor of u_s in G other than u_{s+1} is an external vertex of G. Similarly, every neighbor of u_t in G other than u_{t-1} is an external vertex of G. For an internal vertex v of G, let $N_{outer}(v)$ denote the number of those neighbors of v which are external vertices of G. We now have the following lemma.

Lemma 3. *Let G be a biconnected 2-outerplane graph which satisfies the sufficient condition of Theorem 1. Let $P_{st} = u_s, u_{s+1}, \ldots, u_t$, be the path induced by the internal vertices of G. Let $w \in \{u_s, u_t\}$. Then $N_{outer}(w) \leq 3$.* □

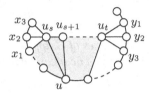

Fig. 2. Illustration of Lemma 3. The shaded region corresponds to the fan of vertex u

Proof. Let us assume that u is a candidate apex of G as illustrated in Fig. 2. Then F_u, the fan of u, will contain exactly one edge connecting u_s with an external vertex x of G where $x \neq u$. Similarly, F_u will contain exactly one edge connecting u_t with an external vertex y of G where $y \neq u$. Clearly, if $N_{outer}(w) > 3$ for either $w = u_s$ or $w = u_t$ as illustrated in Fig. 2, then w will have more than one neighbor outside F_u which violates Condition 2(c) of Theorem 1. Hence, $N_{outer}(w) \leq 3$, for each $w \in \{u_s, u_t\}$.   □

Let G be a biconnected 2-outerplane graph which satisfies the sufficient condition of Theorem 1. Let the path $P_{st} = u_s, u_{s+1}, \ldots, u_t$, be the subgraph of G induced by the internal vertices of G. Then Lemma 3 holds for the end-vertices u_s and u_t of P_{st}. Let x and y be two of those external vertices of G which are neighbors of u_s and u_t respectively. We say that the path $P_{st}^* = x, u_s, u_{s+1}, \ldots, u_t, y$, is a *slicing path of G obtained from P_{st}*. Given a slicing path $P_{st}^* = x, u_s, u_{s+1}, \ldots, u_t, y$, if we traverse P_{st}^* from x to y, then some of the vertices of G will lie on our left hand side, and some will lie on our right hand side. Let G_{st}^L denote the subgraph of G induced by those vertices of G which lie on our left hand side while traversing P_{st}^* from x to y. Similarly, let G_{st}^R denote the subgraph of G induced by those vertices of G which lie on our right hand side while traversing P_{st}^* from x to y. We say that a slicing path $P_{st}^* = x, u_s, u_{s+1}, \ldots, u_t, y$, is a *good slicing path of G* if the following Conditions ($gs1$) and ($gs2$) hold.

($gs1$) There is no vertex v of P_{st}^* such that v has more than one neighbor in G_{st}^L and more than one neighbor in G_{st}^R; and

($gs2$) there are no two vertices u and v of P_{st}^* such that u and v have a common neighbor in G_{st}^L and a common neighbor in G_{st}^R.

The following lemma is immediate from the above definition of a good slicing path of G.

Lemma 4. *Let G be a biconnected 2-outerplane graph which satisfies the sufficient condition of Theorem 1. Let P_{st} be the path induced by the internal vertices of G. Then a good slicing path of G can be obtained from the path P_{st}.*   □

We now have the following lemma.

Lemma 5. *Let G be a biconnected 2-outerplane graph which satisfies the sufficient condition of Theorem 1. Let $P_{st} = u_s, u_{s+1}, \ldots, u_t$, be the path induced by the internal vertices of G. In order to obtain a good slicing path of G from P_{st}, we need to check at most four slicing paths of G obtained from P_{st}.*   □

Before presenting the proof of Lemma 5, we first give the following lemma whose proof is omitted in this extended abstract.

Lemma 6. *Let G be a biconnected 2-outerplane graph which satisfies the sufficient condition of Theorem 1. Let $P_{st} = u_s, u_{s+1}, \ldots, u_t$, be the path induced by the internal vertices of G. Let $N_{outer}(w) = 3$, for each $w \in \{u_s, u_t\}$. Let x_1, x_2, x_3 be the three neighbors of w which are external vertices of G appearing in counter-clockwise order on the external face of G. Then x_2 cannot be a candidate apex of G, but x_1 or x_3 can be a candidate apex of G.*   □

Proof of Lemma 5. Let $w \in \{u_s, u_t\}$. We can obtain slicing paths P_{st}^* from P_{st} as follows. **(1)** If $N_{outer}(w) = 1$, then let x denote the neighbor of w which is an external vertex of G. In this case, we include x in P_{st}^*. **(2)** If $N_{outer}(w) = 3$, then let x_1, x_2, x_3 be the three neighbors of w appearing counter-clockwise on the external face of G. As we have shown in Lemma 6, x_2 cannot be a candidate apex of G, but x_1 or x_3 can be a candidate apex of G. In this case, we include x_2 in P_{st}^*. **(3)** If $N_{outer}(w) = 2$, then let x_1 and x_2 be the two neighbors of w on the external face of G. In this case, we construct two slicing paths from P_{st}, by including x_1 in one and including x_2 in another.

   Hence, our claim holds from Lemma 6 and the above mentioned method to construct slicing paths P_{st}^* from P_{st}.   □

We finally have the following lemma.

Lemma 7. *Let G be a biconnected 2-outerplane graph. Then one can check in linear time whether G satisfies the sufficient condition of Theorem 1 or not, and can compute the set of candidate apices of G in linear time if G satisfies the condition.*

Proof. Our proof is constructive. We first check whether G has at least five external vertices or not. We next check whether G_{in} is a simple path or not. If G_{in} is not a simple path, then Lemma 2 implies that G does not satisfy the sufficient condition of Theorem 1. In the positive case, let P_{st} denote the path induced by the internal vertices of G. Then we check whether Lemma 3 holds for P_{st} or not. If Lemma 3 does not hold for P_{st}, then Lemma 3 implies that G does not satisfy the sufficient condition of Theorem 1. Clearly, the operations in this first step can be performed in linear-time.

   In our next step, we construct slicing paths P_{st}^* of G from P_{st} according to the method outlined in the proof of Lemma 5. It is also implied by Lemma 5 that we will have to construct at most four such slicing paths P_{st}^*. For each P_{st}^*, we check whether it is a good slicing path of G or not. If no good slicing path of G can be obtained from P_{st}, then we get from Lemma 4 that G does not satisfy the sufficient condition of Theorem 1. The checking of this step is independent of the previous step, and can be performed in linear-time. In this phase, we also remember which of the two subgraphs G_{st}^L and G_{st}^R can contain a candidate apex.

In our last step, for each good slicing path P_{st}^*, we traverse all the internal faces of G_{st}^L and G_{st}^R that can contain a candidate apex. Let u be an external vertex of G which appears on each of these faces. All such vertices u will constitute the set of candidate apices. This step can also be implemented in linear time, since we can gather the whole information by traversing each such internal face at most once. Hence the total algorithm has linear time complexity. □

We finally present the main result of this paper in the following theorem.

Theorem 2. *Let G be a biconnected 2-outerplane graph. Then one can check in linear time whether G satisfies the sufficient condition of Theorem 1 or not, and can find a β-drawing of G, where $\beta \in (1, 2)$, in linear time if G satisfies the condition.*

Proof. The claim holds directly from Lemma 7 and the linear-time drawing algorithm presented in [4]. □

4   Conclusion

In this paper, we have given a linear-time algorithm to test whether a biconnected 2-outerplane graph G satisfies the sufficient condition presented in [4], and thus, we have achieved a linear algorithm for computing β-drawings of biconnected 2-outerplane graphs where $\beta \in (1, 2)$. It remains as our future work to obtain efficient algorithms for computing β-drawings of larger classes of graphs.

References

[1] Bose, P., Di Battista, G., Lenhart, W., Liotta, G.: Proximity constraints and representable trees. In: Tamassia, R., Tollis, I(Y.) G. (eds.) GD 1994. LNCS, vol. 894, pp. 340–351. Springer, Heidelberg (1995)

[2] Bose, P., Lenhart, W., Liotta, G.: Characterizing proximity trees. Algorithmica 16, 83–110 (1996)

[3] Tamassia, R.(ed.): Handbook of graph drawing and visualization. CRC Press, Boca Raton, USA(to be published), http://www.cs.brown.edu/~rt/gdhandbook/

[4] Irfan, M.T., Rahman, M.S.: Computing β-drawings of 2-outerplane graphs. In: Kaykobad, M., Rahman, M. S. (eds.) WALCOM. Bangladesh Academy of Sciences (BAS), pp. 46–61 (2007)

[5] Kirkpatrick, D.G., Radke, J.D.: A framework for computational morphology. In: Toussaint, G.T. (ed.) Computational Geometry, pp. 217–248. North-Holland, Amsterdam (1985)

[6] Lenhart, W., Liotta, G.: Proximity drawings of outerplanar graphs. In: North, S.C. (ed.) GD 1996. LNCS, vol. 1190, pp. 286–302. Springer, Heidelberg (1997)

Upward Drawings of Trees on the Minimum Number of Layers
(Extended Abstract)

Md. Jawaherul Alam, Md. Abul Hassan Samee,
Md. Mashfiqui Rabbi, and Md. Saidur Rahman

Department of Computer Science and Engineering,
Bangladesh University of Engineering and Technology (BUET)
jawaherul@yahoo.com, samee@cse.buet.ac.bd,
mashfiqui.rabbi@gmail.com, saidurrahman@cse.buet.ac.bd

Abstract. In a planar straight-line drawing of a tree T on k layers, each vertex is placed on one of k horizontal lines called layers and each edge is drawn as a straight-line segment. A planar straight-line drawing of a rooted tree T on k layers is called an upward drawing of T on k layers if, for each vertex u of T, no child of u is placed on a layer vertically above the layer on which u has been placed. For a tree T having pathwidth h, a linear-time algorithm is known that produces a planar straight-line drawing of T on $\lceil 3h/2 \rceil$ layers. A necessary condition characterizing trees that admit planar straight-line drawings on k layers for a given value of k is also known. However, none of the known algorithms focuses on drawing a tree on the minimum number of layers. Moreover, although an upward drawing is the most useful visualization of a rooted tree, the known algorithms for drawing trees on k layers do not focus on upward drawings. In this paper, we give a linear-time algorithm to compute the minimum number of layers required for an upward drawing of a given rooted tree T. If T is not a rooted tree, then we can select a vertex u of T in linear time such that an upward drawing of T rooted at u would require the minimum number of layers among all other upward drawings of T rooted at the vertices other than u. We also give a linear-time algorithm to obtain an upward drawing of a rooted tree T on the minimum number of layers.

Keywords: Planar Drawing, Straight-line Drawing, k-layer Planar Drawing, Upward Drawing, Minimum Layer Upward Drawing, Trees, Algorithm, Line-labeling.

1 Introduction

A *k-layer planar drawing* of a tree T is a planar drawing of T where each vertex of T is drawn as a point on one of k horizontal lines called layers and each edge of T is drawn as a straight-line segment. Such a drawing of a tree T is also called a *planar straight-line drawing of T on k layers*. A tree T is *k-layer planar* if T admits a planar straight-line drawing on k layers. For example, the tree T_1 in Fig. 1(a) is 1-layer planar since it admits a planar straight-line drawing Γ_1 on a single layer as illustrated in Fig. 1(b). However, an arbitrary tree T may not always admit a k-layer planar drawing for a desired value

S.-i. Nakano and Md. S. Rahman (Eds.): WALCOM 2008, LNCS 4921, pp. 88–99, 2008.

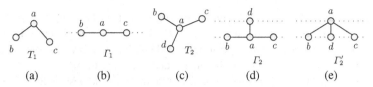

Fig. 1. (a) The tree T_1, (b) a drawing Γ_1 of T_1 on one layer, (c) the tree T_2, (d) a drawing Γ_2 of T_2 on two layers, and (e) another drawing Γ_2' of T_2 on two layers

of k. For example, the tree T_2 in Fig. 1(c) does not admit a k-layer planar drawing for $k = 1$. The reason is as follows. Let l denote the layer in the drawing on which the vertex a of T_2 will be placed. If we want to obtain a planar drawing of T_2, then we can place at most two neighbors of a in T_2 on the layer l. Placing all the three neighbors of a in T_2 on the layer l will violate planarity of the drawing and thus, at least two layers are necessary for a planar straight-line drawing of T_2. Two drawings Γ_2 and Γ_2' of T_2 on two layers are shown in Fig. 1(d) and (e) respectively. Thus, although the tree T_2 admits a k-layer planar drawing for $k = 2$, it does not admit a k-layer planar drawing for $k = 1$. One can easily infer from this simple example that, such problems as to determine whether a given tree T admits a k-layer planar drawing for a given value of k, or to compute the minimum number of layers required for a k-layer planar drawing of T are quite challenging.

Let T be a rooted tree. An *upward drawing of T on k layers* is a k-layer planar drawing of T such that, for each vertex u of T, no child of u is placed on a layer vertically above the layer on which u has been placed [6,7]. A *minimum layer upward drawing* of the rooted tree T is an upward drawing of T on the minimum number of layers. For example, if the vertex a is taken as the root of the tree T_2 in Fig. 1(c), then the drawing Γ_2 of T_2 in Fig. 1(d) is a minimum layer upward drawing of T_2. If a tree T is not a rooted tree, then let T_u be the tree obtained from T by considering a vertex u of T as the root of T. Let v be a vertex of T. Let l_v be the number of layers required for a minimum layer upward drawing of T_v. If for any other vertex w of T, an upward drawing of T_w requires at least l_v layers, then a *minimum layer upward drawing of the tree T* is an upward drawing of T on l_v layers.

A k-layer planar drawing of a tree is a common variant of the well-known "layered drawings" of trees [15,14]. In a *layered drawing* of a tree T, the vertices are drawn on a set of horizontal lines called layers, and the edges are drawn as straight-line segments [14]. Layered drawings have important applications in several areas like VLSI layouts [10], DNA-mapping [16] and information visualization [1]. Layered drawings of trees are usually required to satisfy some constraints arising from the application at hand. One such constraint is to impose bounds on the number of layers [14] and in this regard, it is often sought to know whether a given tree T admits a k-layer planar drawing for a given value of k [14]. However, the solutions for this problem known to date work only for some small values of k [4,2,5]. For example, linear time algorithms have been given in [2,5] for recognizing and drawing trees that are k-layer planar for $k = 2$. For $k > 2$, Felsner *et al* [5] have given necessary conditions for a tree T to be k-layer planar, but these conditions are not sufficient. For a tree T with pathwidth h, a linear-time algorithm has been given in [13] to draw T on $\lceil 3h/2 \rceil$ layers and it has been shown that T cannot be drawn on less than h layers [13]. However, the algorithm presented in [13] cannot ensure that the minimum number of layers have been used in the drawing, neither can it determine whether a given tree T admits a k-layer planar drawing for a desired value of k. Moreover, no algorithm for k-layer planar drawings of trees focuses on producing upward drawings. On the other hand, trees are most often used to model hierarchical data and an upward drawing of a rooted tree is the best means to visualize it. For example, trees representing organization charts, software class hierarchies, phylogenetic evolutions and programming language parsing are always rooted trees and are best visualized through upward drawings. Thus, there is a large gap between the known results for k-layer planar drawings of trees and the practical visualization requirements imposed by trees. In this paper, we give a linear-time algorithm for computing the number of layers required for a minimum layer upward drawing of a rooted tree T. In case T is a tree with no root specified, our algorithm can determine a vertex r of T so that an upward drawing of T rooted at r results in a minimum layers upward drawing of T. Consequently, we can also determine whether a tree T admits an upward drawing on k layers for a desired value of k. We also give a linear-time algorithm that produces a minimum layer upward drawing of T.

Our results presented in this paper are also significant regarding the area bounds for such drawings of rooted trees that are planar, straight-line and also upward. In the remainder of this paper we simply use term "upward drawing of a rooted tree" to denote such a drawing of a rooted tree that is planar, straight-line and also upward. We have shown that our drawing algorithm produces an upward drawing of a rooted tree with an area bound of $O(n \log n)$. This bound matches with the one presented in [3,11]. However, an $O(n \log n)$ area bound is not the best among the known results for upward drawings of rooted trees. Nevertheless, it is notable that, although there are known algorithms for drawing binary trees in $O(n \log \log n)$ area and for drawing general trees in $O(n)$ area [6,7], those algorithms produce polyline drawings and are not aesthetically as appealing as a straight-line drawing. Another algorithm for drawing trees in $O(n)$ area [9,8] fails to produce an upward drawing. Finally, although the algorithm in [12] gives an $O(n \log \log n)$ area upward drawings of trees, the algorithm works only on trees with bounded degrees. Thus, when there is no restriction on vertex-degrees, our algorithm gives the best possible area bound known so far for producing upward drawings of rooted trees.

The rest of this paper is organized as follows. In Section 2, we give some preliminary definitions and present a brief outline of our algorithm. In Section 3, we detail our central concept used in this paper called the "line-labeling" of a tree. Section 4 is the description of our drawing algorithm. Finally, Section 5 is a conclusion.

2   Preliminaries

In this section, we give some definitions and present an outline of our algorithm.

Let $G = (V, E)$ be a simple graph with vertex set V and edge set E. Let (u, v) be an edge of G joining two vertices u and v of G. A vertex u of G is a *neighbor* of another vertex v of G (and vice versa) if and only if G has an edge (u, v). The *degree* of a vertex v in G is the number of neighbors of v in G. A *path* P in G is a sequence $v_0, e_1, v_1, \ldots, e_n, v_n$ of vertices and edges of G such that G contains an edge

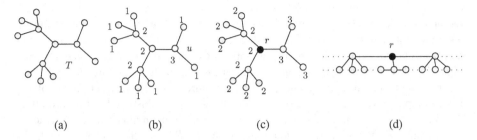

Fig. 2. (a) A tree T, (b) computation of the line-labeling of a vertex u of T, (c) line-labelings of all the vertices of T, and (d) an upward drawing of T on the minimum number of layers

$e_i = (v_{i-1}, v_i)$ for each i, $(1 \leq i \leq n)$ and all the vertices v_i, $(0 \leq i \leq n)$ are distinct. Such a path P is also called a v_0, v_n-path of G. The vertices v_0 and v_n of the path P are called the *end-vertices* of P.

A graph G is *connected* if there exists a u, v-path in G for every pair of vertices $u, v \in V$. A *cycle* in a graph G is a path in G whose end-vertices are the same. A *tree* T is a connected graph which contains no cycle. A vertex u of T having degree one in T is called a *leaf* of T. A vertex u of T having degree greater than one in T is called an *internal vertex* of T. A tree T is called a *rooted tree* if one of the vertices r of T is considered as the *root* of T. In this paper, we use the notation T_r to denote such a rooted tree obtained from T by considering a vertex r of T as the root of T. The *parent* of a vertex v in T_r is the vertex that precedes v in the r, v-path in T_r. All the neighbors of v in T_r other than its parent are called the *children* of v in T_r. The *ancestors* of a vertex v in T_r is the set of all such vertices u of T_r that u is on the r, v-path in T_r. The *descendants* of v in T_r is the set of all such vertices u of T_r that the r, u-path in T_r contains v. A *subtree* of T_r rooted at a vertex v of T_r is the subgraph of T_r induced by the descendants of v in T_r. In this paper, we use the notation $T_r(v)$ to denote the subtree of T_r rooted at a vertex v of T.

A *k-layer planar drawing* of a tree T is a planar drawing where each vertex is placed on one of k horizontal lines called layers and each edge is drawn as a straight-line segment. An *upward drawing of a rooted tree T on k layers* is a k-layer planar drawing of T such that, for each vertex u of T, no child of u is placed on a layer vertically above the layer on which u has been placed. A *minimum layer upward drawing* of T is an upward drawing of T on the minimum number of layers. If T is not a rooted tree, then for a vertex v of T, let l_v be the number of layers required for a minimum layer upward drawing of T_v. If for any other vertex w of T, an upward drawing of T_w requires at least l_v layers, then a *minimum layer upward drawing of the tree T* is an upward drawing of T on l_v layers.

Having been introduced to the necessary graph-theoretic definitions, we now give an outline of our algorithm for producing a minimum layer upward drawing of a tree T. We first outline our approach for the scenario where T is not a rooted tree as illustrated in Fig. 2(a). In this case, we first determine a vertex r of T such that an upward drawing of T rooted at r would require the minimum number of layers among all possible upward drawings of T. Let u be a vertex of T. Let T_u denote the rooted tree obtained from T

by considering u as the root of T. In order to compute the minimum number of layers required for an upward drawing of T_u, we perform a bottom-up traversal of T_u during which we equip every vertex v of T_u with an integer value that represents the minimum number of layers required for any upward drawing of the subtree of T_u rooted at v (See Fig. 2(b)). The integer value assigned to the vertex u is "the line-labeling of u in T" and this represents the minimum number of layers required for any upward drawing of T_u. If we compute the line-labelings of all the vertices of T as illustrated in Fig. 2(c), then the minimum of these line-labelings represents the minimum number of layers required for any upward drawing of T rooted at any vertex of T. Also, the vertex r, which has the minimum line-labeling among all the vertices, represents the root of such a drawing. In the later sections, we have shown that the whole task of computing the line-labelings of all the vertices of T and finding the minimum of these values be done in linear time. Finally, we obtain an upward drawing of T rooted at r on the minimum number of layers in linear time (See Fig. 2(d)). In case T is a rooted tree with a vertex r as the root, we simply compute the line-labeling of r in T and use our drawing algorithm to obtain an upward drawing of T rooted at r on the minimum number of layers.

In the next section, we formally define the concept of line-labeling of the vertices of a tree T and the concept of line-labeling of a tree T. We also provide in the next section the necessary proofs of correctness regarding our claims on line-labeling.

3   Line-Labeling of a Tree

In this section, we detail the concept of the "line-labeling of the vertices" of a tree T and the concept of "line-labeling" of the tree T. We also prove that if the "line-labeling" of a tree T is equal to k, then any upward drawing of T requires at least k layers.

Let T be a tree. Let u be a vertex of T. Prior to defining the line-labeling of the vertex u in T, we first define the line-labeling of u in T with respect to some other vertex r of T. Let T_r be the rooted tree obtained from T by considering the vertex r as the root of T. Then the *line-labeling of a vertex u with respect to the vertex r in T*, which we have denoted by $L_r(u)$ in the remainder of this paper, is defined as follows.

(a) If u is a leaf of T_r, then $L_r(u) = 1$. This scenario is illustrated in Fig. 3(a).
(b) If u is an internal vertex of T_r, then let $u_1, u_2, \ldots, u_p$ be the children of u in T_r. Let k be the maximum among the values $L_r(u_1), L_r(u_2), \ldots, L_r(u_p)$. Then $L_r(u)$ is defined according to any of the three following cases.

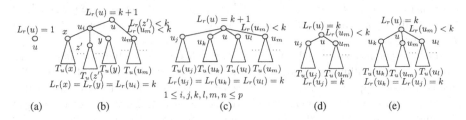

Fig. 3. Definition of the line-labeling $L_r(u)$

(i) If u has at least one child u_i in T_r such that $L_r(u_i) = k$ and if u_i has two children x and y in T_r such that $L_r(x) = k$ and $L_r(y) = k$, then $L_r(u) = k + 1$. This scenario is illustrated in Fig. 3(b)

(ii) If u has no child like u_i in Case (i) above but has three or more children in T_r with line-labeling k with respect to r, then $L_r(u) = k + 1$. This scenario is illustrated in Fig. 3(c).

(iii) Otherwise, u has at most two children in T_r with line-labeling k with respect to r and none of them is like the child u_i of u in Case (i) above. In this case, $L_r(u) = k$. This scenario is illustrated in Fig. 3(d) and (e).

We now define the line-labeling of a vertex u of the tree T. The *line-labeling of a vertex u of the tree T*, which we have denoted by $L(u)$ in the remainder of this paper, is the line-labeling of u with respect to u itself. Thus, according to the notations introduced so far, $L(u) = L_u(u)$.

We now have the following lemma.

Lemma 1. *Let T be a tree. Let u be a vertex of T. Let $L(u)$ be the line-labeling of u. Then any upward drawing of T rooted at u requires at least $L(u)$ number of layers.*

Proof. We can prove this lemma by induction on the number of vertices n of T. Since u is a vertex of T, $n \geq 1$. Hence, as the basis of induction, we take $n = 1$. According to our definition, in this case, $L(u) = 1$. Clearly, any drawing of a single-vertex tree requires at least one layer and thus the claim holds for the basis case.

We now prove the claim for a tree T with the number of vertices $n > 1$. We first assume that for any tree T' with the number of vertices $n' < n$, any upward drawing of T' rooted at a vertex u of T' requires at least $L(u)$ number of layers, where $L(u)$ is the line-labeling of u in T'.

Let u be a vertex of T. Let k be the maximum value of the line-labelings of all the children of u in T. Let T_u be the rooted tree obtained from T by making u the root of T. First, we assume that u has a child v in T_u such that $L_u(v) = k$ and v has two children x and y in T_u such that $L_u(x) = L_u(y) = k$. In this case, $L(u) = k + 1$. The fact that any upward drawing of T rooted at u requires at least $k + 1$ layers can be realized by the following arguments. Let $T_u(x)$ and $T_u(y)$ denote the subtrees of T_u rooted at x and y respectively. From the induction hypothesis, at least k layers are required for any upward drawing of $T_u(x)$ and $T_u(y)$. The drawing of these two subtrees restricts the placement of the vertex v on the topmost of these k layers between x and y. Then u cannot be placed on any of these k layers while retaining both planarity and upwardness of the drawing. Hence, at least $k + 1$ layers are needed for any upward drawing of T rooted at u and the claim holds in this case.

Next, we assume that u has no such child v as mentioned above. Then if u has three (or more) children in T_u with line-labeling k with respect to u, then $L(u) = k + 1$. The fact that any upward drawing of T rooted at the vertex u requires at least $k + 1$ layers can be understood from the following reasonings. Let w, x and y be the three children of u in T_u such that $L_u(w) = L_u(x) = L_u(y) = k$. Let $T_u(w)$, $T_u(x)$ and $T_u(y)$ the subtrees of T_u rooted at w, x and y respectively. From the induction hypothesis, any upward drawing of any of these subtrees requires at least k layers. We note that we cannot place all these three subtrees on the same k layers and then place u on any

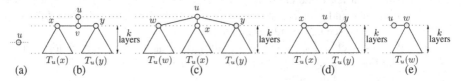

Fig. 4. Drawing of T_u on $L(u)$ number of layers

of these k layers keeping planarity. Therefore, we need a new layer on top of these k layers for placing the vertex u. Hence, at least $k+1$ layers are necessary for any upward drawing of T rooted at u and the claim holds in this case also.

Finally, we assume that u has at most two children in T_u with line-labeling k with respect to u and none of these children of u has two children in T_u with line-labeling k. Then according to our definition, $L(u) = k$. Any upward drawing of T rooted at u requires at least k layers since the upward drawings of some subtrees of T_u require at least k layers. Therefore, the claim holds in this case as well.

This completes the induction step and hence completes the proof.   □

We now define the concept of "line-labeling" of a tree T. The *line-labeling of a tree T*, which we have denoted by L_T in the remainder of this paper, is defined as follows:

(a) If T contains no vertex, then $L_T = 0$.
(b) Otherwise, the line labeling of T is the minimum value of the line-labelings of all the vertices of T. That is, $L_T = \min_{u \in V} L(u)$, where V is the vertex set of T.

We now have the following theorem.

Theorem 1. *If the line-labeling of a tree T is k, then any upward drawing of T rooted at any vertex of T requires at least k layers and the value of k can be computed in linear time.*

Proof. Let L_T denote the line-labeling of the tree T. Let $L_T = k$. Then it follows from the definition of the line-labeling of a tree and Lemma 1 that any upward drawing of T rooted at any vertex of T requires at least k layers.

We now prove that we can compute the line-labeling of T in linear time. Our approach is to compute the line-labelings of all the vertices of T by two linear-time traversals of T and then to compute the minimum of these values.

We first take any vertex r of T and traverse the rooted tree T_r in a bottom-up fashion. During this traversal, we compute the line-labeling $L_r(u)$ of every vertex u of T with respect to r. Let v be a child of u in T_r. The value of $L_r(v)$, computed in this first traversal of T_r would be the same as the value of $L_u(v)$. However, the computation of $L(u)$ depends not only on the value of $L_u(v)$ for each child v of u in T_r, but also on the value of $L_u(w)$, where w is the parent of u in T_r. In order to compute $L_u(w)$ for each vertex u of T, we traverse T_r for a second time, in a top-down fashion.

In the top-down traversal of T_r, for each vertex w, for each of the children u of w in T_r, we compute the value of $L_u(w)$ and use this value to compute $L(u)$. At first, we consider the case when the root r of T_r is traversed. Let $u_1, u_2, \ldots, u_p$ be the p neighbors

$(p > 1)$ of r in T. Let k be the maximum among the values $L_r(u_1), L_r(u_2), \ldots, L_r(u_p)$.
If r has at least four neighbors u_i, u_j, u_k and u_l in T such that $L_r(u_i) = L_r(u_j) = L_r(u_k) = L_r(u_l) = k$, then for any of the neighbors u_m of r in T, r has at least
three children in T_{u_m} with line-labeling k with respect to u_m. Hence, $L_{u_m}(r) = k + 1$
according to condition (b)(i) of the definition of the line-labeling of a vertex. Again,
if r has two neighbors u_i and u_j in T such that $L_r(u_i) = L_r(u_j) = k$ and u_i has
two children x and y in T_r and u_j has two children w and z in T_r such that $L_r(x) = L_r(y) = L_r(w) = L_r(z) = k$, then for each neighbor u_m of r in T, $L_{u_m}(r) = k + 1$
according to condition (b)(ii) of the definition of the line-labeling of a vertex. Hence,
we suppose that r has at most three neighbors in T with line-labeling k with respect
to r and of them, at most one has two children in T_r with line-labeling k with respect
to r.

If none of the neighbors of r with line-labeling k with respect to r has two children
in T_r each with line-labeling k with respect to r, then the computation is accomplished
in the following manner. If r has three neighbors u_i, u_j and u_k such that $L_r(u_i) = L_r(u_j) = L_r(u_k) = k$, then for each of the neighbors u_m with $L_r(u_m) < k$, r has
three children with line-labeling k with respect to v and as such, $L_{u_m}(r) = k + 1$.
For each of the neighbors u_i, u_j and u_k, r has two children having line-labeling k
with respect to itself. Hence, $L_{u_i}(r) = L_{u_j}(r) = L_{u_k}(r) = k$. Again, if r has two
neighbors u_i and u_j such that $L_r(u_i) = L_r(u_j) = k$, from the same line of reasoning,
$L_{u_m}(r) = k$ for each neighbor u_m of r in T. However, for $L_r(u_m) < k$, that is, for u_m,
which is neither u_i nor u_j, r has two children with line-labeling k with respect to u_m.
In the case where r has only a single neighbor u_i with $L_r(u_i) = k$, for each neighbor
u_m of r other than u_i, $L_{u_m}(r) = k$. For u_i, the value of $L_{u_i}(r)$ must be computed from
the values of $L_r(u_m)$ for all such neighbors u_m of r that are not the same as u_i.

Finally, we consider the case when exactly one of the neighbors u_i of r in T with
$L_r(u_i) = k$ has two children x and y in T_r such that $L_r(x) = L_r(y) = k$. In this case,
for every neighbor u_m of r in T other than u_i, $L_{u_m}(r) = k + 1$. For u_i, the value of
$L_{u_i}(r)$ must be computed from the values of $L_r(u_m)$ for all such neighbors u_m of r
that are not the same as u_i.

At this stage, for each of the neighbors v of r, the line-labelings of each of the
neighbors of v in T with respect to v is known. Similarly, when any vertex u of T
is traversed, for each neighbor v of u in T, $L_u(v)$ is known. Therefore, $L(u)$ can be
computed by the definition. Again, the value of $L_v(u)$ for each of the neighbors v of u
in T can be computed in the same way as in the case when the root r is traversed.

We can note that for each vertex u of T, there is at most one neighbor v of u for
which the line-labelings of all the neighbors of u is needed for the computation of
$L_v(u)$. Therefore, the complexity of this top-down traversal is $\sum_{u \in V} deg(u) = O(n)$.
Hence, we can compute the line-labeling of a tree in linear time.   □

In the next section, we give our drawing algorithm for obtaining an upward drawing
of T on exactly k layers where k is the line-labeling of T and thus prove that the
line-labeling of a tree denotes the minimum number of layers required for any upward
drawing of the tree.

4   Upward Drawings on the Minimum Number of Layers

In this section, we give a linear-time algorithm to draw a tree T rooted at a vertex r of T with line-labeling $L(r) = k$ on exactly k layers. We also show that such a drawing of T occupies $O(n \log n)$ area. At the end, we have outlined a linear-time algorithm to obtain an order-preserving and upward drawing of a rooted tree T on the minimum number of layers.

Let T be a tree. Let r be a vertex of T with line-labeling $L(r) = k$. We have proved in the previous section that any upward drawing of T_r requires at least k layers. We now have the following lemma.

Lemma 2. *Let T be a tree. Let r be a vertex of T with line-labeling $L(r) = k$. Then there is an upward drawing of T rooted at r on k layers.*

Prior to giving the proof of Lemma 2, we present the following lemma. We omit the proof of this lemma in this extended abstract.

Lemma 3. *Let T be a tree rooted at a vertex r of T. Let r' be a vertex of T and u be any descendant of the vertex r' in T. Then $L_r(r') \geq L_r(u)$.*

The *skeleton subgraph* of the tree T with respect to the vertex r is the subgraph induced by each vertex u of T such that the line-labeling of u with respect to r in T is equal to the line-labeling of r in T. That is, the *skeleton subgraph* of the tree T with respect to the vertex r is the subgraph induced by all such vertices u of T such that $L_r(u) = L(r)$. In the remainder of this paper, we use the notation $skel(r)$ to denote the skeleton subgraph of a tree T with respect to a vertex r of T. We now have the following lemma whose proof is omitted in this extended abstract.

Lemma 4. *Let T be a tree. Let r be a vertex of T. Then the skeleton subgraph $skel(r)$ of T with respect to r is a path or a single vertex.*

We are now ready to present our proof of Lemma 2.

Proof of Lemma 2. We prove the claim by taking induction on k, i.e., the line-labeling of the vertex r in T.

For the basis case we take $k = 1$, that is $L(r) = 1$. Then, according to Lemma 3, $L(r) = 1$ is the maximum among the line-labelings $L_r(u)$ of all the vertices u of T. Since the line-labeling of a vertex is always positive, for every vertex u of T, $L_r(u) = 1$. Therefore, the skeleton subgraph of T with respect to r is the tree T itself and we get from Lemma 4 that T is a path or a single vertex. Therefore, T can be drawn on one layer. Thus, our claim holds for the basis case.

We now prove the claim for a vertex r of T with line-labeling $L(r) = k > 1$ as the one shown in Fig. 5(a). We assume that the claim holds for any vertex r' in a tree T' with line-labeling $L(r') = k' < k$. That is, a tree T' has an upward drawing rooted at a vertex r' of T' on k' layers, where the line-labeling of r' is $L(r') = k' < k$.

Suppose the line-labeling of the vertex r in the tree T in Fig. 5(a) is $L(r) = k$. In order to obtain an upward drawing of T rooted at r on k layers, we first find the skeleton subgraph $skel(r)$ as illustrated in Fig. 5(b). The skeleton subgraph $skel(r)$ is a path or single vertex according to Lemma 4. We draw $skel(r)$ on the topmost layer as in Fig. 5(c).

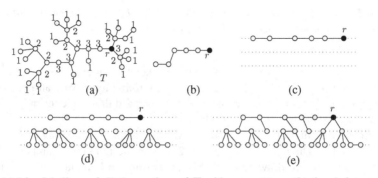

Fig. 5. (a) Line-labelings of all the vertices of T with respect to r, (b) the skeleton subgraph $skel(r)$ of T with respect to r, and (c)–(e) drawing T on $L(r)$ number of layers

If we delete the skeleton subgraph, it leaves the tree with several components, each of which is a tree with the root having line-labeling less than k. From the induction hypothesis, these trees have upward drawings on at most $k-1$ layers. We place the drawings of these trees on the bottommost $k-1$ layers in such a way that for any two of these trees T_1 and T_2 and any pair of vertices u and v in $skel(r)$, if u is placed to the left of v in the drawing of the skeleton subgraph $skel(r)$ and T_1 and T_2 are connected to u and v respectively, then the drawing of T_1 is placed to the left of the drawing of T_2 as illustrated in Fig. 5(d). Then, it is possible to join the edges from the vertices of $skel(r)$ to these trees without violating planarity as illustrated in Fig. 5(e).

Thus, we can obtain an upward drawing of T on k layers where T is rooted at a vertex r with line-labeling $L(r) = k$.   □

The constructive proof of the above lemma gives an algorithm for a k-layer upward drawing of a tree T rooted at a vertex r with line-labeling $L(r) = k$. According to Lemma 1, k represents the minimum number of layers required for any upward drawing of T rooted at r. For the remainder of this paper, we call this algorithm **Draw-min-Layer**. Clearly, the algorithm **Draw-min-Layer** runs in linear time. Also, in the remainder of this paper, we refer to the algorithm outlined in the proof of Theorem 1 for computing the number of layers required for a minimum layer upward drawing of a rooted tree T as the algorithm **Compute-min-Layer**. Note that, if T is not a rooted tree, then the algorithm **Compute-min-Layer** can select a vertex r of T such that a minimum layer upward drawing of T_r results in a minimum layer upward drawing of T.

Now we analyze the area requirement of a drawing of a tree T obtained from the algorithm **Draw-min-Layer**. For this, we establish a relationship between the line-labeling of the vertex r and the number of vertices of T in the following lemma. We omit the proof of this lemma in this extended abstract.

Lemma 5. *Let T be a tree and r be a vertex of T. If $L(r) = k \geq 2$, then T has at least 2^k vertices.*

An immediate consequence of the above lemma is that for any vertex r of an n-vertex tree T, $L(r) = O(\log n)$. We now have the following theorem.

Theorem 2. *Let T be a tree rooted at a vertex r of T. Let n be the number of vertices of T. Then the algorithm* **Draw-min-Layer** *provides a minimum layer upward drawing of T in $O(n \log n)$ area.*

Proof. Let r be the root of the tree T. We find the skeleton subgraph $skel(r)$ of T with respect to r and draw T using the algorithm **Draw-min-Layer**. We can note that the drawing produced by this algorithm maintains the upward drawing convention, that is, no vertex is placed below any of its children in T.

Now, the algorithm draws T in k layers, where k is the line-labeling of r in T. As we have proved in Theorem 1 and Lemma 2, this is the minimum number of layers required for any upward drawing of T. Also, according to Lemma 5, $k = O(\log n)$. Therefore, the height of the drawing is $O(\log n)$. Again, in the drawing, the edges are drawn only between vertices on the same layer or adjacent layers. Then the vertices on a layer can be placed on consecutive x-coordinates keeping planarity. Since the number of vertices on a layer does not exceed n, the width of the drawing is $O(n)$. Hence the area requirement for the upward drawing of T produced by the algorithm is $O(n \log n)$. □

Again for any tree T, the algorithm **Compute-min-Layer** can select a suitable vertex r such that a minimum layer upward drawing of T_r gives a minimum layer upward drawing of T. Therefore, we also have the following theorem.

Theorem 3. *Let T be a tree with n vertices. Then the algorithms* **Compute-min-layer** *and* **Draw-min-Layer** *provides a minimum layer upward drawing of T in $O(n \log n)$ area.*

The drawing we have proposed in this paper for a tree is not order-preserving. The clockwise ordering of the neighbors of a vertex may be shuffled in the drawing. Interestingly, our approach can also be used to obtain a minimum layer drawing of a rooted tree when it is constrained to be order-preserving and upward. Thus we have the following lemma whose proof is omitted in this extended abstract.

Lemma 6. *Let T be a rooted tree. Then a minimum layer upward drawing of T can be computed in linear time where the drawing is constrained to be order-preserving.*

5   Conclusion

In this paper, we have given a linear algorithm to obtain a minimum layer upward drawing of a rooted tree T. If T is not a rooted tree, we have also given a linear algorithm to select a vertex r of T such that a minimum layer upward drawing of T_r results in a minimum layer upward drawing of T. We have shown that our algorithm achieves the best area-bound known so far for upward drawings of arbitrary trees. Almost all the known results on layered drawings of trees are based on the "pathwidth" of trees. We studied layered drawings of trees through a new parameter called the line-labeling of a tree. It remains as our future work to find a relationship between the pathwidth and the line-labeling of a tree. However, we can show that a tree with pathwidth k has line-labeling at least k, which specifies a lower limit of the line-labeling of a tree and thus, a lower limit of the number of layers required for any upward drawing of the tree. It also remains as our future work to obtain planar straight-line drawings of trees that are not necessarily upward but nevertheless requires the minimum number of layers.

References

1. Di Battista, G., Eades, P., Tamassia, R., Tollis, I.G.: Graph Drawing: Algorithms for the Visualization of Graphs. Prentice-Hall, Upper Saddle River, New Jersey (1999)
2. Cornelsen, S., Schank, T., Wagner, D.: Drawing graphs on two and three lines. JGAA 8(2), 161–177 (2004)
3. Crescenzi, P., Di Battista, G., Piperno, A.: A note on optimal area algorithms for upward drawings of binary trees. CGTA 2, 187–200 (1992)
4. Dujmović, V., Fellows, M.R., Hallett, M.T., Kitching, M., Liotta, G., McCartin, C., Nishimura, N., Ragde, P., Rosamond, F.A., Suderman, M., Whitesides, S., Wood, D.R.: On the parameterized complexity of layered graph drawing. In: Meyer auf der Heide, F. (ed.) ESA 2001. LNCS, vol. 2161, pp. 488–499. Springer, Heidelberg (2001)
5. Felsner, S., Liotta, G., Wismath, S.K.: Straight-line drawings on restricted integer grids in two and three dimensions. JGAA 7(4), 363–398 (2003)
6. Garg, A., Goodrich, M.T., Tamassia, R.: Area-efficient upward tree drawings. In: Symposium on Computational Geometry, pp. 359–368 (1993)
7. Garg, A., Goodrich, M.T., Tamassia, R.: Planar upward tree drawings with optimal area. International Journal of Computational Geometry and Applications 6(3), 333–356 (1996)
8. Garg, A., Rusu, A.: Straight-line drawings of general trees with linear area and arbitrary aspect ratio. ICCSA (3), 876–885 (2003)
9. Garg, A., Rusu, A.: Straight-line drawings of binary trees with linear area and arbitrary aspect ratio. JGAA 8(2), 135–160 (2004)
10. Lengauer, T.: Combinatorial Algorithms for Integrated Circuit Layouts. Wiley, New York, USA (1990)
11. Shiloach, Y.: Arrangements of Planar Graphs on the Planar Lattice. PhD thesis, Weizmann Institute of Science (1976)
12. Shin, C.-S., Kim, S.K., Chwa, K.-Y.: Area-efficient algorithms for upward straight-line tree drawings. In: Cai, J.-Y., Wong, C.K. (eds.) COCOON 1996. LNCS, vol. 1090, pp. 106–116. Springer, Heidelberg (1996)
13. Suderman, M.: Pathwidth and layered drawing of trees. International Journal of Computational Geometry and Applications 14(3), 203–225 (2004)
14. Suderman, M.: Proper and planar drawings of graphs on three layers. In: Healy, P., Nikolov, N.S. (eds.) GD 2005. LNCS, vol. 3843, pp. 434–445. Springer, Heidelberg (2006)
15. Warfield, J.N.: Crossing theory and hierarchy mapping. IEEE Transactions on Systems, Man, and Cybernetics 7, 502–523 (1977)
16. Waterman, M.S., Griggs, J.R.: Interval graphs and maps of dna. Bulletin of Mathematical Biology 48, 189–195 (1986)

Guarding Exterior Region of a Simple Polygon

Arindam Karmakar[1], Sasanka Roy[2], and Sandip Das[1]

[1] Indian Statistical Institute, Kolkata - 700 108, India
[2] Tata Consultancy Services, Pune - 411 013, India

Abstract. In this paper, our objective is to locate a position of a guard on the convex hull of a simple polygon P, such that, the farthest point on the boundary of the polygon from the location of the guard, avoiding the interior region of the polygon P, is minimum among all possible locations of the guard on the convex hull. In other words, we try to identify the possible position of guard on the boundary of convex hull such that the maximum distance required to reach a trouble point on the boundary of polygon P avoiding the interior region of P is minimized. The time complexity of our algorithm is $O(n)$ where n is the number of vertices of P.

1   Introduction

Two points a and b are said to be *visible* if the line segment joining them is not obstructed by any other objects. A point p on a given polygon guards the polygon P if every point u of P is visible from point p. Sometimes a single guard may not be able to see the entire polygon. Chvatal [3] established in *Art Gallery theorem* that $\lceil \frac{n}{3} \rceil$ guards are always sufficient and occasionally necessary to cover the interior of a simple polygon with n vertices. Since then a tremendous amount of research on *Art Gallery problem* has been carried out [7,9,11]. The *Fortress Problem*, a variant of Art Gallery problem, considers the minimum number of guards required to cover the exterior of an n-vertex polygon. O'Rourke and Wood [7] proved that $\lceil \frac{n}{2} \rceil$ vertex guards are always sufficient for guarding the exterior of a polygon. Yiu and Choi [4] considered a variation of the problem by allowing each guard to patrol an edge (called an *edgeguard*) of the polygon. Yiu [13] showed that $\lceil \frac{n}{k+1} \rceil$ k-consecutive vertex guards are sometimes necessary and always sufficient to cover the exterior of any n-vertex simple polygons for any fixed $k < n$. Recently, Zylinski [14] used the notion of *cooperative guards* in the Fortress Problem.

Guard placement problems have also been pursued on different interesting variations of the standard notion of visibility [9]. A guard can be considered as a robot that can sense any movement or sound in the boundary of the polygon and quickly reaches the source of problem along the shortest path through the exterior side of the polygon P. Samuel and Toussaint [10] considered the problem of finding two vertices of a simple polygon which maximizes the external shortest path between them in $O(n^2)$ time complexity. This shortest path is denoted as the external geodesic diameter of a simple polygon. Suri [12], proposed an algorithm that computes the farthest neighbors of all the vertices inside a simple

S.-i. Nakano and Md. S. Rahman (Eds.): WALCOM 2008, LNCS 4921, pp. 100–110, 2008.
© Springer-Verlag Berlin Heidelberg 2008

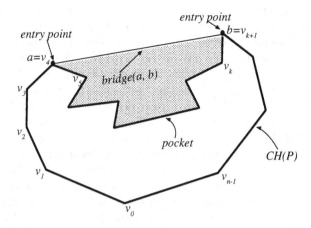

Fig. 1. Pocket, bridge and entry points of a polygon

polygon. Agarwal et. al [1] considered the problem of computing the external farthest neighbors of all the vertices of a simple polygon. Peshkin and Sanderson [8] proposed a linear time algorithm that efficiently finds the externally visible vertices of a simple polygon and the range of the angles from which each is visible. The computation of farthest-site geodesic Voronoi diagram inside a simple polygon was described by Aronov et. al [2].

In this paper, we try to locate a position on the convex hull of a polygon P, such that, the distance of the farthest point(in either direction) on the boundary of the polygon avoiding the interior region of the polygon P from the location of the guard is minimum among all possible location of guard on convex hull. This also gives the size of minimum length rope whose one end is anchored at the optimum location of guard say *anchor* such that other end can touch every point q on the boundary of the polygon where the rope does not pass through the interior of P.

Here we consider P as a simple polygon whose vertices $v_0, v_1, \ldots, v_{n-1}$ are in clockwise order and the edge (v_i, v_{i+1}) is denoted by e_i with length l_i. The sets E and V represent the sets of edges and vertices of the polygon P respectively. The convex hull $CH(P)$ of polygon P is the smallest convex set that includes the set of vertices of polygon P. Note that, all the vertices of P may not be the vertices of convex hull $CH(P)$. The vertices of P, which are not vertices of $CH(P)$ are termed as *notches*. *Bridges* are convex hull edges that connect two non-adjacent vertices of V. *Pockets* are maximal chains of non convex hull edges of P. Therefore each bridge associate with a pocket and the vertices corresponding to a bridge are called the *entry points* of that pocket (see Figure 1).

2   Maximum Distance from Given Source Point

Here, we try to compute the farthest point on the boundary of P from a given point, say p, on $CH(P)$. For a point v on a pocket of P where p is on the bridge of that pocket, the shortest path among these two points avoiding the polygonal region P must be confined in the same pocket. Otherwise the guard located at p must traverse along the convex hull of P either in clockwise direction or in counterclockwise direction. The path of the guard in clockwise direction is termed as *clockwise path*. Similarly, we define the *anti-clockwise path*. Let $\pi_c(v, p)$ and $\pi_a(v, p)$ be respectively the clockwise and anticlockwise shortest paths from p to a point v on boundary of P avoiding interior region of polygon P. $\delta_c(v, p)$ and $\delta_a(v, p)$ denote the lengths of $\pi_c(v, p)$ and $\pi_a(v, p)$ respectively. The minimum among $\delta_c(v, p)$ and $\delta_a(v, p)$ is denoted by $\delta(v, p)$ and the corresponding path is denoted by $\pi(v, p)$. Let $\pi_{opt}(p)$ represents a path which is farthest among all shortest path from p to any point on the boundary of P. The length of the path $\pi_{opt}(p)$ is denoted by $\delta(p)$. Note that, $\frac{\psi}{2} \leq \delta(p) \leq \frac{\phi}{2}$ where ϕ and ψ denote the perimeter of the polygon P and the perimeter of $CH(P)$ respectively. Let the shortest path from p to a point q among all paths in clockwise and anti-clockwise directions be $\Pi = (p, \rho_1, \rho_2, \ldots, \rho_k, q)$ where $\rho_1, \rho_2, \ldots, \rho_k$ are the vertices of P. Then ρ_k is said to be a *clip_vertex* for point q on polygon P. Observe that, any point q on P is visible from its *clip_vertex* ρ_k.

Let e be an edge on a pocket $\wp$ of polygon P and (a, b) be the bridge corresponding to that pocket. Consider p as a point on $CH(P)$ which is not on the bridge (a, b) and q be a point on the edge e. Observe that, the shortest path from p to point q must pass through one of the entry points a or b of pocket $\wp$.

Without loss of generality, we may assume that, the clockwise shortest path from p to b is passing through the vertex a and in that case, shortest path from p to point q on e in pocket $\wp$ in clockwise direction must pass through a and shortest path from p to q in counterclockwise direction must pass through b. The following observation prune our search process.

Observation 1. *For some point $q \in e$, if $\delta_c(q, p)$ is maximum among all points on e then q must be an end point of edge e of polygon P. Similar result holds for $\delta_a(q, p)$ also.*

Lemma 1. *From a point p on convex hull of P, all the distances $\delta(v, p)$, $v \in V$, can be computed in $O(n)$ time.*

Proof: Convex hull vertices of P can be determined in $O(n)$ time using the techniques proposed by McCallum and Avis [6]. For all $v \in CH(P)$, $\delta_c(v, p)$ and $\delta_a(v, p)$ can be computed in $O(n)$ time. Consider a point v on the boundary of P inside a pocket $\wp_i$. If p is not on the bridge of $\wp_i$, the shortest path $\pi(v, p)$ must pass through one of the entry points a_i, b_i of pocket $\wp_i$.

Guibas et al. [5] presented a linear time algorithm for computing the Euclidean shortest paths from a given point to all the vertices inside a simple polygon of n vertices . Considering $\wp_i$ along with it's bridge as a simple polygon, similar technique can be used for locating the shortest paths from the entry points of $\wp_i$ to all the vertices inside it. Note that the union of shortest paths from a entry point say a to all vertices inside $\wp_i$ form a tree, termed as shortest path tree rooted at the entry point a. All these distances can be computed in $O(n_i)$ time where n_i is the number of vertices in $\wp_i$. As a_i and b_i are in $CH(P)$, $\delta(a_i, p)$

and $\delta(b_i, p)$ are already computed. Therefore in $O(n_i)$ time, we can compute $\delta(v, p)$ for all vertex v in $\wp_i$. As the total number of vertices in P is n, shortest path trees for all the pockets rooted at their corresponding entry points can be computed in $O(n)$ time. Hence, the lemma follows.   □

Note that, the farthest point from the guard located at p on the boundary of the polygon P may not be a vertex of P. In such case, it must be on an edge of polygon P. Suppose the farthest point γ is on an edge $e = (v_i, v_{i+1})$ of polygon P and therefore $\delta(\gamma, p) \geq \delta(v, p)$ where v is the vertex farthest from p among the vertex set V. Assume p is not on the bridge whose corresponding pocket contains edge e. In that case, both clockwise and counter clockwise paths from p to γ are different. But note that the lengths of the paths $\pi_c(\gamma, p)$ and $\pi_a(\gamma, p)$ are equal. Without loss of generality, assume that $\delta(v_i, p) = \delta_c(v_i, p)$ and in that case, $\delta(v_{i+1}, p) = \delta_a(v_{i+1}, p)$ and both $\delta_a(v_i, p)$ and $\delta_c(v_{i+1}, p)$ are greater than $\delta(\gamma, p)$.

Consider the point p^* which is on $CH(P)$ and $\delta_a(p^*, p)$ and $\delta_c(p^*, p)$ are both equal to half the perimeter of $CH(P)$. We define p^* as the *image* of point p. Note that the image relation is symmetric. Similarly the *image* of edge $e = (v_i, v_{i+1})$ is the portion of the boundary of $CH(P)$, say e^*, such that each point on e have an *image* on e^* and each point on e^* have *image* on edge e. Note that the chain e^* is bounded by points v_i^* and v_{i+1}^*.

Lemma 2. *If p^* is on edge e of the polygon P, then p^* is the only point on P whose $\delta_c(p^*, p)$ and $\delta_a(p^*, p)$ are same. If p^* is on bridge of a pocket $\wp$ then all the points on edges of the polygon P, whose clockwise and anti-clockwise distances are same, must be inside the pocket $\wp$.*

Proof: For a guard p, $\delta_c(p^*, p)$ is equal to $\delta_a(p^*, p)$ and no other point on the convex hull $CH(P)$ exists whose clockwise and anti-clockwise path from p are of equal length.

The rest part of lemma is proved by contradiction. Assume that p' is a point on some edge say e of pocket $\wp'$ with $\delta_c(p', p) = \delta_a(p', p)$ and the bridge (a, b) of the pocket $\wp'$ is different than the hull edge that contains p^*. Therefore, both the distances from p to a and p to b are less than half of the perimeter of the convex hull $CH(P)$. The shortest distances of a and b from p must be in same direction. Without loss of generality, we can assume that the distances of a and b from p are in clockwise direction. In that case all the points on the boundary of the polygon inside the pocket $\wp'$ have a shorter length from p in clockwise direction. That contradicts our assumption.   □

Lemma 3. *At most one point on edges of polygon P have distinct clockwise and anticlockwise path from guard p with same path length.*

Proof: From lemma 2, we can conclude that the points on edge of the polygon P, having distinct shortest clockwise and anti-clockwise path from p with equal length must be inside the same pocket. Let q and q' are two such points in different edges of that pocket. Note that, if clockwise path from p to q does not intersect with anticlockwise path from p to q', then anti-clockwise path from p

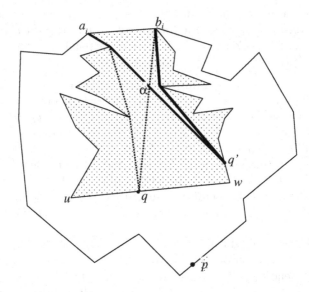

Fig. 2. Illustration of proof of lemma 3

to q must intersects with clockwise path from p to q' (see Figure 2. Let those two paths intersect at point α. $\delta_c(\alpha, p)$ and $\delta_a(\alpha, p)$ must be same. Again, as, clockwise and anticlockwise path from p to q' are disjoint, the anticlockwise path from p to q' is not passing through α. Consider the path from p to α in anticlockwise direction followed by α to q'. This path length is same as the clockwise path from p to q' and hence is equal to shortest anticlockwise path from p to q'. That indicates two different shortest paths from p to q' in anticlockwise direction. That is only possible when α is at q', which implies q and q' are the same points.   □

Theorem 1. *The farthest point on a simple polygon avoiding its internal region from any given location of the convex hull of that polygon can be computed in $O(n)$ time, where n is the number of vertices of the polygon.*

Proof: The farthest point from guard located at p on the boundary of the polygon P may be on an edge of P. In case, the farthest point on the boundary of P from p avoiding its internal region is a vertex, then from Lemma 1, we can recognize it in $O(n)$ time.

Locate the image p^* of point p. If p^* is on some bridge of pocket, say $\wp_i$, then from Lemma 2, farthest point from p is either on some edge or vertex inside $\wp_i$, or on the pocket that contains the point p.

Guibas et al. [5] have shown the algorithm for computing shortest path tree in P considering p as root in time $O(n)$ by using sophisticated data structures for representing, searching and splitting funnels efficiently (Figure 3. For an edge $e' = (u, w)$ of P, let r be the least common ancestor of u and w in the shortest path tree. Let, r_1 and r_2 be the split nodes in path $\Pi(r, u)$ and $\Pi(r, w)$

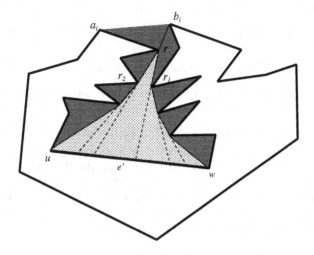

Fig. 3. Funnel splitting

respectively. Then tangent on nodes along the path from r_1 to r_2 through r on the edge e' defines the split nodes on different segments partitioned by the tangents on e'. As the number of nodes in shortest path tree is linear, the total segments formed by the tangents of those nodes on edges of polygon P is of $O(n)$. Suppose a and b are the split nodes for a segment s in clockwise and counter clockwise paths from p respectively, then we can determine the farthest point from p on s in constant time. Hence the lemma follows.   □

3   Locate Guard to Minimize Distance from Its Farthest Point

We characterize below some properties of the optimum guard position for identifying the location of guard on convex hull of the polygon so as to minimize the distance of the farthest point from guard avoiding internal region of the polygon.

Observation 2. *If polygon is convex, then each point on the boundary is the optimal location of guard with distance of its farthest point on polygon boundary avoiding internal region of the polygon is equal to half of perimeter of the polygon.*

From now onwards, we will consider P as a non convex simple polygon. Note that, the number of points that are farthest from location of guard avoiding internal region of polygon may be more than one. Those points may be located on the vertices or edges of the polygon. From lemma 3, we can conclude that, whenever the farthest point f is on an edge of the polygon, then $\delta_c(f, p)$ and $\delta_a(f, p)$ are equal. Let F denotes the set of points on boundary of P which are farthest from optimal location of guard and distance from guard to all points of F are same.

Observation 3. *When the size of set F is one, the guard is either on a bridge of a pocket that contains the farthest point, or the clockwise and anticlockwise paths from guard to $f \in F$ are distinct and $\delta_c(f,p) = \delta_a(f,p)$.*

Again from Lemma 3, we can conclude the following observation.

Observation 4. *If the farthest points from optimal location of guard on $CH(P)$ are not any vertex of the polygon, then the farthest point must be unique.*

Hence we can conclude that, if the cardinality of the set $F = \{f_1, f_2, f_3, \ldots f_k\}$ is greater than one, then there must exist at least one vertex of P in that set. We partition the set F into F_C and F_A depending upon the direction of shortest path from p to $f_i \in F$ in clockwise and anticlockwise directions respectively. If the clockwise and anticlockwise distances of a farthest point are same, place that in both the sets F_C and F_A. Depending upon the organization of the sets F_C and F_A we classify them into following categories (Figure 4):

Type 1: cardinality of set F is one.
Type 2: a vertex of P is common in both the sets F_C and F_A.
Type 3: each set contains at least one vertex of P, but no common vertex in those sets.
Type 4: one set does not contain any vertex of P, but the other one have some vertices of P.

Note that, if the cardinality of the set F is greater than one, then there must exist at least one vertex of P in that set. Surely, we can conclude that partitions F_A and F_B must follow one of the above four types.

Lemma 4. *At most one pocket contains elements from both F_C and F_A. Moreover, if $\wp$ is such a pocket with bridge (a, b), then the location of guard must be in the portion of $CH(P)$ bounded by a^* and b^*, where $\delta_c(a, a^*) = \delta_c(b, b^*) = \frac{\psi}{2}$.*

Observe that, whenever the subsets F_C and F_A follow either of *Type 1* or *Type 2*, then obviously there exist a pocket containing elements from both F_C and F_A and then from Lemma 4, we can estimate the possible location of the guard. In next subsection, we describe an algorithm for locating the optimum position of the guard for the cases *Type 1* and *Type 2*.

3.1   When the Subsets F_C and F_A Are of *Type 1* or *Type 2*

Let the points from both F_C and F_A lie in some pocket $\wp$ with bridge (a, b), where a is on anticlockwise direction of b. Let $v_j(= a), v_{j+1}, \ldots, v_{k-1}, v_k(= b)$ are the points inside the pocket $\wp$ in clockwise order. The optimum location of the guard must be on the image of (a, b) with respect to the convex hull of P. Suppose a point α inside $\wp$ is in both the set F_C and F_A and that may be a vertex or a point on an edge of $\wp$. Let d be the distance of α from p. All the points on the boundary of $\wp$ in clockwise direction of α is reachable from the guard located at p within a distance less than d in anticlockwise direction.

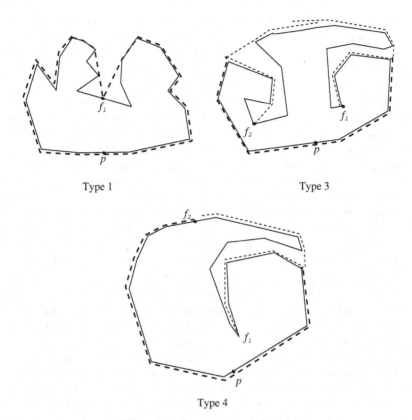

Type 1   Type 3

Type 4

Fig. 4. Farthest point categories

Similarly if the guard moves along clockwise direction, it can reach all the points of $\wp$ in anticlockwise direction of α within a distance less than d. Observe that, $d = \frac{\psi - \delta(a,b)}{2} + \frac{\delta(a,\alpha) + \delta(b,\alpha)}{2}$ where ψ represents the length of the perimeter of $CH(P)$. Again, note that both $\delta_c(a^*, \alpha)$ and $\delta_a(b^*, \alpha)$ are greater than equal to d. In order to locate α inside $\wp$, we follow the following steps.

Step 1: Compute the shortest path tree $T_{\wp,a}$, $T_{\wp,b}$ rooted at a and b respectively for all the vertices inside the pocket $\wp$.

Step 2: For each vertex v in $\wp$, compute a_v and b_v, where a_v is the minimum distance required for reaching all the vertices starting from a to v in clockwise direction and b_v is the minimum distance required for reaching all the vertices starting from b to v in anticlockwise direction.

Step 3: Report the vertex that have minimum $a_v + b_v$ value.

Step 4: Locate the edge (v_i, v_{i+1}) such that $a_{v_i} < b_{v_i}$, $a_{v_{i+1}} > b_{v_{i+1}}$, $a_{v_i} < a_{v_{i+1}}$, and $b_{v_i} > b_{v_{i+1}}$; identify the point q on that edge such that $a_q + b_q$ is minimum among all points on that edge. Note that the time complexity for locating the point q on (v_i, v_{i+1}) is linear in order of the number of vertices in pocket $\wp$.

Step 5: The point α should be either v or q depending upon the lesser values of $a_v + b_v$ and $a_q + b_q$.

Suppose the number of pockets in P is r and consider those r pockets $\{\wp_1, \wp_2, \dots, \wp_r\}$ in clockwise order and a_i and b_i denote the entry points for pocket $\wp_i$ with b_i in clockwise direction of a_i. Images of bridges for all pockets are distinct. Therefore, in order to locate the possible optimal location of guard where subsets F_C and F_A follow either *Type 1* or *Type 2*, we locate points $\alpha_1, \alpha_2, \alpha_3, \dots, \alpha_r$ inside $\wp_1, \wp_2, \dots, \wp_r$ respectively such that α_i is in both the set F_C and F_A whenever members from both F_C and F_A lie inside pocket $\wp_i$. All the points $\alpha_1, \alpha_2, \alpha_3, \dots, \alpha_r$ can be located in $O(n)$ time using the method described above.

Note that, if α_i is in F_A and F_C, guard is located at p_i in $CH(P)$ having distance $d_i = \frac{\delta(a_i, \alpha_i) + \delta(b_i, \alpha_i)}{2} + \frac{\psi - \delta(a_i, b_i)}{2}$ from α_i. Although distance d_i is sufficient to reach every point inside the pocket $\wp_i$ from the guard, but that may not be sufficient to reach other boundary points of polygon P and in that case, farthest point from guard must not be inside that pocket. For eliminating those points from the list of possible locations of guard, we need to solve the query of the following type. For a given point on $CH(P)$ and a given constant ℓ, how many vertices are reachable in clockwise direction (or in anticlockwise direction) from that given point within the path length ℓ. Another query of similar type is to report the minimum distance required from some point for visiting all the vertices in clockwise (anticlockwise) direction up to some given vertex. In order to solve those queries efficiently, we describe below some characterization and preprocessing steps along with some data structures.

Let s_i (s_i') represents the maximum among the distances from a_i (b_i) to all the vertices inside pocket $\wp_i$. A pocket $\wp_i$ *covers up to pocket* $\wp_j$ in clockwise direction implies all the vertices of P from a_i to b_j in clockwise direction are reachable within distance s_i but not able to reach b_{j+1} within that distance, that is from a_i, we are able to reach any point inside pockets $\wp_i, \wp_{i+1}, \dots, \wp_j$ within distance s_i.

Observation 5. *If $\wp_i$ covers up to pocket $\wp_j$ in clockwise direction and $s_{i-1} \geq \delta_c(a_{i-1}, a_i) + s_i$, then $\wp_{i-1}$ also covers up to pocket $\wp_j$ in clockwise direction. If $s_{i-1} < \delta_c(a_{i-1}, a_i) + s_i$ then $\wp_{i-1}$ does not cover pocket $\wp_i$ in clockwise direction.*

From the above observation and using a stack we can compute the amount of cover for pocket $\wp_i$ for all $i = 1, 2, \dots, r$ and therefore we can conclude the following lemma.

Lemma 5. *Total time required for determining the amount of cover for all the pockets $\wp_i$ for $i = 1, 2, \dots, r$ of P in both clockwise and anticlockwise directions is $O(n)$.*

Suppose x is the distance required to visit all the vertices from a_1 to $a_2, a_3, \dots$ etc. in clockwise direction with respect to the image of a_1 i.e, a_1^*. Let the pockets appear along the path $\wp_1, \wp_2, \dots, \wp_m$ and assume that j is the least index such that $\wp_j$ covers $\wp_m$. Hence, x is the maximum value of $\delta_c(a_i, a_j) + s_j$ and $\delta_c(a_1, a_1^*)$. Suppose the pockets which appear along the path from a_2 to a_{2*} in clockwise direction are $\wp_2, \wp_3, \dots, \wp_{m'}$. Obviously $m' \geq m$. Again, note that either $\wp_j$ covers $\wp_{m'}$ or the least index covering $\wp_{m'}$ must be greater m and less than equal to m'.

In case, $\wp_j$ does not cover $\wp_{m'}$, we need to scan $\wp_{m+1}, \wp_{m+2}, \ldots, \wp_{m'}$ in order to locate the least index that covers $r_{m'}$. When $j = 1$, we scan $\wp_2, \wp_3, \ldots, \wp_{m'}$ for locating least index that covers $\wp_{m'}$ and we can compute the minimum distance required to visit all the vertices from a_2 to $a_3, a_4, \ldots$ etc. up to a_2^* in clockwise direction. Iteratively we can compute the minimum distance required from a_i to visit all the vertices a_i^* in clockwise direction for all $i = 1, 2, \ldots, r$ and similarly in anticlockwise direction. Hence we can conclude the following lemma.

Lemma 6. *Let R be the subset of $\{p_1, p_2, \ldots, p_r\}$ such that $p_i \in R$ implies all the points on the boundary of P is reachable from p_i within the distance $d_i = \delta(p_i, \alpha_i)$. The set R can be recognized in $O(n)$ time.*

From the above discussions an lemmata we are in position to conclude the following theorem.

Theorem 2. *If the farthest point from the optimum location of guard follows Type 1 or Type 2, the location of guard can be determined in $O(n)$ time.*

3.2   When the Subsets F_C and F_A Are of *Type 3* or *Type 4*

Here in this case when points from both F_C and F_A lies in same pocket, say $\wp$, with entry points a and b where a is in the anticlockwise direction of b, we locate the edge (v_i, v_{i+1}) in pocket $\wp$ (if it exists) such that $\delta(a, v_i) < \delta(b, v_i)$, $\delta(a, v_{i+1}) > \delta(b, v_{i+1})$, $\delta(a, v_i) < \delta(a, v_{i+1})$ and $\delta(b, v_i) > \delta(b, v_{i+1})$. For the case of *Type 3*, such edge may not exists and even if it exists, we will not get any point q on edge (v_i, v_{i+1}) such that $\delta(a, q) > \delta(a, v_i)$ and $\delta(b, q) > \delta(b, v_{i+1})$. When F_C and F_A follows *Type 4* we will get a point q on edge (v_i, v_{i+1}) such that $\delta(a, q) + \delta(b, q)$ is minimum among all points on that edge and either $\delta(a, q) = \delta(a, v_i)$ and $\delta(b, q) > \delta(b, v_{i+1})$ or $\delta(a, q) > \delta(a, v_i)$ and $\delta(b, q) = \delta(b, v_{i+1})$. The point q can be recognized for each pocket of P and the corresponding location of guard in $O(n)$ time using the same technique described in the previous section.

Consider the case where both F_C and F_A does not contain vertices of P in the same pocket.

Without loss of generality a vertex v_1 in $\wp_1$ is in F_A and assume that the distance of v_1 from entry point a_1 is s_1. Then we have to identify the sequence of vertices in the clockwise direction which are reachable from a_1. Let the last vertex be w. Next we try to compute the minimum distance required to reach all the vertices in anticlockwise direction up to vertex w from a_1. Let the farthest vertex among this set is v_1' and the distance of v_1' be h. Note that h must be less than $\psi - \delta(a_1, b_1) + \delta(b_1, v_1)$. This can be computed in $O(n)$ time. Therefore we can locate the guard whose clockwise farthest point is v_1' and anticlockwise farthest point is v_1. Iteratively we proceed by considering vertex v_2 is in F_A and locate v_2' in similar fashion. Observe that v_2' is either the same vertex v_1' or a vertex in clockwise direction of v_1' and in the anticlockwise direction of v_2'. Hence we can conclude the following theorem.

Theorem 3. *If the farthest point from the optimum location of guard follows Type 3 or Type 4, the location of guard can be determined in $O(n)$ time.*

Here in this version we are omitting the proof.

From the theorems 2 and 3 we can conclude the main result.

Theorem 4. *For a simple polygon P, a guard can be placed on the convex hull of P in O(n) time such that the distance from the guard to its farthest point on the boundary avoiding the interior region of P is minimized.*

References

1. Agarwal, P.K., Aggarwal, A., Aronov, B., Kosaraju, S.R., Schieber, B., Suri, S.: Computing external farthest neighbors for a simple polygon. Discrete Applied Mathematics 31, 97–111 (1991)
2. Aronov, B., Fortune, S., Wilfong, G.: The furthest-site geodesic voronoi diagram. Discrete Computational Geometry 9, 217–255 (1993)
3. Chvatal, V.: A combinatorial theorem in plane geometry. Journal of combinatorial Theory SER B 18, 39–41 (1975)
4. Choi, A.K.O., Yiu, S.M.: Edge guards for the fortress problem. Journal of Geometry 72, 47–64 (2001)
5. Guibas, L., Hershberger, J., Leven, D., Sharir, M., Tarjan, R.E.: Linear Time Algorithms for Visibility and Shortest Path Problems Inside Simple Polygons. Algorithmica 2, 209–233 (1987)
6. McCallum, D., Avis, D.: A linear Algorithm for Finding the Convex Hull of a Simple Polygon. Information Processing Letters 9, 201–206 (1979)
7. O'Rourke, J.: Art Gallery Theorems and Algorithms. Oxford University Press, Oxford (1987)
8. Peshkin, M.A., Sanderson, A.C.: Reachable Grasps on a Polygon: The Convex Rope Algorithm. IEEE Journal of Robotics and Automation RA-2, 53–58 (1986)
9. Sack, J.R., Urrutia, J.: Handbook of computational geometry. Elsevier Science, B. V., Netherlands (2000)
10. Samuel, D., Toussaint, G.T.: Computing the external geodesic diameter of a simple polygon. Computing 44, 1–19 (1990)
11. Shermer, T.: Recent results in Art Galleries. Proceedings of the IEEE, 1384–1399 (1992)
12. Suri, S.: Computing Geodesic Furthest Neighbors in Simple Polygons. Journal of Computer and System Sciences 39, 220–235 (1989)
13. Yiu, S.M.: A generalized fortress problem using k-consecutive vertex guards. Journal of Geometry 39, 220–235 (1989)
14. Zylinski, P.: Cooperative guards in the fortress problem. Balkan Journal of Geometry and Its Applications 72, 188–198 (2004)

Computing Nice Projections of Convex Polyhedra

Md. Ashraful Alam and Masud Hasan

Department of Computer Science and Engineering
Bangladesh University of Engineering and Technology
Dhaka-1000, Bangladesh
{ashrafulalam,masudhasan}@cse.buet.ac.bd

Abstract. In an orthogonal projection of a convex polyhedron P the visibility ratio of a face f (similarly of an edge e) is the ratio of orthogonally projected area of f (length of e) and its actual area (length). In this paper we give algorithms for nice projections of P such that the minimum visibility ratio over all visible faces (over all visible edges) is maximized.

Keywords: Convex polyhedra, nice orthogonal projections, Voronoi diagram, views.

1 Introduction

Polyhedra are 3D solid objects. When we view a polyhedron from a view point, our eyes or camera computes its 2D projection, which can be either orthogonal or perspective based on whether the view point is at infinity or not respectively. Projections are important properties of polyhedra and other 3D objects due to its potential application in the field of computer graphics [10], object reconstruction [4,5], machine vision [1], computational geometry [9,8], and three dimensional graph drawing [7].

Given a polyhedron or other 3D objects such as a set of line segments, it is a well studied problem to compute its "nice" projections based on different criteria for "niceness". McKenna and Seidl [11] studied this problem for convex polyhedra. They presented $O(n^2)$-time algorithms for computing orthogonal projections of a convex polyhedron where the projected area is maximum and minimum. In a similar problem, when a convex polyhedron is orthogonally projected in an arbitrary lower dimension Burger and Gritzmann [6] proved that finding the minimum and maximum volume of the polyhedron is NP-hard, and they gave several approximations algorithms.

Bose et al. [3] studied this problem for line segments in 3D. In their algorithms the criteria for niceness includes minimum crossings among line segments, minimum overlapping among line segments and vertices, and monotonicity of polygonal chains. Eades et al. [7] also studied this problem with similar criteria from the view point of three dimensional graph drawing.

S.-i. Nakano and Md. S. Rahman (Eds.): WALCOM 2008, LNCS 4921, pp. 111–119, 2008.
© Springer-Verlag Berlin Heidelberg 2008

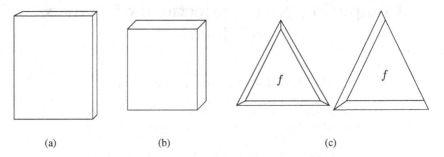

(a)   (b)   (c)

Fig. 1. (a) A maximum area projection, (b) but our algorithm would generate a projection like this. (c) Minimizing the maximum visibility ratio may incur a degenerate projection.

Recently Biedl et al. [2] have studied the problem of computing projections of convex polyhedra such that the silhouette (i.e., the projection boundary) meets certain criteria. They have given several algorithms where a given set of vertices, edges and/or faces appear on the silhouette.

While finding the projections of a polyhedron, it is usually assumed that the degenerate projections are avoided, where a degenerate projection means the view point is coplanar with one or more faces. In the above-mentioned algorithms for computing nice projections of convex polyhedra the degenerate projections are avoided. But in those algorithms it may be possible that the resulting nice projection is almost degenerate. For example, consider the maximum area projection of a convex polyhedron of Figure 1(a). In this projection the big face is visible almost to its full area but the adjacent tiny faces are almost lost. In other words, for some visible faces the visibility ratio, which is the ratio of projected area and actual area, is very small. Similar situation may happen for some edges too. But in some applications such as visual inspection for quality control of manufacturing 3D objects (like toys), a natural expectation is to find a projection where each visible face or edge is viewed in reasonable "amount" so that they can be inspected comfortably for any anomaly (like air pockets).

1.1 The Problem

In this paper, we give algorithms for finding nice orthogonal projections of a convex polyhedron such that in a particular "view" each of the visible faces and edges is viewed as much as possible. More formally, we give algorithms to find orthogonal projections such that within a particular view the minimum visibility ratio over all visible faces (similarly over all visible edges) is maximized. For example, for the polyhedron of Figure 1(a), our algorithm would prefer a projection like that in Figure 1(b) as a nice projection of faces.

We give separate algorithms for faces and edges. Moreover, for edges we consider line segments in 3D and edges of a polyhedron. We consider convex polyhedra and orthogonal projections only. (So from now on by a polyhedron we mean it to be convex and by a projection we mean it to be an orthogonal projection.)

For nice projections of faces and edges in a view $\mathcal{V}$ of P we give $O(|\mathcal{V}| \log |\mathcal{V}| + |C|)$-time algorithms, where $|\mathcal{V}|$ is the number of faces visible in $\mathcal{V}$ and $|C|$ is the size of the "view cone" of $\mathcal{V}$ and can be as small as or less than $|\mathcal{V}|$. Over all possible views of P, our algorithms take a total of $O(n^2 \log n)$ time, where n is the number of vertices of P. For a set of line segments E in 3D, we give an $O(|E| \log |E|)$-time algorithm.

While delving the case of maximizing the minimum visibility ratio, it is natural to ask "Is it possible to minimize the maximum visibility ratio?" Yes, it is possible, and more interestingly, it is possible by using the same technique that we will use for maximizing the minimum ratio. (By the end of this paper it will be clear to the reader.) However, such a nice projection may incur a degenerate projection, which is against the motivation in this paper, and that's why in this paper we do not explore that criteria. For example, in Figure 1(c) the top face f is almost at right angles with the side faces and for that the maximum visibility ratio is due to f. To minimize this ratio we can rotate f as long as an adjacent face does not degenerate. Similar may be the case with lines.

1.2   Outline of the Paper

Rest of the paper is organized as follows. In Section 2 we define the visibility ratio and give other preliminaries. Then Sections 3 and 4 give algorithms for nice projections of faces and edges respectively. Finally, Section 5 concludes the paper with some future work.

2   Preliminaries

A *convex polyhedron* P is the bounded intersection of a finite number of half-spaces. A *face/edge/vertex* of P is the maximal connected set of points which belong to exactly one/exactly two/at least three planes that support these half spaces. By Euler's theorem [14] every convex polyhedron with n vertices has $\Theta(n)$ edges and $\Theta(n)$ faces.

We represent an (orthogonal) view direction d as a unit vector pointing from the origin to the view point (at infinity).

The *visibility ratio* r_f of a visible face f of P with respect to d is the ratio of the projected area of f from d and its actual area. More formally, if θ is the angle between d and the outward normal of f and if $\theta_f = 90° - \theta$, then

$$r_f = \frac{|f| \cos \theta}{|f|} = \sin \theta_f,$$

where $|f|$ means the area of face f.

Let s be a unit sphere centered at the origin. Each point of s uniquely represents a view direction and let p be the point on s for d. The *translated plane* of f is the plane that is parallel to f and passes through the origin, and the intersection of this plane with s gives a great circle and let it be g. Since $\theta_f \leq 90°$ (and

$\sin \theta_f \propto \theta_f$ for $90° \leq \theta_f \leq 0°$), the ratio r_f with respect to d can alternatively be defined as the geodesic distance of p from g on s.

Similarly, the *visibility ratio* r_e of an edge e in 3D with respect to d is the ratio of the projected length of e from d and its actual length. If the acute angle between d and e (assuming e is translated to the origin) is θ_e, then r_e is:

$$r_e = \frac{|e| \sin \theta_e}{|e|} = \sin \theta_e,$$

where $|e|$ means the length of line e.

The *translated line* of an edge e is the line that is parallel to e and passes through the origin, and the intersection of this line with s gives two antipodal points. The ratio r_e with respect to d can alternatively be defined as the minimum geodesic distance of p on s from these two antipodal points.

Among the two half spaces of the translated plane of f one that contains all view directions from which f is visible is called the *positive* half space of f, and the other one is called the *negative* half space of f.

Consider the set of translated planes for all faces (visible or invisible) of P. Arrangement of these planes divides the 3D space into cones called the *view cones*. Within a particular view cone, all view directions have a unique combinatorial projection of P called a *view* of P. It is known that a convex polyhedron with n vertices has $\Theta(n^2)$ different views [13].

Consider a view $\mathcal{V}$ of P. The intersection of the view cone of $\mathcal{V}$ with s gives a spherical convex polygon called the *view polygon* of $\mathcal{V}$. Let this view polygon be C. Each point within C now represents a view direction for $\mathcal{V}$.

3   Nice Projection of Faces

Let the set of faces visible in $\mathcal{V}$ be F. Consider the positive half spaces for all faces in F. Intersection of these half spaces gives a cone called the *face cone* of $\mathcal{V}$. Intersection of this face cone with s gives another spherical polygon called the *face polygon* of $\mathcal{V}$. Let this face polygon be D.

Lemma 1. $C \subseteq D$.

Proof. Both C and D are connected set. A view direction outside D does not see at least one face of F. A view direction within D sees all faces of F, but it may also see one or more faces that are not in F. Let f' be such a face. Let h be the negative half space of f'. Then $C \subseteq D \cap h$. Note that if f' does not exist, then $C = D$.   □

Let G_F be the set of great circles of s that are due to the translated planes of the faces of F. Among G_F let G_D be the set of those who contribute an edge in the boundary of D. Then $G_D \subseteq G_F$. Of course, all faces in F may not contribute to the face cone of $\mathcal{V}$. For example, in Figure 2 five faces are visible but the corresponding face polygon does not contain any piece of the great circle that is due to the shaded top face. We now have the following obvious lemma.

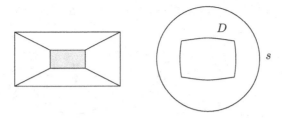

Fig. 2. The great circle corresponding to the top face of the polyhedron does not contribute to the boundary of the corresponding face polygon D

Lemma 2. *From any point within D the closest among all great circles in G_F must be one in G_D.*

After the above two lemmas, our problem of finding an optimum projection for $\mathcal{V}$ now reduces to the problem of finding a point p inside C such that the minimum distance of p from the great circles in G_D is maximized. In what follows in this section we describe how to do that.

3.1   The Optimum Projection

We first consider a special case where F has only one face f. Let y be the point on s that represents the outward normal of f. y is called the *normal point* of f. Then optimum p is y, if y is within C. Otherwise optimum p is a point on C that is closest to y, which can be found trivially in $O(|C|)$ time (by considering only the corner points of C or by considering for each edge x of C a point z such that the segment yz is perpendicular to x.)

Now we come to the general case. *Medial axis* of a polygon Q is the generalized voronoi diagram of Q where the voronoi sites are the vertices and edges of Q. There exists several $O(n \log n)$-time algorithms for computing medial axis on a sphere where the sites are points and segments of great circles [12].

We take the medial axis of D on s and let it be M. Since D is convex and its edges are segments of great circles the edges of M are also segments of great circles. We take the intersection of M and C. This intersection will partition C into small pieces called the *voronoi pieces*. See Figure 3(a).

Consider a particular voronoi piece v. Let the voronoi site of v, which is an edge of D, be a segment of the great circle g. For any point p in v, g is the closest among all great circles in G_D. The optimum value of p within v is the point of v that is furthest from g. We call this point the *candidate point* of v and its geodesic distance from g as its *value*. Finding a candidate point is not obvious and needs some exploration.

Let e_v be a boundary edge of v. We will find the point of e_v that is furthest from g, we again call this point the *candidate point* of e_v. Then the candidate point of v is the maximum among them for all boundary edges of v. Remember that e_v is also a segment of a great circle. Let y be the middle point of the half circle that contains e_v and is bounded by g. If y is in e_v, then the candidate point of e_v is y. Otherwise it is a corner point of e_v. See Figure 3(b).

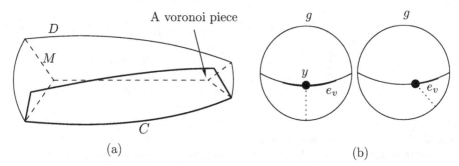

Fig. 3. (a) Intersection of C and M inside D. (b) Finding the candidate point (the filled circle) of e_v from g.

Finally, the resulting p is the maximum among all candidate points.

We now see the time complexity of our algorithm. Each edge of M intersects C into at most two points. So the total number of intersection points of M and C is $O(|M|+|C|)$. Thus the intersection of M and C has a total size of $O(|M|+|C|)$. Moreover, this intersection can be found in $O(|M|+|C|)$ time by first finding one intersection point and then always moving to the other adjacent voronoi region of the currently intersected voronoi edge of M. Since the voronoi pieces are the connected regions of C due to this intersection, $\sum |v| = O(|M|+|C|)$.

The candidate point of e_v can be found in constant time. Thus the candidate point of v can be found in $O(|v|)$ time. So over all v finding optimum p takes $O(|M|+|C|)$ time. Computing M can be done in $O(|D|\log|D|)$ time [12]. Thus the total time taken by the algorithm is $O(|D|\log|D|+|M|+|C|)$. Since $|M|$ is $O(|D|)$ and $O(|D|)$ is $O(|V|)$, where $O(|V|)$ is the number of faces visible in V, this time complexity becomes $O(|V|\log|V|+|C|)$.

The following theorem summarizes the above result.

Theorem 1. *For a view V of P the projection in which the minimum visibility ratio of the visible faces is maximized can be found in $O(|V|\log|V|+|C|)$ time, where $|V|$ is the number of faces visible in V and $|C|$ is the size of the view cone of V.*

Corollary 1. *Optimum projections for all views of P can be found in $O(n^2\log n)$ time.*

Proof. Remember that all of $O(n^2)$ views of P are the result of the arrangement of $O(n)$ planes parallel to $O(n)$ faces of P. So the complexity of the arrangement is also $O(n^2)$. Between any two adjacent view cones the only difference is the common face of the cones. So over all views $\sum_V O(|V|)$ and $\sum_V O(|C|)$ is $O(n^2)$ and thus the total time is $\sum_V O(|V|\log|V|) = O(n^2\log n)$.   □

4   Nice Projection of Line Segments

Let E be the set of line segments. We take the corresponding translated lines for E and then take their intersections with s. Thus we get $|E|$ pairs of antipodal points, whose set we denote by S.

4.1   Line Segments in 3D

Here we do not have any constraint of views and the problem of finding an optimum projection reduces to finding a point p on s such that the minimum distance of p on s from the points of S is maximized.

To do that we take the voronoi diagram V of S on s. If E contains only one edge, then V is simply a great circle and p is any point of V. For E containing more than one edge, V contains at least two voronoi vertices and in that case p is clearly one of those voronoi vertices. By the algorithm of [12] we can find V in $O(|E|\log|E|)$ time. Since the size of V is $O(|E|)$, total time required for finding an optimum p is $O(|E|\log|E|)$.

Theorem 2. *Given a set of line segments E, a projection in which the minimum visibility ratio over all segments is maximized can be found in $O(|E|\log|E|)$ time.*

4.2   Line Segments of a Polyhedron

Let $\mathcal{V}$ be a particular view of P with visible edge set E, view polygon C, and the corresponding set of antipodal points S. Then the problem of finding an optimum projection reduces to the the problem of finding a point p within C such that the minimum distance from p on s to the points of S is maximized.

Since in a view of a polyhedron it is not possible to see only one edge, V is not simply a great circle and has voronoi vertices. We take the intersection of V and C. As before, for each voronoi piece v (created by the intersection) we find the candidate point of v. Since the voronoi edges of V are segments of great circles this point is a corner point of v. Resulting p is the maximum among all candidate points in C.

As before, time complexity of the above algorithm is $O(|\mathcal{V}|\log|\mathcal{V}| + |C|)$.

Theorem 3. *For a view $\mathcal{V}$ of P the projection in which the minimum visibility ratio of a line is maximized can be found in $O(|\mathcal{V}|\log|\mathcal{V}| + |C|)$ time, where $|\mathcal{V}|$ is the number of faces visible in $\mathcal{V}$ and $|C|$ is the size of the view cone of $\mathcal{V}$. Moreover, for all views of P these projections can be found in $O(n^2\log n)$ time.*

5   Conclusion

In this paper, we studied several nice projections of convex polyhedra. We have shown how to find an orthogonal projection for a particular view such that the minimum visibility ratio over the visible faces is maximized. For lines in 3D we have shown a similar result.

An alternative approach of computing nice projection of faces would be as follows. Remember that r_f for a face f increases if the angle θ between the view direction and the normal of f decreases. So finding nice projections of faces is equivalent to finding p on s such that its maximum geodesic distance from the normal points is as small as possible. This can be done by considering the furthest site voronoi diagram M' of the normal points, taking the intersection of C and M', then computing the candidate point for each voronoi piece as the point whose distance is the minimum from the corresponding normal point, and finally finding an optimum p as the *minimum* among all candidate points. However, we found this approach less convenient and avoided for clarity's shake.

As indicated at the begining of the paper, now it should be clear that our algorithm can be easily modified, in conjunction with the above idea of using furthest site voronoi diagram, to find nice projections where the maximum visibility ratio of the faces or edges is minimized. The time complexity should also remain the same.

Finally, we find the following open problem interesting for future work: Is there any way to find a projection such that sum of the projected length of the visible edges is maximum or minimum?

References

1. Bhattacharya, P., Rosenfeld, A.: Polygon in three dimensions. Journal of Visual Communication and Image Representation 5(2), 139–147 (1994)
2. Biedl, T., Hasan, M., Lopez-Ortiz, A.: Efficient view point selection for silhouettes of convex polyhedra. In: Fiala, J., Koubek, V., Kratochvíl, J. (eds.) MFCS 2004. LNCS, vol. 3153, pp. 735–747. Springer, Heidelberg (2004)
3. Bose, P., Gomez, F., Ramos, P., Toussaint, G.: Drawing nice projections of objects in space. Journal of Visual Communication and Image Representation 10(2), 155–172 (1999)
4. Bottino, A., Jaulin, L., Laurentini, A.: Reconstructing 3D objects from silhouettes with unknown view points: The case of planar orthographic views. In: Proceedings of the 8th Iberoamerican Congress on Pattern Recognition, pp. 153–162 (2003)
5. Bottino, A., Laurentini, A.: Introducing a new problem: Shape from silhouette when the relative position of view point is unknown. IEEE Transactions on Pattern Analysis and Machine Intelligence 25(11), 1484–1493 (2003)
6. Burger, T., Gritzmann, P.: Finding optimal shadows of polytopes. Journal of Discrete Comput Geom 24, 219–239 (2000)
7. Eades, P., Houle, M.E., Webber, R.: Finding the best viewpoints for three dimensional graph drawings. In: 5th International Symposium on Graph Drawing, Rome, Italy, pp. 87–98 (September 1998)
8. Gomez, F., Hurtado, F., Selleres, J., Toussaint, G.: Perspective projections and removal of degeneracies. In: Proceedings of the Tenth Canadian Conference on Computational Geometry, pp. 100–101 (1998)
9. Gomez, F., Ramaswami, S., Toussaint, G.: On removing non-degeneracy assumptions in computational geometry. In: Proceedings of the 3rd Italian Conference on Algorithms and Complexity (1997)
10. Kamada, T., Kawai, S.: A simple method for computing general position in displaying three-dimensional objects. Computer Vision, Graphics and Image Processing 41, 43–56 (1988)

11. McKenna, M., Seidel, R.: Finding the optimal shadows of a convex polytope. In: Proceedings of the 1st Annual ACM Symposium of Computational Geometry, pp. 24–28 (1985)
12. Na, H., Lee, C., Cheong, O.: Voronoi diagrams on the sphere. Computational Geometry: Theory and Application 23(2), 183–194 (2002)
13. Plantinga, H., Dyer, C.R.: Visibility, occlusion and the aspect graph. Int. J. Comput. Vision 5(2), 137–160 (1990)
14. West, D.B.: Introduction to Graph Theory, 2nd edn. Pearson Education, London (2002)

A Compact Encoding of Plane Triangulations with Efficient Query Supports

(Extended Abstract)

Katsuhisa Yamanaka and Shin-ichi Nakano

Department of Computer Science, Gunma University,
1-5-1 Tenjin-cho Kiryu, Gunma, 376-8515 Japan
yamanaka@nakano-lab.cs.gunma-u.ac.jp, nakano@cs.gunma-u.ac.jp

Abstract. In this paper we give a coding scheme for plane triangulations. The coding scheme is very simple, and needs only $6n$ bits for each plane triangulation with n vertices. Also with additional $o(n)$ bits it supports adjacency, degree and clockwise neighbour queries in constant time. Our scheme is based on a realizer of a plane triangulation.

The best known algorithm needs only $4.35n + o(n)$ bits for each plane triangulation, however, within $o(n)$ bits it needs to store a complete list of all possible triangulations having at most $(\log n)/4$ nodes, while our algorithm is simple and does not need such a list. The second best known algorithm needs $2m + (5 + 1/k)n + o(m + n)$ bits for each (general) plane graph with m edges and $7n + o(n)$ bits for each plane triangulation, while our algorithm needs only $6n + o(n)$ bits for each plane triangulation.

1 Introduction

Given a class C of graphs how many bits are needed to encode a graph $G \in C$ into a binary string S_G so that S_G can be decoded to reconstruct G? If C contains n_C graphs, then for any coding scheme the average length of S_G is at least $\log n_C$ bits, which is called *the information-theoretically optimal bound*.

By using any generating algorithm, we can encode the k-th generated graph into the binary representation of k, and attain the optimal bound. However such method may need exponential time for encoding and decoding.

On the other hand, for many applications, efficient running time for encoding and decoding is required. Thus for various classes of graphs many coding schemes with efficient running time have been proposed. Moreover, some of those coding schemes support several graph operations in constant time. See [4,5,6,7,8,9,10,11,13,14,19].

In this paper we consider the problem for plane triangulations. We wish to design a simple scheme to encode a given plane triangulation G into a binary string S_G so that (1) S_G can be efficiently decoded to reconstruct G, (2) the length of S_G is short, and (3) S_G supports several graph operations in constant time.

The following results are known for the problem. Let m be the number of edges in a graph, and n the number of vertices. Note that for planar triangulation $m = 3n - 6$ holds.

S.-i. Nakano and Md. S. Rahman (Eds.): WALCOM 2008, LNCS 4921, pp. 120–131, 2008.

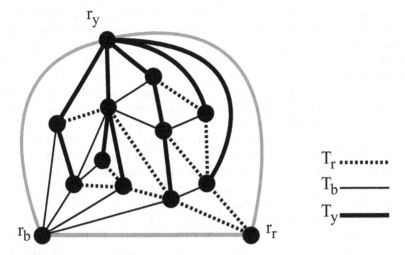

r_y

r_b

r_r

T_r ··········

T_b ——

T_y ▬▬

Fig. 1. An example of a realizer

[20] shows that the information-theoretically optimal bound is $1.08m$ bits for plane triangulations. However this coding scheme needs exponential time for encoding.

For schemes without any query support the following results are known. [19] gives a scheme to encode a general planar graph into asymptotically $4m$ bits. [11] gives schemes to encode a general planar graph into $m \log 12 = 3.58m$ bits, a triconnected planar graph into $3m$ bits, and a plane triangulation into $(3 + \log 3)m/3 = 1.53m$ bits. [8] gives a scheme based on "the canonical ordering" to encode a plane triangulation into $4m/3 - 1$ bits. [15] gives a scheme based on a bijection with a class of trees to encode a plane triangulation into $4m/3$ bits. [3] gives a scheme to encode into $3.37m/3$ bits. [9] gives a scheme to encode a triconnected planar graph and plane triangulation with information-theoretically optimal bounds, respectively.

For schemes with query support the following results are known. [10] gives a scheme to encode trees achieving the information-theoretically optimal bound to within a lower order term, and still supporting some natural query operations quickly. [13,14] gives a scheme to encode a planar graph into $2m + 8n + o(n)$ bits with supporting adjacency and degree query in constant time. [4] gives a scheme to encode a planar graph into $2m + (5 + 1/k)n + o(n)$ bits, where k is any constant, and a plane triangulation into $7n + o(n)$ bits. [5,6] gives a scheme to encode a planar graph into $2m + 2n + o(n)$ bits. [1] gives the best known scheme which needs only $4.35n + o(n)$ bits for each plane triangulation. However, since within $o(n)$ bits it need to store a complete list of all possible triangulations having at most $(\log n)/4$ vertices, it is theoretically nice but an implementation is not easy. [2] gives a scheme to encode a plane triangulation on the topological sphere into $3.24n + o(n)$ bits.

In this paper we design a coding scheme for plane triangulations. Our scheme is very simple and easy to implement.

The class of plane triangulations is an important class of graphs, since the standard representation for fine 3D models, called triangle meshes, consists of a huge amount of vertex coordinate data and a huge size of connectivity data[17]. If the triangle mesh is homeomorphic to a sphere then the connectivity data is a plane triangulation.

We give a simple coding scheme for plane triangulations. The coding scheme needs only $6n$ bits for each plane triangulation, and with additional $o(n)$ bits it supports adjacency and degree queries in $O(1)$ time. Given a vertex u and its neighbour v, many plane graph algorithms need to find the "next" neighbour of u succeeding v in clockwise order, because with this query one can trace a face, and it is one of basic operation for plane graph algorithms. Our coding scheme also finds such a neighbour in $O(1)$ time. Our algorithm is based on a realizer[18] (See an example in Fig. 1.) of a plane triangulation.

The rest of the paper is organized as follows. Section 2 gives some definitions. Section 3 introduces a realizer of a plane triangulation. Section 4 presents our coding scheme. In Section 5 we explain query support. Finally Section 6 is a conclusion.

2   Preliminaries

In this section we give some definitions.

Let $G = (V, E)$ be a connected graph with vertex set V and edge set E. We denote $n = |V|$ and $m = |E|$. An edge connecting vertices x and y is denoted by (x, y). The *degree* of a vertex v, denoted by $d(v)$, is the number of neighbours of v in G.

A graph is *planar* if it can be embedded in the plane so that no two edges intersect geometrically except at a vertex to which they are both incident. A *plane* graph is a planar graph with a fixed planar embedding. A plane graph divides the plane into connected regions called *faces*. The unbounded face is called *the outer face*, and other faces are called *inner faces*. We regard *the contour* of a face as the clockwise cycle formed by the vertices and edges on the boundary of the face. We denote the contour of the outer face of plane graph G by $C_o(G)$. A vertex is *an outer vertex* if it is on $C_o(G)$, and *an inner vertex* otherwise. An edge is *an outer edge* if it is on $C_o(G)$, and *an inner edge* otherwise. A plane graph is called *a plane triangulation* if each face has exactly three edges on its contour. By Euler's Formula: $n - m + f = 2$, where f is the number of faces, one can show $m = 3n - 6$ for any plane triangulation.

3   Realizer

In this section we briefly introduce *a realizer*[18] of a plane triangulation.

Let G be a plane triangulation with three outer vertices r_r, r_b, r_y. We can assume that r_r, r_b, r_y appear on $C_o(G)$ in clockwise order. Those vertices are called *red root*, *blue root* and *yellow root*, respectively. We denote by V_I the set of inner vertices of G.

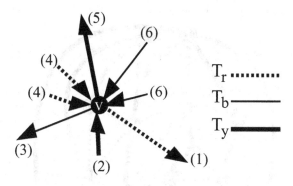

Fig. 2. Edges around an inner vertex v

A *realizer* R of G is a partition of the inner edges of G into three edge-distinct trees T_r, T_b, T_y satisfying the following conditions (c1) and (c2). See an example in Fig. 1.

(c1) For each $i \in \{r, b, y\}$, T_i is a tree with vertex set $V_I \cup \{r_i\}$.
(c2) For each $i \in \{r, b, y\}$, we regard r_i as the root of T_i, and orient each edge in T_i from a child to its parent. Then at each $v \in V_I$ the edges incident to v appear in clockwise order as follows. See Fig. 2.

(1) exactly one edge in T_r leaving from v.
(2) (zero or more) edges in T_y entering into v.
(3) exactly one edge in T_b leaving from v.
(4) (zero or more) edges in T_r entering into v.
(5) exactly one edge in T_y leaving from v.
(6) (zero or more) edges in T_b entering into v.

Let G be a plane triangulation, and $R = \{T_r, T_b, T_y\}$ be a realizer of G. Again for each $i \in \{r, b, y\}$ we regard r_i as the root of T_i, and orient each edge in T_i from a child to its parent.

Then $T = T_y \cup \{(r_y, r_b), (r_y, r_r)\}$ is a spanning tree of G with root r_y. By preorder traversal of T we assign an integer $i(v)$ for each vertex v. See an example in Fig. 3. Note that $i(r_y) = 1, i(r_b) = 2, i(r_r) = n$ always holds.

We have the following lemma.

Lemma 1. *(a) If $e = (u, v)$ is an edge in T_r and orient from u to v, then $i(u) < i(v)$.*
(b) If $e = (u, v)$ is an edge in T_b and orient from u to v, then $i(u) > i(v)$.

Proof. (a) Assume otherwise for the contradiction. Now there is an edge $e = (u, v)$ in T_r and orient from u to v, but $i(u) > i(v)$.

For each $i \in \{r, b, y\}$ let P_i be the path in T_i starting at v and ending at the root r_i. Then by those three paths we partite the plane graph into three regions as follows. Region $\overline{R_r}$: The region inside of $P_y \cup P_b \cup \{(r_b, r_y)\}$. Region $\overline{R_b}$: The region inside of $P_r \cup P_y \cup \{(r_y, r_r)\}$. Region $\overline{R_y}$: The region inside of $P_b \cup P_r \cup \{(r_r, r_b)\}$.

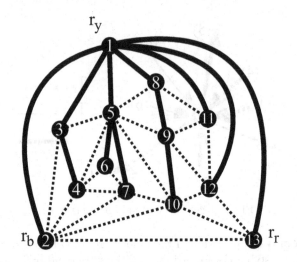

Fig. 3. The spanning tree and remaining edges

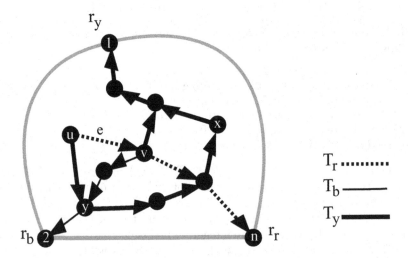

Fig. 4. Illustration for Lemma 1

By the condition (c2) of the realizer, vertex u is in $\overline{R_r}$.

By assumption $i(u) > i(v)$ above, the path P in T_y starting at u and ending at the root r_y must contain at least one vertex in $\overline{R_y} \cup \overline{R_b}$. See Fig. 4. Thus P must cross P_b from $\overline{R_r}$ to $\overline{R_y}$. However, at the crossing point, say vertex y, the condition (c2) of the realizer does not hold. A contradiction.

(b) Similar to (a). Omitted. □

Let $\overline{G_T}$ be the graph derived from G by deleting all edges in the spanning tree T. If $\overline{G_T}$ has an edge (u, v) with $i(u) > i(v)$ then we say v is a *smaller* neighbour of u and u is a *larger* neighbour of v.

We have the following lemma. See Fig. 3.

Lemma 2. *(a) Each inner vertex v has at least one smaller neighbour and at least one larger neighbour.*
(b) r_r has at least one smaller neighbour and no larger neighbour.
(c) r_b has no smaller neighbour and at least one larger neighbour.
(b) r_y has neither smaller nor larger neighbour.

Proof. (a)Immediate from Lemma 1 and the condition (c2) of the realizer.

Intuitively each inner vertex has one outgoing edge in T_b connecting to one smaller neighbour, and one outgoing edge in T_r connecting to one larger neighbour. See Fig. 3.

(b)(c)(d) Omitted. □

4 Coding

In this section we give our coding scheme for plane triangulations.

Let G be a plane triangulation with a realizer $R = \{T_r, T_b, T_y\}$. Let $T = T_y \cup \{(r_y, r_b), (r_y, r_r)\}$ be a spanning tree of G with root r_y. Assume that by preorder traversal of T each vertex v has an integer label i, as explained in the previous section. Again $\overline{G_T}$ be the graph derived from G by deleting all edges in T. See Fig. 3.

We first encode T into string S_1, then the rest of the graph $\overline{G_T}$ into string S_2. See an example in Fig. 5(a) and (c).

In S_1 each vertex except the root corresponds to a pair of matching parentheses, and if vertex p is the parent of vertex c then the matching parentheses corresponding to p immediately enclose the matching parentheses corresponding to c. See Fig. 5(a).

S_2 consists of $|S_1| - 2$ blocks. See Fig. 5(b). Blocks are hatched alternately to show their boundary. Each block consists of one or more (square) brackets. Each matching brackets corresponds to an edge in $\overline{G_T}$. See Fig. 5(c). Each parenthesis (except for the first and the last one) in S_1 has a corresponding block in S_2. Each open parenthesis "(" in S_1 (except for the first one) has a corresponding block, denoted by $s(v)$, consisting of some "]" 's, and each close parenthesis ")" in S_1 (except for the last one) has a corresponding block, denoted by $l(v)$, consisting of some "[" 's. The length of $s(v)$ is the number of smaller neighbours of v. Thus $s(v)$ consists of $|s(v)|$ of consecutive "]" 's. Similarly, the length of $l(v)$ is the number of larger neighbours of v, and $l(v)$ consists of $|l(v)|$ of consecutive "[" 's.

Since $|s(v)| \geq 1$ always holds, we can encode the block $s(v)$ as $|s(v)| - 1$ consecutive 0's followed by one 1. See Fig. 5(d). Similarly we encode the block $l(v)$ as $|l(v)| - 1$ consecutive 0's followed by one 1. By the encoding above we can easily recognize the boundary of each block. Note that each block always ends with 1, and 1 is always the end of some block.

Now we explain how to encode given G into S_1 and S_2.

First we encode T as follows. Given a (ordered) trees T we traverse T starting at the root with depth first manner. If we go down an edge then we code it with 0, and if we go up an edge then we code it with 1. Let S_1 be the resulting bit string. The length of S_1 is $2(n - 1)$ bits. By regarding the 0 as the open parenthesis "(" and the 1 as the close parenthesis ")", we can regard S_1 as a sequence of balanced parentheses. In S_1 each vertex v except the root r_y correspond to a pair of matching parentheses. Moreover if $i(v) = k$, then v corresponds to the $(k - 1)$-th "(" and its matching ")". Note that the root r_y has no corresponding "(" and ")".

Next we encode $\overline{G_T}$ as follows. We first copy S_1 above into S_2, and then replace each "(" and ")" by some "]" 's, and "[" 's as follows.

Let $i(v) = k$ and $|s(v)|$ be the number of smaller neighbours of v. If $k \neq 1, 2$ then we replace the $(k - 1)$-th "(" by consecutive $|s(v)| - 1$ zeros followed by one "1". Similarly, let $|l(v)|$ be the number of larger neighbours of v. If $k \neq 1, n$ then we replace the ")" which matches the $(k - 1)$-th "(" by consecutive $|l(v)| - 1$ zeros followed by one "1". Note that $|s(v)| \geq 1$ and $|l(v)| \geq 1$ always hold for inner vertex v by Lemma 2.

The idea above is similar to [5], however by utilizing the claim of Lemma 2 we can save two bits for S_2 at each inner vertex.

Now estimate the length of $S_1 + S_2$. We have $|S_1| = 2(n - 1)$ and $|S_2| = 2(3n - 6 - (n - 1)) = 4n - 10$. Thus $|S_1 + S_2| = 2(n - 1) + 4n - 10 = 6n - 12(= 2m)$.

For example the code in Fig. 5 has length $|S_1| + |S_2| = 24 + 42 = 66$ bits.

We have the following lemma.

Lemma 3. *Given a triangulation G we can encode G into $S_1 + S_2$ in $O(n)$ time, where $|S_1 + S_2| = 6n - 12$.*

5   Query

In this section we give an efficient algorithm to answer an adjacency and degree queries with a help of an additional string S_A of $o(n)$ bits. We can construct S_A in $O(n)$ time. We also give an algorithm to answer "the clockwise neighbour" query.

We first define several basic operations. Using those basic operations, we can solve each adjacency, degree and clockwise neighbour queries in constant time.

Given a bitstring, $rank(p)$, the rank of the bit at position p is the number of 1's up to and including the position p, and $select(i)$ is the position of the i-th 1 in the bitstring.

Given a sequence of balanced parentheses, the following operations are defined. Operation $findclose(p)$ computes the position of the close parenthesis that matches the open parenthesis at position p. Operation $findopen(p)$ computes the position of the open parenthesis that matches the close parenthesis at position p. Given an open parenthesis at position p, assume q is the position of p's matching close parenthesis, then $enclose(p)$ is the position of the open parenthesis which immediately encloses the pair, p and q, of the matching parentheses. Operation $wrapped(p)$ computes the number of the positions c_i of open parentheses such

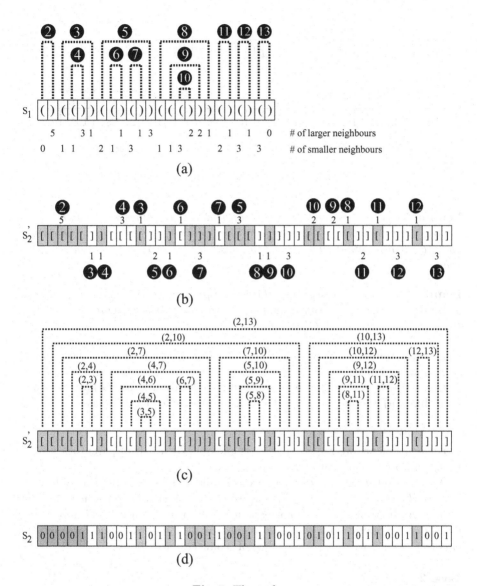

Fig. 5. The code

that $enclose(c_i)=p$. Intuitively $wrapped(p)$ is the number of matching parenthesis pairs which are immediately enclosed by the given matching parenthesis pair. The following lemmas are known.

Lemma 4. *[13,14] Given a bitstring of length $2n$, using $o(n)$ auxiliary bits, we can perform the operations rank(p), select(i), in constant time. One can construct the $o(n)$ auxiliary bits in $O(n)$ time.*

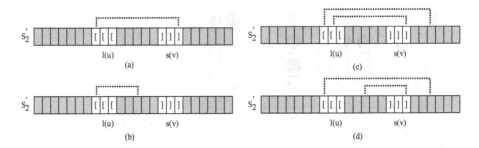

Fig. 6. Illustration for the adjacency query

Lemma 5. *[13,14] Given a sequence of balanced parentheses of length 2n, using o(n) auxiliary bits, we can perform the operations findclose(p), findopen(p), enclose(p) in constant time. One can construct the o(n) auxiliary bits in O(n) time.*

Lemma 6. *[5] Given a sequence of balanced parentheses of length 2n, using o(n) auxiliary bits, we can perform wrapped(p) in constant time. One can construct the o(n) auxiliary bits in O(n) time.*

Then using the basic operations above we can solve an adjacency query in constant time as follows.

Given two integers a and b we are going to decide whether G has edge (u, v) such that $i(u) = a$ and $i(v) = b$. We consider the following two cases.

Case 1: $(u, v) \in T$.

For convenience we regard that S_1 is enclosed by a pair of parentheses corresponding to r_y for operation *enclose()*.

Assume that $a < b$. (The other case is similar.) Then $(u, v) \in T$ iff $select(a - 1) = enclose(select(b - 1))$ in S_1 and we can check this in constant time. Note that since the root r_y has no corresponding "(" thus we need "-1" above. Also note that for operation *select()* we regard S_1 as a bitstring, and for operation *enclose()* we regard S_1 as a sequence of balanced parentheses.

Case 2: $(u, v) \in \overline{G_T}$.

Assume that $a < b$. (The other case is similar.) Then $(u, v) \in \overline{G_T}$ iff some "[" in $l(u)$ matches some "]" in $s(v)$. We can check this as follows.

We can recognize the block $l(u)$ in S_2 as follows. First $q = findclose(select(a-1))$ is the position of ")" corresponding to u in S_1. The block corresponding to $l(u)$ starts at position $s_u = select(q - 2) + 1$ and ends at $e_u = select(q - 1)$ in S_2. Note that S_2 has no block corresponding to $s(2)$, thus we need "-1" above. Similarly we can recognize the block $s(v)$, and assume that the block starts at position s_v and ends at e_v.

If $findclose(s_u)$ is located among the block $s(v)$, as shown in Fig. 6(a), then $(u, v) \in \overline{G_T}$. Otherwise, if $findclose(s_u) < s_v$, then $(u, v) \notin \overline{G_T}$. See Fig. 6(b).

Otherwise, $findclose(s_u) > e_v$ always holds. If $findopen(e_v)$ is located among the block $l(u)$, then $(u, v) \in \overline{G_T}$. See Fig. 6(c). Otherwise $findopen(e_v) > e_u$ always holds, and $(u, v) \notin \overline{G_T}$. See Fig. 6(d). Thus we can decide whether $(u, v) \in \overline{G_T}$ in constant time.

Also we can solve a degree query in constant time as follows. Given a vertex v we first count the neighbours in T, then the neighbours in $\overline{G_T}$. The sum of them is the degree.

First we count the neighbours in T as follows.

If $i(v) = 1$, then the number n_T of neighbours in T is the number of matching parenthesis pairs which are not enclosed by any matching parenthesis pairs in S_1. For convenience we regard that S_1 is enclosed by a pair of parentheses corresponding to r_y, and compute n_T by so-called "$wrapped(select(0))$". Note that if $i(v) = 1$ then v is the root and has no parent in T.

Otherwise, the number is $1 + wrapped(select(i(v)))$.

Then we count the neighbours in $\overline{G_T}$ as follows. If we can recognize the blocks $s(v)$ and $l(v)$ then the number is $|s(v)| + |l(v)|$.

If $i(v) = 1$ then $|s(v)| + |l(v)| = 0$. If $i(v) = 2$ then $|s(v)| = 0$. If $i(v) = n$ then $|l(v)| = 0$. Otherwise, we can recognize $s(v)$ and $l(v)$ as above, and compute the number in constant time.

Thus we can compute the degree of a given vertex in constant time.

Given two vertices u and its neighbour v with $i(u) = a$ and $i(v) = b$, many plane graph algorithm need to find the neighbour of u succeeding v in clockwise (or counterclockwise) order, since with this query we can (1) trace the boundary of a face, (2) list up the edges around a vertex in clockwise order, and (3) reconstruct G. The neighbour is called *the clockwise neighbour* of u *with respect to* v, and denoted by $cn(u, v)$. We can compute $cn(u, v)$ in constant time, as follows.

Assume that $a > b$. (The other case is similar.) Let $e = (u, cn(u, v))$ be the edge between u and $cn(u, v)$. We have two cases.

Case 1: $(u, v) \in T$.

Then v is the parent of u in T. If $i(u) = n$, then $u = r_r$, $v = r_y$ and $cn(u, v) = r_b$. Otherwise e corresponds to the first "[" in $l(u)$ and its matching "]". With a similar method for adjacency query above we can find the block $l(u)$ and then $cn(u, v)$ in constant time.

Case 2: $(u, v) \in \overline{G_T}$.

In this case e corresponds to some "[" in $l(v)$ and some "]" in $s(u)$.

Assume that e corresponds to the x-th "[" in block $l(v)$ and y-th "]" in $s(u)$. Now we have the following lemma.

Lemma 7. *Either* $x = 1$ *or* $y = |s(u)|$ *holds.*

Proof. Otherwise, $l(v)$ and $s(u)$ has one more matching parenthesis pair "[" and "]", which immediately encloses the matching parenthesis pair corresponding to e. This means G has one more edge between u and v. This contradicts the fact that G has no multi-edge.   □

We have the following theorem.

Theorem 1. *Given $S_1 + S_2$, one can construct an additional string S_A of $o(n)$ bits in $O(n)$ time. Then one can compute adjacency, degree, and clockwise neighbour queries in $O(1)$ time, and decode G in $O(n)$ time.*

6   Conclusion

In this paper we designed a coding scheme for plane triangulations. The coding scheme is simple and needs only $6n$ bits for each plane triangulation. Also with additional $o(n)$ bits it supports adjacency, degree, and clockwise neighbour queries in constant time for each. With a help of recent paper [12], we can also support "a generalized clockwise neighbour query," which finds the clockwise k-th neighbour of u with respect to v in constant time, using $o(n)$ auxiliary bits.

References

1. Aleardi, L.C., Devillers, O., Schaeffer, G.: Succinct representation of triangulations with a boundary. In: Dehne, F., López-Ortiz, A., Sack, J.-R. (eds.) WADS 2005. LNCS, vol. 3608, pp. 134–145. Springer, Heidelberg (2005)
2. Aleardi, L.C., Devillers, O., Schaeffer, G.: Optimal Succinct Representations of Planar Maps. In: Proc. of SCG 2006, pp. 309–318 (2006)
3. Bonichon, N., Gavoille, C., Hanusse, N.: An Information-Theoretic Upper Bound of Planar Graphs Using Triangulation. In: Alt, H., Habib, M. (eds.) STACS 2003. LNCS, vol. 2607, pp. 499–510. Springer, Heidelberg (2003)
4. Chuang, R.C.N., Garg, A., He, X., Kao, M.Y., Lu, H.I.: Compact Encodings of Planar Graphs via Canonical Orderings and Multiple Parentheses. In: Larsen, K.G., Skyum, S., Winskel, G. (eds.) ICALP 1998. LNCS, vol. 1443, pp. 118–129. Springer, Heidelberg (1998)
5. Chiang, Y.T., Lin, C.C., Lu, H.I.: Orderly Spanning Trees with Applications to Graph Encoding and Graph Drawing. In: Proc. of 12th SODA, pp. 506–515 (2001)
6. Chiang, Y.T., Lin, C.C., Lu, H.I.: Orderly Spanning Trees with Applications. SIAM J. Comput. 34, 924–945 (2005)
7. Clark, D.R.: Compact Pat Trees, PhD thesis, University of Waterloo (1998)
8. He, X., Kao, M.Y., Lu, H.I.: Linear-Time Succinct Encodings of Planar Graphs via Canonical Orderings. SIAM J. Discrete Math. 12, 317–325 (1999)
9. He, X., Kao, M.Y., Lu, H.I.: A Fast General Methodology for Information-Theoretically Optimal Encodings of Graphs. SIAM J. Comput. 30, 838–846 (2000)
10. Jacobson, G.: Space-efficient Static Trees and Graphs. In: Proc. of 30th FOCS, pp. 549–554 (1989)
11. Keeler, K., Westbrook, J.: Short Encodings of Planar Graphs and Maps. Discrete Appl. Math. 58, 239–252 (1995)
12. Lu, H.I., Yeh, C.C.: Balanced Parentheses Strike Back. ACM Transactions on Algorithms (to appear)
13. Munro, J.I., Raman, V.: Succinct Representation of Balanced Parentheses, Static Trees and Planar graphs. In: Proc. of 38th FOCS, pp. 118–126 (1997)
14. Munro, J.I., Raman, V.: Succinct Representation of Balanced Parentheses and Static Trees. SIAM J. Comput. 31, 762–776 (2001)

15. Poulalhon, D., Schaeffer, G.: Optimal Coding and Sampling of Triangulations. In: Baeten, J.C.M., Lenstra, J.K., Parrow, J., Woeginger, G.J. (eds.) ICALP 2003. LNCS, vol. 2719, pp. 1080–1094. Springer, Heidelberg (2003)
16. Rosen, K.H. (eds.): Handbook of Discrete and Combinatorial Mathematics. CRC Press, Boca Raton (2000)
17. Rossignac, J.: Edgebreaker: Connectivity compression for triangle meshes. IEEE Trans. on Visualization and Computer Graphics 5, 47–61 (1999)
18. Schnyder, W.: Embedding Planar Graphs in the Grid. In: Proc. of 1st SODA, pp. 138–147 (1990)
19. Turan, G.: Succinct Representations of Graphs. Discrete Appl. Math. 8, 289–294 (1984)
20. Tutte, W.: A Census of Planar Triangulations. Canad. J. Math. 14, 21–38 (1962)

Four-Connected Spanning Subgraphs
of Doughnut Graphs
(Extended Abstract)

Md. Rezaul Karim[1] and Md. Saidur Rahman[2]

[1] Dept. of Computer Science and Engineering, Bangladesh University of Engineering
and Technology(BUET), Dhaka-1000, Bangladesh. Also Dept. of Computer Science
and Engineering, University of Dhaka, Dhaka-1000, Bangladesh
rkarim@univdhaka.edu
[2] Dept. of Computer Science and Engineering, Bangladesh University of Engineering
and Technology(BUET), Dhaka-1000, Bangladesh
saidurrahman@cse.buet.ac.bd

Abstract. The class doughnut graphs is a subclass of 5-connected planar graphs. In a planar embedding of a doughnut graph of n vertices there are two vertex-disjoint faces each having exactly $n/4$ vertices and each of all the other faces has exactly three vertices. Recently the class of doughnut graphs is introduced to show that a graph in this class admits a straight-line grid drawing with linear area and hence any spanning subgraph of a doughnut graph also admits a straight-line grid drawing with linear area. But recognition of a spanning subgraph of a doughnut graph is a non-trivial problem, since recognition of a spanning subgraph of a given graph is an NP-complete problem in general. In this paper, we establish a necessary and sufficient condition for a 4-connected planar graph G to be a spanning subgraph of a doughnut graph. We also give a linear-time algorithm to augment a 4-connected planar graph G to a doughnut graph if G satisfies the necessary and sufficient condition.

Keywords: Planar Graph, Doughnut Graph, Straight-Line Drawing, Grid Drawing, Linear Area Drawing.

1 Introduction

Recently automatic aesthetic drawings of graphs have created intense interest due to their broad applications in computer networks, VLSI layout, information visualization etc., and as a consequence a number of drawing styles have come out [NR04]. The most typical and widely studied drawing style is the "straight-line drawing" of a planar graph. A *straight-line drawing* of a planar graph G is a drawing of G such that each vertex is drawn as a point and each edge is drawn as a straight-line segment without edge crossings. A *straight-line grid drawing* of a planar graph G is a straight-line drawing of G on an integer grid such that each vertex is drawn as a grid point as illustrated in Fig. 1(c). Wagner [Wag36], Fary [Far48] and Stein [Ste51] independently proved

S.-i. Nakano and Md. S. Rahman (Eds.): WALCOM 2008, LNCS 4921, pp. 132–143, 2008.
© Springer-Verlag Berlin Heidelberg 2008

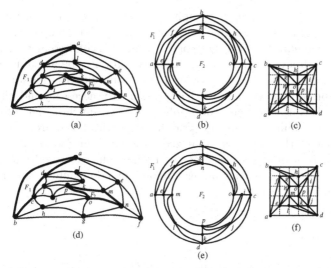

Fig. 1. (a) A doughnut graph G, (b) a doughnut embedding of G, (c) a straight-line grid drawing of G with linear area, (d) a spanning subgraph G' of G, (e) an embedding of G' where face F_1 is embedded as the outerface, and (f) a straight-line grid drawing of G' with area $O(n)$

that every planar graph G has a straight-line drawing. In 1990, de Fraysseix *et al.* [FPP90] proved that both the upper bound and the lower bound of area requirement for a straight-line grid drawing of any planar graph of $n \geq 3$ vertices is $O(n^2)$.

For some restricted classes of graphs, more compact straight-line grid drawings are known. Garg and Rusu showed that a binary tree of n vertices has a planar straight-line grid drawing with area $O(n)$ [GR03b]. Although trees admit straight-line grid drawings with linear area, it is generally thought that triangulations may require a grid of quadratic size. Hence finding nontrivial classes of planar graphs of n vertices richer than trees that admit straight-line grid drawings with area $o(n^2)$ is posted as an open problem in [BEGKLM03]. Garg and Rusu showed that an outerplanar graph with n vertices and the maximum degree d has a planar straight-line drawing with area $O(dn^{1.48})$ [GR03a]. Di Battista and Frati showed that a "balanced" outerplanar graph of n vertices has a straight-line grid drawing with area $O(n)$ and a general outerplanar graph of n vertices has a straight-line grid drawing with area $O(n^{1.48})$ [DF05]. Recently Karim and Rahman introduced a class of graphs, called doughnut graphs, which admits a straight-line grid drawing with linear area [KR07]. They also provide a linear-time algorithm to find such a drawing.

Let $G = (V, E)$ be a planar graph. A subgraph G' is a *spanning subgraph* of G if $V(G') = V(G)$ and $E(G') \subseteq E(G)$. Since a doughnut graph admits a straight-line grid drawing with linear area, one can easily observe that any spanning subgraph of a doughnut graph also admits a straight-line grid drawing with linear area. Fig. 1(f) illustrates a straight-line grid drawing of a graph G' in

Fig. 1(d) where G' is a spanning subgraph of a doughnut graph G in Fig. 1(a). Recognition of a spanning subgraph of a given graph is an NP-complete problem.

In this paper, we narrow down the scope to a subclass of 4-connected planar graphs and establish a necessary and sufficient condition to be a spanning subgraph of a doughnut graph. We also provide a linear-time algorithm to augment a 4-connected graph G to a doughnut graph if G satisfies the necessary and sufficient condition. This gives us a new class of graphs which is a subclass of 4-connected planar graphs that admits straight-line grid drawings with linear area.

The remainder of the paper is organized as follows. In Section 2, we give some definitions. Section 3 provides a necessary and sufficient condition for a 4-connected planar graph to be a spanning subgraph of a doughnut graph. Finally Section 4 concludes the paper.

2   Preliminaries

In this section we give some definitions.

Let $G = (V, E)$ be a connected simple graph with vertex set V and edge set E. Throughout the paper, we denote by n the number of vertices in G, that is, $n = |V|$, and denote by m the number of edges in G, that is, $m = |E|$. An edge joining vertices u and v is denoted by (u, v). The degree of a vertex v, denoted by $d(v)$, is the number of edges incident to v in G. We denote by $\Delta(G)$ the maximum of the degrees of all vertices in G. G is called r-regular if every vertex of G has degree r. We call a vertex v a *neighbor* of a vertex u in G if G has an edge (u, v). The *connectivity* $\kappa(G)$ of a graph G is the minimum number of vertices whose removal results in a disconnected graph or a single-vertex graph K_1. G is called k-*connected* if $\kappa(G) \geq k$. We call a vertex of G a cut-vertex of G if its removal results in a disconnected or single-vertex graph. For $W \subseteq V$, we denote by $G - W$ the graph obtained from G by deleting all vertices in W and all edges incident to them. A *cut-set* of G is a set $S \subseteq V(G)$ such that $G - S$ has more than one component or $G - S$ is a single vertex graph. A *path* in G is an ordered list of distinct vertices $v_1, v_2, ..., v_q \in V$ such that $(v_{i-1}, v_i) \in E$ for all $2 \leq i \leq q$. The *length* of a path is the number of edges on the path. We call a path P is *even* if the number of edges on P is even. We call a path P is *odd* if the number of edges on P is odd.

A graph is *planar* if it can be embedded in the plane so that no two edges intersect geometrically except at a vertex to which they are both incident. A *plane graph* is a planar graph with a fixed embedding. A plane graph G divides the plane into connected regions called *faces*. The unbounded region is called the *outer face*. Let $v_1, v_2, ..., v_l$ be all the vertices in a clockwise order on the contour of a face f in G. We often denote f by $f(v_1, v_2, ..., v_l)$. For a face f in G we denote by $V(f)$ the set of vertices of G on the boundary of face f. We call two faces F_1 and F_2 are *vertex-disjoint* if $V(F_1) \bigcap V(F_2) = \emptyset$. Let f be a face in a plane graph G with $n \geq 3$. If the boundary of f has exactly three edges then we call f a *triangle face* or a *triangulated face*. One can divide a face f of p $(p > 3)$ vertices into $p - 2$ triangulated faces by adding $p - 3$ extra edges. The operation above is called *triangulating a face*. If every face of a graph is triangulated, then the graph is called a *triangulated plane graph*.

A *maximal planar* graph is one to which no edge can be added without losing planarity. Thus in any embedding of a maximal planar graph G with $n \geq 3$, the boundary of every face of G is a triangle, and hence an embedding of a maximal planar graph is often called a triangulated plane graph. It can be derived from the Euler's formula for planar graphs that if G is a maximal planar graph with n vertices and m edges then $m = 3n$ - 6, for more details see [NR04]. We call a face a *quadrangle face* if the face has exactly four vertices.

An *isomorphism* from a simple graph G to a simple graph H is a bijection $f : V(G) \rightarrow V(H)$ such that $(u, v) \in E(G)$ if and only if $(f(u), f(v)) \in E(H)$. Let G_1 and G_2 are two graphs. The *subgraph isomorphism problem* asks to determine whether G_2 contains a subgraph isomorphic to G_1. Subgraph isomorphism is known to be an NP-complete problem [GJ79]. It is not difficult to prove that recognition of a spanning subgraph of a graph is an NP-complete problem using a transformation from subgraph isomorphism problem.

Let G be a 5-connected planar graph, let Γ be any planar embedding of G and let p be an integer such that $p \geq 3$. We call G a p-*doughnut* graph if the following Conditions (d_1) and (d_2) hold: (d_1) Γ has two vertex-disjoint faces each of which has exactly p vertices, and all the other faces of Γ has exactly three vertices; and (d_2) G has the minimum number of vertices satisfying Condition (d_1).

In general, we call a p-doughnut graph for $p \geq 3$ a *doughnut graph*. The following result is known for doughnut graphs [KR07].

Lemma 1. *Let G be a p-doughnut graph. Then G is 5-regular and has exactly $4p$ vertices.*

For a cycle C in a plane graph G, we denote by $G(C)$ the plane subgraph of G inside C excluding C. Let C_1, C_2 and C_3 be three vertex-disjoint cycles in G such that $V(C_1) \cup V(C_2) \cup V(C_3) = V(G)$. Then we call an embedding Γ of G a *doughnut embedding* of G if C_1 is the outer face and C_3 is an inner face of Γ, $G(C_1)$ contains C_2 and $G(C_2)$ contains C_3. We call C_1 the *outer cycle*, C_2 the *middle cycle* and C_3 the *inner cycle* of Γ. Fig. 1(b) illustrates the doughnut embedding of the doughnut graph in Fig. 1(a). The following result regarding doughnut embedding is known for doughnut graphs [KR07].

Lemma 2. *A p-doughnut graph always has a doughnut embedding.*

The algorithm in [KR07] for finding a straight-line grid drawing of a doughnut graph first finds the doughnut embedding of a doughnut graph. Then the three vertex-disjoint cycles are drawn on three nested rectangles in such a way that each edge can be drawn as a straight-line segment without any edge crossings. Fig. 1(c) illustrates a straight-line grid drawing of a doughnut embedding of a doughnut graph in Fig. 1(b). The algorithm takes linear time as stated in the following lemma [KR07].

Lemma 3. *A doughnut graph G of n vertices has a straight-line grid drawing on a grid of area $O(n)$. Furthermore, the drawing of G can be found in linear time.*

3   Characterization

In this section, we give a necessary and sufficient condition for a 4-connected planar graph to be a spanning subgraph of a doughnut graph. The following theorem is the main result of our paper.

Theorem 1. *Let G be a 4-connected planar graph with $4p$ vertices where $p > 4$ and let $\Delta(G) \leq 5$. Let Γ be a planar embedding of G. Assume that Γ has exactly two vertex disjoint faces F_1 and F_2 each of which has exactly p vertices. Then G is a spanning subgraph of a p-doughnut graph if and only if Γ holds the following conditions.*

(a) G has no edge (x, y) such that $x \in V(F_1)$ and $y \in V(F_2)$.
(b) Every face f of Γ has at least one vertex $v \in \{V(F_1) \cup V(F_2)\}$.
(c) For any vertex $x \notin \{V(F_1) \cup V(F_2)\}$, total number of neighbors of x on faces F_1 and F_2 are at most three.
(d) Every face f of Γ except the faces F_1 and F_2 has either three or four vertices.
(e) For any x-y path P such that $V(P) \cap \{V(F_1) \cup V(F_2)\} = \emptyset$ the following conditions hold.
   (i) If the vertex x has exactly two neighbors on face $F_1(F_2)$ and P is even, then the vertex y has at most two neighbors on face $F_1(F_2)$ and at most one neighbor on face $F_2(F_1)$.
   (ii) If the vertex x has exactly two neighbors on face $F_1(F_2)$ and P is odd, then the vertex y has at most one neighbor on face $F_1(F_2)$ and at most two neighbors on face $F_2(F_1)$.

Since G is a 4-connected planar graph, the following fact holds.

Fact 4. *Let G be a 4-connected planar graph with $4p$ vertices. Let Γ and Γ' be any two planar embeddings of G. Then any facial cycle of Γ is a facial cycle of Γ'.*

Fact 4 implies that decomposition of a 4-connected graph G into its facial cycles is unique. (Generally speaking, Fact 4 holds for any 3-connected planar graph.) Throughout the paper we often mention faces of G without mentioning its planar embedding where description of the faces are valid for any planar embedding of G, since $\kappa(G) \geq 4$ for every graph G considered in this paper.

Before proving the necessity of Theorem 1, we have the following fact.

Fact 5. *Let G be a 4-connected planar graph with $4p$ vertices where $p > 4$ and let $\Delta(G) \leq 5$. Assume that G has exactly two vertex disjoint faces F_1 and F_2 each of which has exactly p vertices. If G is a spanning subgraph of a doughnut graph then G can be augmented to a 5-connected 5-regular graph G' through triangulation of all the non-triangulated faces of G except the faces F_1 and F_2.*

We now prove the necessity of Theorem 1.

Proof for Necessity of Theorem 1
Assume that G is a spanning subgraph of a p-doughnut graph. Then by Lemma 1 G has $4p$ vertices. Clearly $\Delta(G) \leq 5$ and G satisfies the conditions (a), (b) and

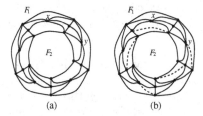

Fig. 2. (a) G has an even x-y path P such that x has exactly two neighbors on face F_1 and y has exactly two neighbors on face F_2 and $V(P) \cap \{V(F_1) \cup V(F_2)\} = \emptyset$; and (b) G' is the resulting graph after triangulations of all non-triangulated faces of G except F_1 and F_2 where vertex y has degree exactly six

(c), otherwise G would not be a spanning subgraph of a doughnut graph. Hence it is sufficient to prove the necessity of Conditions (d) and (e).

(**d**) Since the graph G is simple and four connected, G does not have any face of two or less vertices. Then every face of G has three or more vertices. We now show that G has no face of more than four vertices. Assume for a contradiction that G has a face f of more than four vertices. Any polygon of q vertices can be triangulated by adding $q - 3$ extra edges. These extra edges increase the degrees of $q - 2$ vertices and the sum of the degrees will be increased by $2(q - 3)$. Using pigeon hole principle one can easily observe that there is a vertex among the $q(> 4)$ vertices whose degree will be raised by at least 2 after triangulation of the polygon. Then G' would have a vertex of degree six or more where G' is a graph obtained after triangulation of f. Hence we cannot augment G to a 5-regular graph through triangulation of all the non-triangulated faces of G other than the faces F_1 and F_2. Therefore G can not be a spanning subgraph of a doughnut graph by Fact 5, a contradiction. Hence each face f of G except F_1 and F_2 has either three or four vertices.

(**e**) Assume that G is a spanning subgraph of a doughnut graph. Assume for a contradiction that graph G has an even x-y path P such that $V(P) \cap \{V(F_1) \cup V(F_2)\} = \emptyset$, vertex x has exactly two neighbors on face F_1 and vertex y has more than one neighbor on face F_2, as illustrated in Fig. 2(a). Then one can't augment G to a 5-regular graph by triangulating the non-triangulated faces except F_1 and F_2 since a vertex of degree more than five will appear in such a triangulation as illustrated in Fig. 2(b), where a vertex y of degree exactly six appears in the triangulation. Hence G can not be a spanning subgraph of a doughnut graph by Fact 5, a contradiction. Similarly we can prove that the graph G does not have an odd x-y path P such that vertex x has exactly two neighbors on F_1 and vertex y has more than one neighbor on F_1 and $V(P) \cap \{V(F_1) \cup V(F_2)\} = \emptyset$.

$\mathcal{Q.E.D.}$

In the remaining of this section we give a constructive proof for the sufficiency of Theorem 1. Assume that G satisfies the conditions of Theorem 1. We now have the following lemma.

Lemma 6. *Let G be a 4-connected planar graph satisfying the conditions in Theorem 1. Assume that all the faces of G except F_1 and F_2 are triangulated. Then G is a doughnut graph.*

Proof. To prove the claim, we have to prove that (i) G is a 5-connected graph, (ii) G has two vertex disjoint faces each of which has exactly p, $p > 4$ vertices, and all the other faces of G has exactly three vertices, and (iii) G has the minimum number of vertices satisfying properties (i) and (ii).

(i) We first prove that G is a 5-regular graph. Every face of G is a triangle except F_1 and F_2. Furthermore each of F_1 and F_2 has exactly p, $p > 4$ vertices. Then G has $3(4p) - 6 - 2(p - 3) = 10p$ edges. Since none of the vertices of G has degree more than five and G has exactly $10p$ edges, each vertex of G has degree exactly five. We now prove that vertices of G lie on three vertex-disjoint cycles C_1, C_2 and C_3 such that cycles C_1, C_2, C_3 contain exactly p, $2p$ and p vertices, respectively. We take an embedding Γ of G such that F_1 is embedded as the outer face and F_2 is embedded as an inner face. We take the contour of face F_1 as cycle C_1 and contour of face F_2 as cycle C_3. Then each of C_1 and C_2 contain exactly p, $p > 4$ vertices. Since G satisfies Conditions (a), (b) and (c) of Theorem 1 and all the faces of G except F_1 and F_2 are triangulated, the rest $2p$ vertices of G form a cycle in Γ. We take this cycle as C_2. $G(C_2)$ contains C_3 since G satisfies Condition (b) in Theorem 1. Clearly C_1, C_2 and C_3 are vertex-disjoint and cycles C_1, C_2, C_3 contain exactly p, $2p$ and p vertices, respectively. We now prove that G is 5-connected. Assume for a contradiction that G has a cut-set of less than five vertices. In such a case, G would have a vertex of degree less than five or G would have a face of four or more vertices except F_1 and F_2, a contradiction. Hence G is 5-connected.

(ii) The proof of this part is obvious since G has two vertex disjoint faces each of which has exactly p vertices and all the other faces of G has exactly three vertices.

(iii) The number of vertices of G is $4p$. It is the minimum number of vertices required to construct a graph that satisfies the properties (i) and (ii) [KR07] .

Therefore G is a doughnut graph. $\qquad\qquad\qquad\qquad\qquad\mathcal{Q.E.D.}$

We thus assume that G has a non-triangulated face f except faces F_1 and F_2. By Condition (d) of Theorem 1, f has exactly four vertices. We call a quadrangle face f of G an α-face if f contains at least one vertex from each of the faces F_1 and F_2 as illustrated in Fig. 3, where $f_1(a, b, c, d)$ is an α-face. We call a quadrangle face f of G a β-face if f contains at least one vertex either from F_1 or from F_2 but not from both the faces as illustrated in Fig. 3, where $f_2(p, q, r, s)$ is a β-face. We call a triangulation of an α-face f of G a *valid triangulation* of f if no edge is added between any two vertices x, $y \in V(f)$ such that $x \in V(F_1)$ and $y \in V(F_2)$. Faces $f_1(a, b, c, d)$ and $f_2(p, q, r, s)$ in Fig. 4(a) are two α-faces and Fig. 4(b) illustrates the valid triangulations of f_1 and f_2.

We now have the following fact for an α-face f.

Fact 7. *Let G be a 4-connected planar graph satisfying the conditions in Theorem 1. Let f be an α-face in G. Then f admits unique valid triangulation and*

Fig. 3. $f_1(a, b, c, d)$ is an α-face and $f_2(p, q, r, s)$ is a β-face

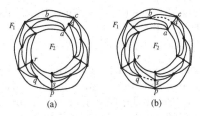

Fig. 4. (a) $f_1(a, b, c, d)$ and $f_2(p, q, r, s)$ are two α-faces, and (b) valid triangulations of f_1 and f_2

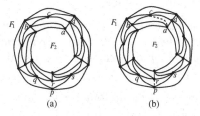

Fig. 5. (a)$f_1(a, b, c, d)$ and $f_2(p, q, r, s)$ are two β_1-faces, and (b) valid triangulations of f_1 and f_2

the triangulation is obtained by adding an edge between two vertices those are not on F_1 and F_2.

We call a β-face a *β_1-face* if the face contains exactly one vertex either from F_1 or from F_2 as illustrated in Fig. 5(a), where $f_1(a, b, c, d)$ and $f_2(p, q, r, s)$ are two β_1-faces. Otherwise we call a β-face a *β_2-face* as illustrated in Fig. 6(a), where face $f_1(a, b, c, d)$ is a β_2-face. Since G is 4-connected, a β_2-face f of G has exactly two vertices either from F_1 or from F_2. We call a vertex of a β_1-face f a *middle* vertex of f if the vertex is in the middle position among the three consecutive vertices other than the vertex on F_1 or F_2. In Fig. 5(a) vertex c of face f_1 and vertex r of face f_2 are the middle vertices. We call a triangulation of a β_1-face f of G a *valid triangulation* of f if no edge is added between any two vertices $x, y \in V(f)$ such that $x, y \notin V(F_1) \cup V(F_2)$ as illustrated in Fig. 5(b), where valid triangulations of faces f_1 and f_2 in Fig. 5(a) are illustrated.

We now have the following fact for a β_1-face f.

Fact 8. *Let G be a 4-connected planar graph satisfying the conditions in Theorem 1. Let f be a β_1-face of G. Then f admits unique valid triangulation and*

*the triangulation is obtained by adding an edge between the vertex on F_1 or F_2
and the middle vertex.*

We call a triangulation of a β_2-face f of G a *valid triangulation* of f if the
triangulation is obtained by adding an edge (a, c) where $a \in V(F_1)(V(F_2))$,
$c \notin \{V(F_1) \cup V(F_1)\}$ and one of the following conditions holds.

(i) G has an even x-c path P such that x has exactly two neighbors on $F_1(F_2)$
and $V(P) \cap \{V(F_1) \cup V(F_2)\} = \emptyset$.
(ii) G has an odd x-c path P such that x has exactly two neighbors on $F_2(F_1)$
and $V(P) \cap \{V(F_1) \cup V(F_2)\} = \emptyset$.
(iii) G satisfies none of the Conditions (i) and (ii).

The graph G in Fig. 6(a) satisfies Condition (i) and the face $f_1(a, b, c, d)$ in G
is a β_2-face, and Fig. 6(d) illustrates the valid triangulation of f_1. The graph G
in Fig. 6(b) satisfies Condition (ii) and the face $f_2(p, q, r, s)$ in G is a β_2-face,
and Fig. 6(e) illustrates the valid triangulation of f_2. The graph G in Fig. 6(c)
satisfies Condition (iii) and the face $f_3(w, x, y, z)$ in G is a β_2-face, and Fig 6(f)
illustrates the valid triangulation of f_3.

We now have the following fact for a β_2-face f.

Fact 9. *Let G be a 4-connected planar graph satisfying the conditions in Theorem 1. Let f be a β_2-face of G. Then f admits unique valid triangulation and
the triangulation is obtained by adding an edge between a vertex on face F_1 or
F_2 and a vertex $x \notin V(F_1) \cup V(F_2)$.*

We now need a definition. Let v_1, v_2, v_3, v_4 are four vertices in a clockwise order
on the contour of a quadrangle face f. We call a vertex v_i on the contour of

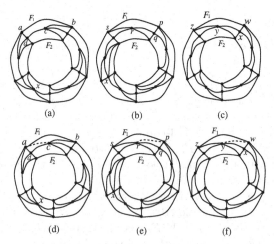

Fig. 6. Illustration for valid triangulation of β_2-face; (a) $f_1(a, b, c, d)$ is a β_2-face and
G satisfies condition (i), (b) $f_2(p, q, r, s)$ is a β_2-face and G satisfies condition(ii), (c)
$f_3(w, x, y, z)$ is a β_2-face and G satisfies condition (iii), and (d), (e), (f) illustrates the
valid triangulations of f_1, f_2 and f_3, respectively

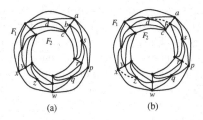

Fig. 7. (a) G has an α-face $f_1(a,b,c,d)$, a β_1-face $f_2(p,q,r,s)$ and a β_2-face $f_3(w,x,y,z)$, and (b) valid triangulation of faces f_1, f_2 and f_3

f a *good vertex* if v_i is one of the end vertex of an edge which is added for a valid triangulation of f as mentioned in Facts 7, 8 and 9. Fig. 7(a) illustrates an example where $f_1(a,b,c,d)$ is an α-face, $f_2(p,q,r,s)$ is a β_1-face and $f_3(w,x,y,z)$ is a β_2 face and Fig. 7(b) illustrates the valid triangulations of faces f_1, f_2 and f_3 where b and d are good vertices of f_1, p and r are good vertices of f_2, and x and z are good vertices of f_3. We now have the following facts regarding good vertices on α- and β_1-faces.

Fact 10. *For any α-face f of G, a good vertex of f is contained in none of F_1 and F_2 and has neighbors on both F_1 and F_2.*

Fact 11. *For any β_1-face f of G, one of the good vertices is on $F_1(F_2)$ and the other good vertex is the middle vertex of f having neighbors only on $F_2(F_1)$.*

We now have the following Lemmas 12 and 13 whose proofs are omitted in this extended abstract. Condition (e) of Theorem 1 is effectively used in the proof of Lemmas 12 and 13.

Lemma 12. *Let G be a 4-connected planar graph satisfying the conditions in Theorem 1. Then any quadrangle face f of G admits unique valid triangulation such that after triangulation $d(v) \le 5$ holds for any vertex v in the resulting graph.*

Lemma 13. *Let G be a 4-connected planar graph satisfying the conditions in Theorem 1. Also assume that G has quadrangle faces. Then no two quadrangle faces f_1 and f_2 have a common vertex which is a good vertex for both the faces f_1 and f_2.*

We are now ready to give a proof for the sufficiency of the Theorem 1.

Proof for Sufficiency of Theorem 1

Assume that the graph G satisfies all the conditions of Theorem 1. If all the faces of G except F_1 and F_2 are triangulated, then by Lemma 6 G is a doughnut graph. Otherwise, we triangulate each quadrangle face of G, using its valid triangulation. Let G' be the resulting graph. For each vertex v in G' $d(v) \le 5$, since according to Lemma 12 degree of each vertex of the graph remains five or less after valid triangulation of each quadrangle face and by Lemma 13 no two quadrangle faces of G have a common vertex which is a good vertex for both the faces. Since G satisfies the conditions in Theorem 1, G' is obtained from G using valid triangulations of quadrangle faces and $d(v) \leq 5$ for each vertex v in G', G' satisfies the conditions in Theorem 1. Hence G' is a doughnut graph by Lemma 6. Therefore G is a spanning subgraph of a doughnut graph. $\mathcal{Q.E.D.}$

We now have the following lemma.

Lemma 14. *Let G be a 4-connected planar graph satisfying the conditions in Theorem 1. Then G can be augmented to a doughnut graph in linear time.*

Proof. We first embed G such that F_1 is embedded as the outer face and F_2 is embedded as an inner face. We then triangulate each of the quadrangle faces of G using its valid triangulation if G has quadrangle faces. Let G' be the resulting graph. As shown in the sufficiency proof of Theorem 1, G' is a doughnut graph. One can easily find all quadrangle faces of G and perform their valid triangulations in linear time, hence G' can be obtained in linear time. $\mathcal{Q.E.D.}$

In Theorem 1 we have given a necessary and sufficient condition for a 4-connected planar graph to be a spanning subgraph of a doughnut graph. As described in the proof of Lemma 14 we have provided a linear-time algorithm to augment a 4-connected planar graph G to a doughnut graph if G satisfies the conditions in Theorem 1. We have thus identified a subclass of 4-connected planar graphs that admits straight-line grid drawings with linear area as stated in the following theorem.

Theorem 2. *Let G be a 4-connected planar graph satisfying the conditions in Theorem 1. Then G admits a straight-line grid drawing on a grid of area $O(n)$. Furthermore, the drawing of G can be found in linear time.*

Proof. Using the method described in the proof of Lemma 14, we augment G to a doughnut graph G' by adding dummy edges (if required) in linear time. By Lemma 3, G' admits a straight-line grid drawing on a grid of area $O(n)$. We finally obtain a drawing of G from the drawing of G' by deleting the dummy edges (if any) from the drawing of G'. By Lemma 14, G can be augmented to a doughnut graph in linear time and by Lemma 3, straight-line grid drawing of a doughnut graph can be found in linear time. Moreover, the dummy edges can also be deleted from the drawing of a doughnut graph in linear time. Hence the drawing of G can be found in linear time. $\mathcal{Q.E.D.}$

4   Conclusion

In this paper, we established a necessary and sufficient condition for a 4-connected planar graph G to be a spanning subgraph of a doughnut graph. We also gave a linear-time algorithm to augment a 4-connected planar graph G to a doughnut graph if G satisfies the necessary and sufficient condition. By introducing the necessary and sufficient condition, in fact, we have identified a subclass of 4-connected planar graphs that admits straight-line grid drawings with linear area. Recognition of a three-connected spanning subgraph of a doughnut graph and it's augmentation to a doughnut graph looks a non-trivial problem and it is left as an open problem.

References

[BEGKLM03] Brandenburg, F., Eppstein, D., Goodrich, M.T., Kobourov, S., Liotta, G., Mutzel, P.: Selected open problems in graph drawing. In: Liotta, G. (ed.) GD 2003. LNCS, vol. 2912, pp. 515–539. Springer, Heidelberg (2004)

[DF05] Di Battista, G., Frati, F.: Small area drawings of outerplanar graphs. In: Healy, P., Nikolov, N.S. (eds.) GD 2005. LNCS, vol. 3843, pp. 89–100. Springer, Heidelberg (2006)

[Far48] Fary, I.: On Straight line representation of planar graphs. Acta Sci. Math. Szeged 11, 229–233 (1948)

[FPP90] de Fraysseix, H., Pach, J., Pollack, R.: How to draw a planar graph on a grid. Combinatorica 10, 41–51 (1990)

[GJ79] Garey, M.R., Johnson, D.S.: Computers and Intractability: A guide to the theory of NP-completeness. W. H. Freeman and Company, New York (1979)

[GR03a] Garg, A., Rusu, A.: Area-efficient drawings of outerplanar graphs. In: Liotta, G. (ed.) GD 2003. LNCS, vol. 2912, pp. 129–134. Springer, Heidelberg (2004)

[GR03b] Garg, A., Rusu, A.: A more practical algorithm for drawing binary trees in linear area with arbitrary aspect ratio. In: Liotta, G. (ed.) GD 2003. LNCS, vol. 2912, pp. 159–165. Springer, Heidelberg (2004)

[KR07] Karim, M.R., Rahman, M.S.: Straight-line grid drawings of planar graphs with linear area. In: Hong, S., Ma, K. (eds.) Proc. of Asia-Pacific Symposium on Visualisation 2007, pp. 109–112. IEEE, Los Alamitos (2007)

[NR04] Nishizeki, T., Rahman, M.S.: Planar Graph Drawing. World Scientific, Singapore (2004)

[Ste51] Stein, K.S.: Convex maps. Proc. Amer Math. Soc. 2, 464–466 (1951)

[Wag36] Wagner, K.: Bemerkugen zum veierfarben problem. Jasresber. Deutsch. Math-Verien. 46, 26–32 (1936)

Exact Algorithms for Maximum Acyclic Subgraph on a Superclass of Cubic Graphs

Henning Fernau and Daniel Raible

Univ.Trier, FB 4—Abteilung Informatik, 54286 Trier, Germany
{fernau,raible}@uni-trier.de

Abstract. Finding a maximum acyclic subgraph is on the list of problems that seem to be hard to tackle from a parameterized perspective. We develop two quite efficient algorithms (one is exact, the other parameterized) for $(1, n)$-graphs, a class containing cubic graphs. The running times are $\mathcal{O}^*(1.1871^m)$ and $\mathcal{O}^*(1.212^k)$, respectively, determined by an amortized analysis via a non-standard measure.

1 Introduction and Definitions

Our Problem. The FEEDBACK ARCSET PROBLEM FAS, is on the list of 21 problems that was presented by R.M. Karp [10] in 1972 exhibiting the first $\mathcal{NP}$-complete problems. It has numerous applications [7], ranging from program verification, VLSI and other network applications to graph drawing, where in particular the re-orientation of arcs in the first phase of the Sugiyama approach to hierarchical layered graph drawing is equivalent to FAS, see [2,17]. More formally, we consider the dual of FAS, namely the following problem:

MAXIMUM ACYCLIC SUBGRAPH MAS
Given a directed graph $G(V, A)$, and the parameter k.
We ask: Is there a subset $A' \subseteq A$, with $|A'| \geq k$, which is acyclic?

In this paper, we deal with finding exact and parameterized algorithms for MAS. Mostly, we focus on a class of graphs that, to our knowledge, has not been previously described in the literature. Let us call a directed graph $G = (V, E)$ $(1, n)$-*graph* if, for each vertex $v \in V$, its indegree $d^+(v)$ obeys $d^+(v) \leq 1$ or its outdegree $d^-(v)$ satisfies $d^-(v) \leq 1$ (i.e, $\forall v \in V : \min\{d^+(v), d^-(v)\} \leq 1$.). In particular, graphs of maximum degree three are $(1, n)$-graphs. Notice that MAS, restricted to cubic graphs, is still $\mathcal{NP}$-complete. For some applications from graph drawing (e.g., laying out "binary decision diagrams" where vertices correspond to yes/no decisions) even the latter restriction is not so severe at all. Having a closer look at the famous paper of I. Nassi and B. Shneiderman [13] where they introduce structograms to aid structured programming (and restricting the use of GOTOs), it can be seen that the resulting class of flowchart graphs is that of $(1, n)$-graphs.

Cubic graphs also have been discussed in relation to approximation algorithms: A. Newman [14] showed a factor $\frac{12}{11}$-approximation. Having a closer look

S.-i. Nakano and Md. S. Rahman (Eds.): WALCOM 2008, LNCS 4921, pp. 144–156, 2008.
© Springer-Verlag Berlin Heidelberg 2008

at her algorithm reveals that it also works for $(1,n)$-graphs with the same approximation factor. This largely improves on the general situation, where only a factor of 2 is known [2]. We point out that finding a minimum feedback arc set (in general graphs) is known to possess a factor $\log n \log\log n$-approximation, see [7], and hence shows an approximability behavior much worse than MAS.

Our Framework: Parameterized Complexity. A *parameterized problem* P is a subset of $\Sigma^* \times \mathbb{N}$, where Σ is a fixed alphabet and $\mathbb{N}$ is the set of all non-negative integers. Therefore, each instance of the parameterized problem P is a pair (I, k), where the second component k is called the *parameter*. The language $L(P)$ is the set of all YES-instances of P. We say that the parameterized problem P is *fixed-parameter tractable* [5] if there is an algorithm that decides whether an input (I, k) is a member of $L(P)$ in time $f(k)|I|^c$, where c is a fixed constant and $f(k)$ is a function independent of the overall input length $|I|$. We will also write $\mathcal{O}^*(f(k))$ for this run-time bound. Equivalently, one can define the class of fixed-parameter tractable problems as follows: strive to find a polynomial-time transformation that, given an instance (I, k), produces another instance (I', k') of the same problem, where $|I'|$ and k' are bounded by some function $g(k)$; in this case, (I', k') is also called a *(problem) kernel*.

Discussion of Related Results. MAS on general directed graphs can be solved in time $\mathcal{O}^*(2^k)$ and $\mathcal{O}^*(2^n)$, shown by V. Raman and S. Saurabh in [15], with n the number of vertices. Most recently, parallel to our work, I. Razgon, J. Chen *et al.* showed during a workshop [12] that FAS $\in \mathcal{FPT}$, relying on results from J. Chen et al [4]. In contrast to MAS, it still admits a fairly vast run time of $\mathcal{O}^*(8^k k!)$. Likewise, I. Razgon [16] provided an exact (non-parameterized) $\mathcal{O}^*(1.9977^n)$-algorithm for FEEDBACK VERTEX SET (FVS), which translates to a FAS-algorithm with the same base, but measured in m.

The complexity picture changes when one considers undirected graphs. The task of removing a minimum number of edges to obtain an acyclic graph can be accomplished in polynomial time, basically by finding a spanning forest. The task of removing a minimum number of vertices to obtain an acyclic graph is (again) $\mathcal{NP}$-complete, but can be approximated to a factor of two, see V. Bafna *et al.* [1], and is known to be solvable in $\mathcal{O}^*(5^k)$ with J. Chen *et al.* [3] being the currently leading party in a run time race. Also, exact algorithms have been derived for this problem by F. V. Fomin *et al.* [8].

Our Contributions. Our main technical contribution is to derive a parameterized $\mathcal{O}^*(1.212^k)$-algorithm for MAS on $(1,n)$-graphs. On cubic graphs the run time reduces to $\mathcal{O}^*(1.1960^k)$ via a novel combinatorial observation. We also derive an exact algorithm for MAS on $(1,n)$-graphs and as a by-products two other for DIRECTED FEEDBACK VERTEX SET on cubic and planar graphs with running times $\mathcal{O}^*(1.1871^m)$, $\mathcal{O}^*(1.282^n)$ and $\mathcal{O}^*(1.986^n)$, respectively. Besides being a nice combinatorial problem on its own right, we think that our contribution is also interesting from the more general perspective of a development of tools for constructing efficient parameterized algorithms. Namely, the algorithm we present is of a quite simple overall structure, similar in simplicity as, e.g., the recently presented algorithms for HITTING SET [6]. But the analysis is quite intricate and seems to offer a novel way of amortized search tree analysis that might be applicable in other situations in parameterized algorithmics, as well. It is also one of the fairly rare applications of the "measure & conquer" paradigm [9] in parameterized algorithmics. Due to lack of space, some proofs had to be omitted.

Fixing Terminology. We consider directed multigraphs $G(V, A)$ in the course of our algorithm, where V is the vertex set and A the arc set. From A to V we have two kinds of mappings: For $a \in A$, $init(a)$ denotes the vertex at the tip of the arc a and $ter(a)$ the end. We distinguish between two kinds of arc-neighborhoods of a vertex v which are $E^+(v) := \{a \in A \mid ter(a) = v\}$ and $E^-(v) := \{a \in A \mid init(a) = v\}$. We have an in- and outdegree of a vertex, that is $d^+(v) := |E^+(v)|$ and $d^-(v) := |E^-(v)|$. We set $E(v) := E^+(v) \cup E^-(v)$ and $d(v) := |E(v)|$ called the degree of v. We also define a neighborhood for arcs a $N_A(a) := \{a_1, a_2 \in A \mid ter(a_1) = init(a), ter(a) = init(a_2)\}$ and for $A' \subseteq A$ we set $N_A(A') := \bigcup_{a' \in A'} N_A(a')$. For $V' \subseteq V$ we set $A(V') := \{a \in A \mid \exists u, v \in V', init(a) = u, ter(a) = v\}$. We call an arc (u, v) a *fork* if $d^-(v) \geq 2$ (but $d^+(v) = 1$) and a *join* if $d^+(u) \geq 2$ (but $d^-(u) = 1$). With $\mathcal{MAS}$, we refer to a set of arcs, which is acyclic and is a partial solution. An *undirected cycle* is an acyclic arc set, which is a cycle in the underlying undirected graph.

Kernels Via Approximation. We first link approximability and parameterized algorithmics by a simple but interesting observation. We call a maximization problem P a *set maximization problem* if its task, given instance I, is to identify a subset S (satisfying additional requirements) of a *ground set* M of maximum cardinality. The natural parameterized problem related to P (denoted by P_{par}) is to find, given (I, k) such a subset S of cardinality at least k.

Proposition 1. *If a set maximization problem P has a c-approximation, where the ratio is measured with respect to the whole ground set M that is part of the input I whose size is measured in terms of the cardinality $|M|$ of M, then P_{par}, parameterized by k, has a kernel of size upper-bounded by ck.*

Both above mentioned approximations exhibit the required properties of Proposition 1. This entails a $2k$-kernel for the general case of MAS, as well as a $\frac{12}{11}k$-kernel when restricted to $(1, n)$-graphs, based on A. Newman's result [14].

2   Preprocessing and Reduction Rules

Firstly, we can assume that our instance $G(V, A)$ forms a strongly connected component. Every arc not in such a component can be taken into a solution, and two solutions of two such components can be simply joined.

Preprocessing. In [7,14] a set of preprocessing rules is already mentioned:

Pre-1: For every $v \in V$ with $d^+(v) = 0$ or $d^-(v) = 0$, delete v and $E(v)$, take $E(v)$ into $\mathcal{MAS}$ and decrement k by $|E(v)|$.

Pre-2: For every $v \in V$ with $E(v) = \{(i,v),(v,o)\}$, $v \neq i$ and $v \neq o$, delete v and $E(v)$ and introduce a new arc (i,o). Decrease k by one.

Pre-3: Remove any loop.

Any preprocessing rule, which applies, will be carried out exhaustively. Afterwards the resulting graph has no vertices of degree less than three.

Definition 1. *An arc g is an α-arc if it is a fork and a join.*

We need the next lemma, which is a sharpened version of [14, Lemma 2.1] and follows the same lines of reasoning.

Lemma 1. *Any two non-arc-disjoint cycles in a $(1,n)$-graph with minimum degree at least 3 share an α-arc.*

We partition A in A_α containing all α-arcs and $A_{\bar{\alpha}} := A \setminus A_\alpha$. By Lemma 1, the cycles in $G[A_{\bar{\alpha}}]$ must be arc-disjoint. This justifies the next preprocessing rule.

Pre-4: In G delete the arc set of every cycle C contained in $G[A_{\bar{\alpha}}]$. For an arbitrary $a \in C$ adjoin $C \setminus \{a\}$ to $\mathcal{MAS}$ and decrease k by $|C| - 1$.

After exhaustively applying Preprocess() (shown in Figure 1), every cycle has an α-arc. For $v \in V$ with $E^+(v) = \{a_1, \ldots, a_s\}$ ($E^+(v) = \{c\}$, resp.) and $E^-(v) = \{c\}$ ($E^-(v) = \{a_1, \ldots, a_s\}$), it is always better to delete c than one of $a_1, \ldots, a_s$. Therefore, we adjoin $a_1, \ldots, a_s$ to $\mathcal{MAS}$, adjusting k accordingly. Having applied this rule on every vertex, we adjoined $A_{\bar{\alpha}}$ to $\mathcal{MAS}$, and the remaining arcs are exactly A_α.

So, the next task is to find $S \subseteq A_\alpha$ with $|\mathcal{MAS} \cup S| \geq k$ so that $G[\mathcal{MAS} \cup S]$ is acyclic. We have to branch on the α-arcs, deciding whether we take them into $\mathcal{MAS}$ or if we delete them. α-arcs which we take into $\mathcal{MAS}$ will be called *red*. For the purpose of measuring the complexity of the algorithm, we will deal with two parameters k and k', where k measures the size of the partial solution and k' will be used for purposes of run-time estimation: We do not account the arcs in $A_{\bar{\alpha}}$ immediately into k'. For every branching on an α-arc, we count only a portion of them into k'. More precisely, upon first seeing an arc $b \in A_{\bar{\alpha}}$ within the neighborhood $N_A(g)$ of an α-arc g we branch on, we will count b only by an amount of ω, where $0 < \omega < 0.5$ will be determined later. So, we will have two weighting functions w_k and $w_{k'}$ for k and k' with $w_k(a) \in \{0,1\}$ and $w_{k'}(a) \in \{0,(1-\omega),1\}$ for $a \in A$, indicating each how much of the arc has not been counted into k, or k' respectively, yet. In the course of the algorithm, we always have $w_k(a) \leq w_{k'}(a)$. In the very beginning, we have $w_k(a) = w_{k'}(a) = 1$ for all $a \in A$. For a set $A' \subseteq A$, we define $w_{k'}(A') := \sum_{a' \in A'} w_{k'}(a')$ and $w_k(A')$ accordingly. Observe that for $a \in A$ we have $a \in \mathcal{MAS}$ iff $w_k(a) = 0$.

The preprocessing rules, together with the mentioned kernel of $\frac{12}{11}k$ arcs, gives us another simple brute-force algorithm for MAS: Within the kernel with m arcs,

Procedure:	Procedure:
Preprocess($\mathcal{MAS}$,$G(V,A)$,k):	Reduce($\mathcal{MAS}$,$G(V,A)$),k,k',w_k,$w_{k'}$):
1: **repeat**	1: **repeat**
2: cont ← **false**	2: cont ← **false**
3: apply **Pre-1 - Pre-3** exhaustively.	3: **for** i=1 **to** 6 **do**
	4: apply **RR-i** exhaustively.
4: **if Pre-4** applies **then**	5: **if RR-i** applied **then**
5: cont ← **true**	6: cont ← **true**
6: **until** cont=false	7: **until** cont=false
7: **return** ($\mathcal{MAS}$,$G(V,A)$,k,1)	8: **return** ($\mathcal{MAS}$,$G(V,A)$,k,k',w_k,$w_{k'}$)

Fig. 1. The procedures Preprocess() and Reduce()

there could be at most $m/3$ arcs that are α-arcs. It is obviously sufficient to test all possible $2^{m/3} \leq 2^{(4/11)k} \approx 1.288^k$ many possibilities of choosing α-arcs into the (potential) feedback arc set.

Reduction Rules. There is a set of reduction rules from [14], which we adapted and modified to deal with weighted arcs. Also, we define a (linear time checkable) predicate *contractible* for all $a \in A$.

$$contractible(a) = \begin{cases} 0 : w_k(a) = 1, \exists \text{ cycle } C \text{ with } a \in C \text{ and } w_k(C \setminus \{a\}) = 0 \\ 1 : \text{else} \end{cases}$$

The meaning of this predicate is the following: if $contractible(a) = 0$, then a is the only remaining arc of some cycle, which is not already determined to be put into $\mathcal{MAS}$. Thus, a has to be deleted. In the following, **RR-(i-1)** is always carried out exhaustively before **RR-i**.

RR-1: For $v \in V$ with $d^+(v) = 0$ or $d^-(v) = 0$, take $E(v)$ into $\mathcal{MAS}$, delete v and $E(v)$ and decrease k by $w_k(E(v))$ and k' by $w_{k'}(E(v))$.

RR-2: For $v \in V$ with $E(v) = \{a,b\}$ let $z = \arg\max\{w_{k'}(a), w_{k'}(b)\}$ and $y \in E(v) \setminus \{z\}$. If $contractible(y) = 1$, then contract y, decrement k by $w_k(y)$, k' by $w_{k'}(y)$. If y was red, then z becomes red.

RR-3: If for $g \in A$, we have $contractible(g) = 0$, then delete g.

We point out that due to **RR-2** also non-α-arcs can become red. But it is still true for a α-arc a that $a \in \mathcal{MAS}$ iff a is red. Let $A_\alpha^U := \{a \in A_\alpha \mid a \text{ is non-red}\}$. We classify the arcs of A_α^U in *thin α-arcs*, which are contained in exactly one cycle, and *thick α-arcs*, which are contained in at least two cycles. Because G is strongly connected, there are no other α-arcs. We can distinguish them as follows: For every α-arc g, find the smallest cycle C_g which contains g via BFS. If g is contained in a second cycle C_g', then there is an arc $a \in C_g$ with $a \notin C_g'$. So for all $a \in C_g$, remove a and restart BFS, possibly finding a second cycle.

RR-4: If $g \in A_\alpha^U$ is thin and $contractible(g) = 1$, then take g into $\mathcal{MAS}$ and decrease k by $w_k(g)$, k' by $w_{k'}(g)$ and set $w_k(g) \leftarrow 0$, $w_{k'}(g) \leftarrow 0$.

Algorithm 1. A parameterized algorithm for MAXIMUM ACYCLIC SUBGRAPH on $(1, n)$-graphs

---

1: $(\mathcal{MAS},G(V, A),k,w_k) \leftarrow$ Preprocess($\emptyset$,$G(V, A)$,k).
2: $\mathcal{MAS} \leftarrow A_{\bar\alpha} \cup \mathcal{MAS}, k' \leftarrow k, \; k \leftarrow k - w_k(A_{\bar\alpha}), \; w_k(A_{\bar\alpha}) \leftarrow 0$
3: Sol3MAS($\mathcal{MAS}$,$G(V, A)$,k,k',$w_{k'}$,w_k)

Procedure: Sol3MAS($\mathcal{MAS}$,$G(V, A)$,k,k',w_k,$w_{k'}$):

1: $(\mathcal{MAS},G(V, A),k,k',w_k,w_{k'}) \leftarrow$ Reduce($\mathcal{MAS}$,$G(V, A)$,k,k',w_k,$w_{k'}$)
2: **if** $k \leq 0$ **then**
3:     **return** YES
4: **else if** there is a component C with at most 9 arcs **then**
5:     Test all possible solutions for C.
6: **else if** there is a α-arc $g \in A_\alpha^U$ **then**
7:     **if not** Sol3MAS($\mathcal{MAS}$,$G[A \setminus \{g\}]$,k,k',w_k,$w_{k'}$) **then**
8:         $k \leftarrow k-1, k' \leftarrow k'-w_{k'}(g), w_k(g) \leftarrow w_{k'}(g) \leftarrow 0, \mathcal{MAS} \leftarrow \mathcal{MAS} \cup N_A(g) \cup \{g\}$.
9:         **for all** $a \in N_A(g)$ **do**
10:            Adjust $w_{k'}$, see Figure 2.
11:         **return** Sol3MAS($\mathcal{MAS}$,$G(V, A)$,k,k',w_k,$w_{k'}$)
12:     **else**
13:         **return** YES
14: **else**
15:     **return** NO

RR-5: If $a, b \in A$ form an undirected 2-cycle then let $z = \arg\min\{w_{k'}(a), w_{k'}(b)\}$, decrease k by $w_k(z)$, k' by $w_{k'}(z)$, take z into $\mathcal{MAS}$ and delete z.

RR-6: Having $(u, v), (v, w), (u, w) \in A$ (an undirected 3-cycle), decrease k by $w_k((u, w))$, k' by $w_{k'}((u, w))$, take (u, w) into $\mathcal{MAS}$ and delete (u, w).

Lemma 2. a)*The reduction rules are sound.* b)*After the application of Reduce(), see Figure 1, we are left with a $(1, n)$-graph with only thick α-arcs and no directed or undirected 2- or 3-cycle.*

3 The Algorithm and Its Analysis

3.1 The Algorithm

We are ready now to state our main Algorithm 1; observe that the handling of the second parameter k' is only needed for the run-time analysis and could be avoided when implementing the algorithm. Therefore, the branching structure of the algorithm is quite simple, as expressed in the following:

Lemma 3. *Branching in Alg. 1 either puts a selected α-arc g into $\mathcal{MAS}$, or it deletes g. Only if arcs are deleted, reduction rules will be triggered in the subsequent recursive call. This can can be also due to triggering* **RR-3** *after putting g into $\mathcal{MAS}$.*

Adjust $w_{k'}$:

1: **if** $w_{k'}(a) = 1$ **then**
2:    **if** $\exists b \in (N_A(a) \setminus (N_A(g) \cup \{g\}))$ with
      $w_{k'}(b) = 0$ **then**
3:       $k' \leftarrow k' - 1, w_{k'}(a) \leftarrow 0,$
         $k \leftarrow k - w_k(a), w_k(a) \leftarrow 0$ (case a.)
4:    **else**
5:       $k' \leftarrow k' - \omega, w_{k'}(a) \leftarrow (1 - \omega),$
         $k \leftarrow k - w_k(a), w_k(a) \leftarrow 0$ (case b.)
6: **else**
7:    $k' \leftarrow k' - (1 - \omega), w_{k'}(a) \leftarrow 0.$ (case c.)

Fig. 2. In case a. we set $w_{k'}(a) = 0$, because there will not be any other neighboring non-red α-arc of a. In case b., this might not be the case, so we count only a portion of ω. In case c., we will prove that $w_{k'}(a) = (1 - \omega)$ and that there will be no other non-red neighboring α-arc of a, see Theorem 1.5.

3.2 Analysis

Basic Combinatorial Observations. While running the algorithm, $k \leq k'$. Now, substitute in line 2 of Sol3MAS of Algorithm 1 k by k'. If we run the algorithm, it will create a search tree $T_{k'}$. The search tree T_k of the original algorithm must be contained in $T_{k'}$, because $k \leq k'$. If $|T_{k'}| \leq c^{k'}$, then it follows that also $|T_k| \leq c^{k'} = c^k$, because in the very beginning, $k = k'$. So in the following, we will state the different recurrences derived from Algorithm 1 in terms of k'. For a good estimate, we have to calculate an optimal value for ω.

Theorem 1. *In every node of the search tree, after applying Reduce(), we have*

1. *For all $a = (u, v) \in A_{\bar{\alpha}}$ with $w_{k'}(a) = (1 - \omega)$, there exists a red fork (u', u) or a red join (v, v').*
2. *For all non-red $a = (u, v) \in A_{\bar{\alpha}}$ with $w_{k'}(a) = 0$, we find a red fork (u', u) and a red join (v, v'). We will also say that a is protected (by the red arcs).*
3. *For all red arcs $d = (u, v)$ with $w_{k'}(d) = 0$, if we have only non-red arcs in $E(u) \setminus \{d\}$ ($E(v) \setminus \{d\}$, resp.), then d is a join (d is a fork, resp.).*
4. *For each red arc $d = (u, v)$ with $w_{k'}(d) = 0$ that is not a join (fork, resp.), if there is at least one red arc in $E(u) \setminus \{d\}$ (in $E(v) \setminus \{d\}$, resp.), then there is a red fork (red join, resp.) in $G[E(u)]$ ($G[E(v)]$, resp.).*
5. *For all $g \in A_{\alpha}^U$ and for all $a \in N_A(g)$, we have: $w_{k'}(a) > 0$.*

Proof. We use induction on the depth of the search tree. Clearly, all claims are trivially true for the original graph instance, i.e., the root node. Notice that each claim has the form $\forall a \in A : X(a) \implies Y(a)$. Here, X and Y express local situations affecting a. Therefore, we have to analyze how $X(a)$ could have been created by branching. According to Lemma 3, we have to discuss what happens (1) if a certain α-arc had been put into $\mathcal{MAS}$ and (2) if reduction rules were

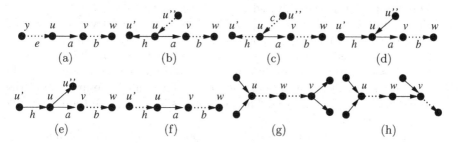

Fig. 3. Dotted lines indicate red arcs

triggered. As a third point, we must consider the possibility that $X(a)$ is true both in the currently observed search tree node s and in its predecessor, but that $Y(a)$ was possibly affected upon entering s.

1. *(sketch)* Such an arc (u, v) can be created in two ways: (1) By taking a neighboring α-arc g into $\mathcal{MAS}$. Then g is a join and a fork. (2) By applying **RR-2** on arcs $(u, z), (z, v)$. W.l.o.g. $w_{k'}((u, z)) = (1 - \omega)$. Then by induction there is a fork (u', u). How can this situation be affected by the reduction rules? Any **RR-2**-application leaves it unchanged as red arcs are 'dominant'. If w.l.o.g. (u', u) is a red fork then this situation may be changed by deletions in the case where $d(u)$ becomes two. However, (u, v) disappears by a subsequent **RR-2**-application.

2. We will actually prove points 2. through 4. by a parallel induction. To improve readability of our main argument, we refrain from giving all possible details how the employment of **RR-2** may affect (but not drastically change) the situation in particular. How can $a = (u, v) \in A_{\bar{\alpha}}$ with $w_{k'}(a) = 0$ have been created? Firstly, it could be due to a **RR2**-contraction with a non-red arc t with $w_{k'}(t) = 0$. But then a was not protected, which is a contradiction to the induction hypothesis. Secondly, it could be due to branching on a neighboring α-arc b, say $b = (v, w)$ with b a join, in two different ways: (1) either we branched at b at a point of time when $w_{k'}(a) = (1 - \omega)$ (case c. of Procedure "Adjust"), or (2) we branched at b when $w_{k'}(a) = 1$ (case a.)

In case (1), there must have been another red arc e incident to a by item 1 of our property list, see Figure 3(a). e is not incident to v, since it is a fork. Hence, $e = (y, u)$. This displays the two required red arcs (namely b and e) in this case. In case (2), a was created by case a. of Procedure "Adjust." Obviously, b is red after branching. Since we have branched according to case a., there is another arc h incident with a (but not with b) such that $w_{k'}(h) = 0$. There are four subcases to be considered:
(a) $h = (u, u')$ is not red, see Figure 3(b). By induction (item 2.), there must be a red fork arc (u'', u). Hence, a is protected.
(b) $h = (u, u')$ is red, see Figure 3(c). Consider all other arcs incident to u. Since we are dealing with reduced instances and by the $(1, n)$-property, there must be exactly one of the form $c = (u'', u)$, since otherwise u would be a

sink. Suppose h is the only red arc in $E(u)$. Then this contradicts item 3. Now, suppose c is not red. Then there is a red arc $(u, \bar{u})$. By item 4. c must be also red, a contradiction, Therefore, c is the red fork, which protects a.

(c) $h = (u', u)$ is not red. All other arcs incident to u could be of the form (u'', u), see Figure 3(d). Since h must be protected, by induction, a should be red, contradicting our assumption on a. Thus, all these arcs are of the form (u, u''), see Figure 3(e). This contradicts item 2., since there is no red join protecting h.

(d) $h = (u', u)$ is red, see Figure 3(f). Suppose h is not a fork. Then all other arcs beside a and h are of the form $(\tilde{u}, u)$. If none of them is red we have a contradiction concerning item 3. If one of them is red then by item 4. a is red, which contradicts our assumptions. Hence, a will be protected.

3. How could d have been created? If it had been created by branching, then there are two cases: (1) d was put into $\mathcal{MAS}$; (2) d was neighbor of an arc b which we put into $\mathcal{MAS}$.

   In case (1), the claim is obviously true. In case (2), let, w.l.o.g., u be the common neighbor of b and d. After putting b into $\mathcal{MAS}$, there will be a red arc (namely b), incident to u, so that there could be only non-red arcs incident with v that have the claimed property by induction. If d has been created by reduction rules, it must have been through **RR-2**. So, there have been (w.l.o.g.) two arcs (u, w) and (w, v) with $w_{k'}$-weights zero. One of them must be red. W.l.o.g., assume that (u, w) is red. If (w, v) is red, see Figure 3(g), then the claim holds by induction. If (w, v) is not red, see Figure 3(h), then (w, v) must be protected due to item 2. Hence, the premise is falsified for vertex v.

4. We again discuss the possibilities that may create a red d with $w_{k'}(d) = 0$. If d was created by taking it into $\mathcal{MAS}$ during branching, then d would be both fork and join in contrast to our assumptions.

   If we branch in the neighborhood of d, then the claim could be easily verified directly. Finally, d could be obtained from merging two arcs $e = (u, w), f = (w, v)$ with $w_{k'}(e) = w_{k'}(f) = 0$. If both e and f are red, the claim follows by induction. If only f is red and e is non-red, then there is a red fork, which protects e by item 2. Again, by induction the claim follows. The case where only e is red is symmetric.

5. Assume the contrary. Discuss a neighbor arc a of g with $w_{k'}(a) = 0$. If a is not red, then g must be red due to item 2., contradicting $g \in A_\alpha^U$. If a is red, then discuss another arc b that is incident to the common endpoint of a and g. If there is no red b, then the situation contradicts item 3. So, there is a red b. This picture contradicts item 4. $\qquad\Box$

Estimating the Running Time for Maximum Degree 3 Graphs. In Algorithm 1, depending in which case of Figure 2 we end up, we decrement k' by a different amount for each arc $a \in N_A(g)$ in the case that we put g into $\mathcal{MAS}$. We can be sure that we may decrement k' by at least $(1 - \omega)$ for each neighbor $a \in N_A(g)$ due to the last property of the Theorem 1. If we do not put g into $\mathcal{MAS}$, we delete g and $N_A(g)$ immediately afterwards by **RR-1**, decrementing

Table 1. Summarizes by which amount k' can be decreased for $a \in N_A(g)$, subject to if we take g into $\mathcal{MAS}$ or delete g and to the case applying to a

α-arc g	$a.$	$b.$	$c.$	$b'.$
$\mathcal{MAS}$	1	ω	$(1 - \omega)$	ω
Deletion	1	$(2 - \omega)$	$(1 - \omega)$	1

k' accordingly (by $w_{k'}(N_A(g))$). Moreover, if case $b.$ applies to $a \in N_A(g)$, we know that the two arcs $d, e \in (N_A(a) \setminus (N_A(g) \cup g\}))$ obey $w_{k'}(d)w_{k'}(e) > 0$ (observe that we do not have triangles). By deleting a, no matter whether **RR-1** or **RR-2** applies to d and e (this depends on the direction of the arcs) we can decrement k' by an extra amount of at least $(1 - \omega)$, cf. the handling of k' by these reduction rules. This is true even if $V(d), V(e) \subset V(N_A(g))$. Note that if $V(N_A(N_A(g))) \subseteq V(N_A(g))$, then $A(V(N_A(g)))$ is a component of 9 arcs (which are handled separately). Let i denote the number of arcs $a \in N_A(g)$ for which case $a.$ applies. In the analogous sense j stands for the case $b.$ and q for $c.$ For every positive integer solution of $i + j + q = 4$, we can state a total of 15 recursions $T_1, \ldots, T_{15}$ according to Table 1 depending on ω (ignoring the last column for the moment). For every T_i and for a fixed ω, we can calculate a constant $c_i(\omega)$ such that $T_i[k] \in \mathcal{O}^*(c_i(\omega)^k)$. We want to find a ω with subject to minimize $\max\{c_1(\omega), \ldots, c_{15}(\omega)\}$. We numerically obtained $\omega = 0.1687$ so that $\max\{c_1(\omega), \ldots, c_{15}(\omega)\}$ evaluates to 1.201. The dominating cases are when $i = 0, j = 0, q = 4$ (T_5) and $i = 0, j = 4, q = 0$ (T_{15}). We conclude that MAS on graphs G with $\Delta(G) \leq 3$ can be solved in $\mathcal{O}^*(1.201^k)$. Measuring the run time in terms of $m := |A|$ the same way is also possible. Observe that if we delete an α-arc, we can decrement m by one more. By adjusting $T_1, \ldots, T_{15}$ according to this and by choosing $\omega = 0.2016$, we derive an upper bound of $\mathcal{O}^*(1.1798^m)$. Note that if we run this exact algorithm on the $\frac{12}{11}k$-kernel, we finish in $\mathcal{O}^*(1.1977^k)$, being slightly better than the pure parameterized algorithm.

We will obtain a better bound for the search tree by a precedence rule, aiming to improve recurrence T_5. If we branch on an α-arc g according to this recurrence, for all $a \in N_A(g)$ we have $w_{k'}(a) \geq (1 - \omega)$. Such α-arcs will be called α_5-arcs. We add the following rule: branch on α_5-arcs with least priority. Let $l := |A_\alpha^U|$.

Lemma 4. *Branching on an α_5-arc, we can assume:* $\left\lfloor \frac{1}{5-4\omega}k' \right\rfloor \leq l < \left\lceil \frac{1}{4-4\omega}k' \right\rceil$.

Employing this lemma, we can find a good combinatorial estimate for a brute-force search at the end of the algorithm. This allows us to conclude:

Theorem 2. MAS *is solvable in time* $\mathcal{O}^*(1.1798^m)$ *and* $\mathcal{O}^*(1.1960^k)$ *on maximum-degree-3-graphs* .

Corollary 1. FEEDBACK VERTEX SET *on cubic graphs is solvable in* $\mathcal{O}^*(1.282^n)$.

Proof. We argue that MAS and FAS are equivalent for graphs of degree at most three as follows. Namely, if A is a feedback arc set, then we can remove instead the set S of vertices the arcs in A are pointing to in order to obtain a directed feedback vertex set with $|S| \leq |A|$. Conversely, if S is a directed feedback vertex set, then we can assume that each vertex $v \in S$ has one ingoing and two outgoing arcs or two ingoing and one outgoing arc; in the first case, let a_v be the ingoing arc, and in the second case, let a_v be the outgoing arc. Then, $A = \{a_v \mid v \in S\}$ is a feedback arc set with $|A| \leq |S|$. With $m \leq \frac{3}{2}n$ the claim follows.   □

Estimating the Running Time for $(1, n)$-Graphs. There is a difference to maximum degree 3 graphs, namely the entry for case $b.$ in case of deletion in Table 1. For $a \in N_A(g)$ it might be the case that $|N_A(a) \setminus (N_A(g) \cup \{g\})| \geq 3$, so that when we delete g and afterwards a by **RR-1** that whether **RR-1** nor **RR-2** applies (due to the lack of sources, sinks or degree two vertices). We call this case b'. Remember, case $b.$ refers to the same setting but with $|N_A(a) \setminus (N_A(g) \cup \{g\})| = 2$. Thus the mentioned entry should be 1 for b'. As long as $|N_A(g)| \geq 6$ the reduction in k' is great enough for the modified table, but for the other cases we must argue more detailed. We introduce two more reduction rules, the first already mentioned in [14], and add items **a)-c)** to Algorithm 1.

RR-7 Contract and adjoin to $\mathcal{MAS}$ any $(u, v) \in A$ with $d^+(u) = d^+(v) = 1$ $(d^-(u) = d^-(v) = 1$, resp.). If (u, v) was red the unique arc $a := (x, u)$ $((v, y)$, resp.) will be red. Decrease k' by $\min\{w_{k'}((u, v)), w_{k'}(a)\}$ and set $w_{k'}(a) \leftarrow \max\{w_{k'}((u, v)), w_{k'}(a)\}$. Proceed similarly with (v, y).
RR-8 For a red $g' \in A_\alpha$ with $w_{k'}(g') > 0$, set $k' \leftarrow k' - w_{k'}(g')$ and $w_{k'}(g') \leftarrow 0$.

a) After Reduce(), first apply **RR-7** and then **RR-8** exhaustively.
b) Prefer α-arcs g such that $|N_A(g)|$ is maximal for branching.
c) Forced to branch on $g \in A_\alpha^U$ with $|N_A(g)| = 5$, choose an α-arc with the least occurrences of case b'.

Lemma 5. **RR-7** *and* **RR-8** *are sound and do not violate Theorem 1.*

Lemma 6. *We can omit branching on arcs* $g \in A_\alpha^U$ *with* **a)** $|N_A(g)| = 5$ *and 5 occurrences of case* b' *or* **b)** $|N_A(g)| = 4$ *with an occurrence of case* b'.

Let x, y, z denote the occurrences of cases $a., b'$ and c. To upperbound the branchings according to α-arcs g with $|N_A(g)| \geq 6$, we put up all recurrences resulting from integer solutions of $x + y + z = 6$. Note that we also use the right column of Table 1. To upperbound branchings with $|N_A(g)| = 5$ we put up all recurrences obtained from integer solutions of $x + y + z = 5$, except when $x = z = 0$ and $y = 5$ due to Lemma 6.a). Additionally we have to cover the case where we have 4 occurrences of case b' and one of case b. $(T[k] \leq T[k - 5w] + T[k - (6 - w)])$. To upperbound the case where $|N_A(g)| = 4$ the recurrences derived from Table 1 for the integer solutions of $x + y + z = 4$ suffice due to Lemma 6.b).

Theorem 3. *On* $(1, n)$-*graphs with* m *arcs,* MAS *is solvable in time* $\mathcal{O}^*(1.1871^m)$ *($w = 0.2392$) and* $\mathcal{O}^*(1.212^k)$ *($w = 0.21689$), respectively.*

Corollary 2. *We solve* FEEDBACK VERTEX SET *on planar graphs in* $\mathcal{O}^*(1.986^n)$.

Proof. Reduce [7] FVS to FAS. The resulting graph G' has no more than $4n$ arcs. Then run Algorithm 1 on G'. □

4   Reparameterization

M. Mahajan, V. Raman and S. Sidkar [11] have discussed a rather general setup for re-parameterization of problems according to a "guaranteed value." In order to use their framework, we only need to exhibit a family of example graphs where Newman's approximation bound for MAS is sharp. Consider $G_r(V_r, A_r)$, $r \geq 2$, with $V_r = \{(i, j) \mid 0 \leq i < r, 0 \leq j \leq 7\}$, and A_r contains two types of arcs:
1. $((i, j), (i, (j + 1) \bmod 8)$ for $0 \leq i \leq r$ and $0 \leq j \leq 7$.

2a. $((i, j), ((i + 1) \bmod r, (1 - j) \bmod 8)$ for $0 \leq i < r$ and $j = 1, 2$.

2b. $(((i + 1) \bmod r, (1 - j) \bmod 8, (i, j))$ for $0 \leq i < r$ and $j = 3, 4$.

For $r = 2$ we find an example to the right. G_r is cubic with $|V_r| = 8r$ and $|A_r| = 12r$. Its α-arcs are $((i, 0), (i, 1))$ and $((i, 4), (i, 5))$ for $0 \leq i < r$.

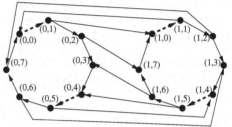

Since we have to destroy all 'rings' as described by the arcs from 1., any feasible solution to these instances require r arcs to go into the feedback arc set. Also r arcs suffice, namely $((0, 4), (0, 5))$ and $((i, 0), (i, 1))$ for $0 < i < r$, giving the 'tight example' as required in [11] to conclude:

Corollary 3. *For any $\epsilon > 0$, the following question is not fixed-parameter tractable unless $\mathcal{P} = \mathcal{NP}$: Given a cubic directed graph $G(V, E)$ and a parameter k, does G possess an acyclic subgraph with at least $\left(\frac{11}{12} + \epsilon\right) |E| + k$ many arcs ?*

References

1. Bafna, V., Berman, P., Fujito, T.: A 2-approximation algorithm for the undirected feedback vertex set problem. SIAM Journal of Discrete Mathematics 12, 289–297 (1999)
2. Berger, B., Shor, P.W.: Approximation algorithms for the maximum acyclic subgraph problem. In: SODA, pp. 236–243 (1990)
3. Chen, J., Fomin, F.V., Liu, Y., Lu, S., Villanger, Y.: Improved algorithms for the feedback vertex set problems. In: WADS, pp. 422–433 (2007)
4. Chen, J., Liu, Y., Lu, S.: An improved parameterized algorithm for the minimum node multiway cut problem. In: WADS, pp. 495–506 (2007)
5. Downey, R.G., Fellows, M.R.: Parameterized Complexity. Springer, Heidelberg (1999)
6. Fernau, H.: Parameterized algorithms for HITTING SET: The weighted case. In: Calamoneri, T., Finocchi, I., Italiano, G.F. (eds.) CIAC 2006. LNCS, vol. 3998, pp. 332–343. Springer, Heidelberg (2006)
7. Festa, P., Pardalos, P.M., Resende, M.G.C.: Feedback set problems. In: Handbook of Combinatorial Optimization, vol. Supplement Volume A, pp. 209–258. Kluwer Academic Publishers, Dordrecht (1999)

8. Fomin, F.V., Gaspers, S., Pyatkin, A.V.: Finding a minimum feedback set in time $\mathcal{O}(1.7548^n)$. In: Bodlaender, H.L., Langston, M.A. (eds.) IWPEC 2006. LNCS, vol. 4169, pp. 184–191. Springer, Heidelberg (2006)
9. Fomin, F.V., Grandoni, F., Kratsch, D.: Measure and conquer: Domination – a case study. In: Caires, L., Italiano, G.F., Monteiro, L., Palamidessi, C., Yung, M. (eds.) ICALP 2005. LNCS, vol. 3580, pp. 191–203. Springer, Heidelberg (2005)
10. Karp, R.M.: Reducibility among combinatorial problems. In: Complexity of Computer Computations, Plenum Press, New York (1972)
11. Mahajan, M., Raman, V., Sikdar, S.: Parameterizing MAXNP problems above guaranteed values. In: Bodlaender, H.L., Langston, M.A. (eds.) IWPEC 2006. LNCS, vol. 4169, pp. 38–49. Springer, Heidelberg (2006)
12. Seminar 07281: Structure Theory and FPT Algorithmics for Graphs, Digraphs and Hypergraphs, Dagstuhl, Germany, July 8-13, 2007, Proceedings (to appear, 2007), http://drops.dagstuhl.de/portals/index.php?semnr=07281, http://uk.arxiv.org/pdf/0707.0282.pdf (for a pre-version)
13. Nassi, I., Shneiderman, B.: Flowchart techniques for structured programming. ACM SIGPLAN Notices 12 (1973)
14. Newman, A.: The maximum acyclic subgraph problem and degree-3 graphs. In: Goemans, M.X., Jansen, K., Rolim, J.D.P., Trevisan, L. (eds.) RANDOM 2001 and APPROX 2001. LNCS, vol. 2129, pp. 147–158. Springer, Heidelberg (2001)
15. Raman, V., Saurabh, S.: Improved fixed parameter tractable algorithms for two edge problems: MAXCUT and MAXDAG. Inf. Process. Lett. 104(2), 65–72 (2007)
16. Razgon, I.: Computing minimum directed feedback vertex set in $O(1.9977^n)$. In: ICTCS, vol. 6581, pp. 70–81. World Scientific, Singapore (2007), www.worldscibooks.com/compsci/6581.html
17. Sugiyama, K., Tagawa, S., Toda, M.: Methods for visual understanding of hierarchical system structures. IEEE Trans. Systems Man Cybernet. 11(2), 109–125 (1981)

Linear-Time 3-Approximation Algorithm for the r-Star Covering Problem

Andrzej Lingas[1], Agnieszka Wasylewicz[2], and Paweł Żyliński[3,*]

[1] Department of Computer Science, Lund University, 221 00 Lund, Sweden
Andrzej.Lingas@cs.lth.se
[2] Department of Mathematics, University of Oslo, N-0371 Oslo, Norway
agniesew@math.uio.no
[3] Institute of Computer Science, University of Gdańsk, 80-952 Gdańsk, Poland
pz@inf.univ.gda.pl

Abstract. The problem of finding the minimum r-star cover of orthogonal polygons had been open for many years, until 2004 when Ch. Worman and J. M. Keil proved it to be polynomial tractable (Polygon decomposition and the orthogonal art gallery problem, *IJCGA* 17(2) (2007), 105-138). However, their algorithm is not practical as it has $\tilde{O}(n^{17})$ time complexity, where $\tilde{O}()$ hides a polylogarithmic factor. Herein, we present a linear-time 3-approximation algorithm based upon the novel partition of a polygon into so-called $[w]$-star-shaped orthogonal polygons.

1 Introduction

One of the main topics of computational geometry is efficient decomposition of polygonal objects into simpler polygons, such as triangles, rectangles, convex polygons, star-shaped polygons, etc. In general, we distinguish between two main types of decompositions: *partitions* and *covers*. A decomposition is called a *partition* if the object is decomposed into non-overlapping pieces. Otherwise, if the pieces are allowed to overlap, then we call the decomposition a *cover*, and a collection of polygons $S = \{P_1, P_2, \ldots, P_k\}$ is said to *cover* a polygon P if the union of all polygons in S is P.

The problems of covering polygons with simpler components have received considerable attention in the literature [3,12,13,24]. Most of these problems are NP-hard, even if the polygon to be covered has no holes [1,5,23]. Because of the difficulty of the general problems, one has focussed on orthogonalized versions of the above problems, and several algorithmic results concerning coverings with the minimum number of rectangles or the minimum number of orthogonally convex polygons have been obtained [4,7,18,19,22]. Let us recall that an orthogonal polygon is *convex* if its intersection with every horizontal and vertical line segment is either empty or a single line segment.

Two types of visibility related to covering with the so-called star-shaped orthogonal polygons have been studied in the literature [4]. Namely, two points

---

* Supported by The Visby Programme Scholarship 01224/2007.

S.-i. Nakano and Md. S. Rahman (Eds.): WALCOM 2008, LNCS 4921, pp. 157–168, 2008.
© Springer-Verlag Berlin Heidelberg 2008

Table 1. The r-star cover problem in orthogonal polygons

Class		Complexity	
0		$O(1)$	
1		$O(n)$ [8,16]	
2	2a	$O(n)$ [8,16]	
	2b	2-approximation, $O(n)$ [This paper]	$\tilde{O}(n^{17})$ [25,26]
3		2-approximation, $O(n)$ [This paper]	
4		3-approximation, $O(n)$ [This paper]	

of an orthogonal polygon P are *s-visible* if there exists an orthogonally convex polygon in P that contains these two points. Similarly, we say that two points are *r-visible* if there exists a rectangle that contains these two points. An *s-star-shaped* orthogonal polygon contains a point p such that for every point q in the polygon, there is an orthogonally convex polygon containing p and q; an *r-star-shaped* orthogonal polygon is defined analogously. Thus, an *s-star cover* is a cover with s-star-shaped polygons, and an *r-star cover* is a cover with r-star-shaped polygons.

For covering with r-stars, Keil [11] has provided $O(n^2)$ algorithm for optimally covering horizontally convex orthogonal polygons. This quadratic time bound was later improved to $O(n)$ in [8], which has been recently simplified by the authors [16]. For the general so-called class 2, class 3, and general simple orthogonal polygons (see Preliminaries for the definition), the problem of covering with the minimum r-star orthogonal polygons had remained open for many years. And just in 2004, Worman and Keil [25,26] proved that the minimum r-star cover problem in simple orthogonal polygons can be solved in $\tilde{O}(n^{17})$-time complexity, where $\tilde{O}()$ hides a polylogarithmic factor. Similarly as other algorithms for convex or star covering problems, the result in [25,26] uses a graph theory approach based upon properties of region visibility graphs [4,20,21]. Since the time complexity of the aforementioned algorithm is far from practical, there is need to design faster exact or approximation algorithms with constant performance ratio for this problem (if possible).

Our Contributions. We present a linear-time 3-approximation algorithm for the minimum r-star cover problem for simple orthogonal polygons (Section 5). Our algorithm starts from the partition of a polygon into histograms introduced by Levcopoulos in [15]. Then, it groups the histograms into components called $[w]$-star-shaped polygons such that in any minimum r-star cover of the original polygon, an element of the coverage can overlap with at most three of the components. This yields the approximation factor 3. And, for the class of orthogonal polygons the components constitute, we derive an optimal linear-time algorithm for finding the minimum number of r-star-shaped orthogonal polygons covering the polygon (Section 4). Finally, we show that our approach yields a linear-time 2-approximation algorithm for the classes 2 and 3 of orthogonal polygons (see Preliminaries for the definition). Table 1 summarizes our results in the context of the known ones.

Additionally (Section 3), to familiarize the reader with the problem and to provide some intuition, we present a simpler 5-approximation algorithm first.

2 Preliminaries

In this section we introduce several classes of polygons which are considered through the paper.

Histograms. A *histogram* H is an orthogonal polygon having a distinguished edge, called its *base*, such that the length of the base of H is equal to the total length of all other edges of H which are parallel to the base (Edelsbrunner *et al.* [6]); see Fig. 1 for an example.

An orthogonal polygon can be partitioned into histograms in linear time, as was shown by Levcopoulos [15]. Let P be a simple orthogonal polygon, and let e be an edge of P. $\mathcal{HIST}(P, e)$ is the set of segments, partitioning P into histograms, which is recursively defined as follows. Let H be the maximal histogram lying within P with the base e. If $H = P$, then $\mathcal{HIST}(P, e)$ is the empty set. Otherwise, let $P_1, P_2, \ldots, P_k$ be the set of subpolygons of P into which P is partitioned by H, with $P_1 = H$. Let s_i, $2 \le i \le k$, be the line segment where P_i touches P_1. The set $\mathcal{HIST}(P, e)$ is defined to be the union

$$\bigcup_{2 \le i \le k} (\{s_i\} \cup \mathcal{HIST}(P_i, s_i)).$$

Theorem 1. [15] $\mathcal{HIST}(P, e)$ *can be computed in linear time.*

Class k Polygons. For a compass direction $D \in \{N, E, S, W\}$, an edge of a polygon P which lies on the D side of P and has both endpoints reflex is said to be a *dent-D*. Analogously, an edge of P which lies on the D side of P and has both endpoints concave is called an *antident-D*. The *class k* of orthogonal polygons,

Fig. 1. A histogram

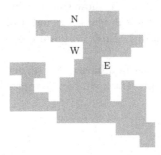

Fig. 2. A polygon with 3 dent orientations

$k = 0, 1, 2, 3, 4$, has dents of k different orientations {N,W,S,E}; this classification of orthogonal polygons is due to Culberson and Reckhow [4]. Thus for example, the class 0 of orthogonal polygons does not have dents, and histograms are in the class 1. The class 2 of orthogonal polygons is subdivided into the class 2a, where the two dent orientations are opposite, and the class 2b, where the two dent orientations are orthogonal to one another; thus the class 2a is the class of orthogonal monotone polygons. We also recognize the class 3 (Fig. 2) and the class 4 polygons which are all orthogonal polygons.

[w]-Star-Shaped Polygons. A polygonal chain is called *orthogonal* if each of its segments is parallel to one of the coordinate axes. An orthogonal chain is called a *staircase* if, while traversing from one end of the chain to the other, one moves in at most two of the four compass directions: N, E, S, and W. Thus there are four types of staircases: those oriented NW or SE, and those oriented NE or SW. Following these definitions, we introduce the class of [w]-*star-shaped* orthogonal polygons.

Definition 1. *Let* $w \in$ {NE, SW, NW, SE}. *An orthogonal polygon P is said to be* [w]-star-shaped *if there is a vertex v in P such that for every point q in P there is a staircase from v to q that contains no points outside of P and has the orientation w.*

An example of [NW]-star-shaped orthogonal polygon is shown in Fig. 3. Note that any [w]-star-shaped orthogonal polygon is s-star-shaped as well.

3   A 5-Approximation Algorithm for Orthogonal Polygons

In this section we present a simple 5-approximation algorithm for covering of a polygon with the minimum number of r-star-shaped polygons. Our algorithm combines the linear-time method of partitioning an orthogonal polygon into histograms, due to Levcopoulos [15], with a simple linear-time algorithm for covering of a histogram with the minimum number of r-star-shaped polygons [8,16].

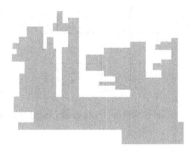

Fig. 3. An [NW]-star-shaped polygon

Algorithm 1

Input. An orthogonal polygon P with n vertices.

Output. A covering of P with r-star-shaped orthogonal polygons.

1. Partition P into orthogonal histograms using the method of Levcopoulos [15].
2. Cover each of the resulting histograms with the minimum number of r-star-shaped orthogonal polygons by running the algorithm of Gewali et al. [8] (or that of Lingas et al. [16] which is its improved variant).

Theorem 2. Algorithm 1 *runs in linear time and yields a 5-approximation for the problem of covering a simple orthogonal polygon with the minimum number of r-star-shaped orthogonal polygons.*

Proof. The linear-time complexity of Algorithm 1 immediately follows from the linear-time complexity of Levcopoulos' method and the linear-time complexity of Gewali et al.'s algorithm [8] (or Lingas et al.'s algorithm as well [16]).

To prove the approximation factor 5, consider a minimum cover OPT of P with r-star orthogonal polygons. Let H be any histogram in the partition of P into histograms. Note that the cardinality of a minimum covering of H r-star orthogonal polygons is not greater than the number of r-star orthogonal polygons in OPT that have a non-empty intersection with H. On the other hand, the partition into histograms has a rooted tree structure with alternating levels of histograms with horizontal and vertical bases [15]. It follows that any r-star orthogonal polygon in OPT can have a non-empty intersection with at most five histograms in the partition — see Fig. 4. Consequently, we conclude that Algorithm 1 produces a covering of cardinality at most five time larger than that of OPT. $\square$

In Section 5, we extend the above approach to obtain a 3-approximation algorithm. The idea is to group the histograms into $[w]$-star-shaped orthogonal polygons such that in any minimum r-star cover of the original polygon, an element of the r-star coverage can overlap with at most three of the components.

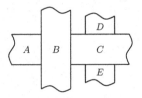

Fig. 4. The possible histogram configuration

In order to proceed with the complete description of the 3-approximation algorithm, we need to derive a linear-time algorithm for the r-star cover problem in the class of $[w]$-star-shaped orthogonal polygons, which we shall discuss in the next section.

4   Optimal Algorithm for $[w]$-Star-Shaped Polygons

Recall that the problem of finding the minimum number of r-star-shaped orthogonal polygons covering an orthogonal polygon P can be defined as the problem of finding the minimum set S of points such that each point of P is r-visible to an element of S. From now on, we shall use the latter convention, and we will say that the set S covers the polygon. Of course, by using such a convention, we then show how to translate the resulting set of covering points into the appropriate r-star partition (i.e., the complete description of covering polygons); due to space limits, details are omitted.

Let P be a $[w]$-star-shaped orthogonal polygon, we may assume w.l.o.g $w = $ NW. The idea of our algorithm for the minimum r-star cover of polygon P relies on the following observation (we omit the proof as for the purpose of the correctness proof of the algorithm, a more general proposition is derived).

Lemma 1. *Let an antident-W e_W be connected by an NW-staircase, whose segments are edges of P, with an antident-N e_N in P. There exists a minimum r-star cover of P which includes the point p whose X coordinate is equal to that of the right-hand endpoint of e_N and whose Y coordinate is equal to that of the lowest endpoint of e_W.*

Consequently, we can determine a minimum r-star cover of P by finding an antident pair (e_W, e_N), then computing the point p, cutting off the region of P which is (r-star) covered by p, and augmenting point p with the union of the recursive solutions for the remaining parts of P.

Theorem 3. *The minimum r-star cover problem for an n-vertex $[w]$-star-shaped orthogonal polygon can be computed in $O(n)$ time and $O(n)$ storage.*

Due to space limits, we omit the complete formal description of the algorithm and we only provide some intuition and an example illustrating our method. Consider the $[NW]$-star-shaped orthogonal polygon P that is shown in Fig. 5(a). Starting

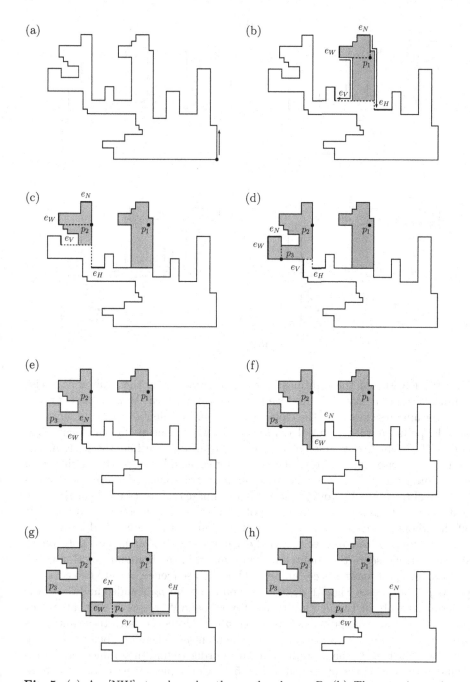

Fig. 5. (a) An [NW]-star-shaped orthogonal polygon P. (b) The covering point p_1 whose X coordinate is equal to that of the right-hand endpoint of e_N and whose Y coordinate is equal to that of the lowest endpoint of e_W. The gray region is cut off.

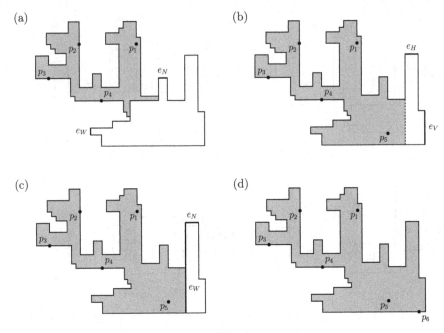

Fig. 6.

from the lowest and the rightmost vertex, we traverse, in the counterclockwise manner, the boundary of P until we find the first antident-W e_W. Next, for e_W we determine its closest, in the clockwise manner, antident-N e_N. By Lemma 1, we add point p_1 to the initial empty set S of covering points, and cut off a region of P which is visible to p_1: starting from e_W, we find the first, in the counterclockwise manner, vertical edge e_V not covered by p_1. Next, starting from e_N, we find the first, in the clockwise manner, horizontal edge e_H not covered by p_1, and the endpoints of e_V and e_H determinate our cut (see Fig. 5(b)). Note that in order to avoid the remaining polygon being disconnected, we do not cut off the whole region covered by p_1 — this region we will be handled later.

Next, starting from e_H, we repeat the same procedure for the new polygon P. We again traverse, in the counterclockwise manner, the boundary of P until we find the first pair antidents e_W and e_N. Thus, in a sequence, we add two covering points p_2 and p_3 (see Fig. 5(c-d)). But, when the point p_3 is added and the cut is generated, we obtain two new antidents of type N and W, respectively, that are not present in the original polygon P (see Fig. 5(e)). Moreover, both these antidents are covered by our temporary set S, i.e., the point p_2. Therefore, before applying Lemma 1, we have to make additional cut in order to find (new) uncovered antidents — the new cut is presented in Fig. 5(f). Consequently, we obtain the temporary covering set $S = \{p_1, p_2, p_3, p_4\}$ (see Fig. 5(g)). However, once again this is the case when in the resulting polygon P, the first antident-W e_W is covered — now by point p_1 — thus we have to modify cut (see Fig. 5(h)-6(a)). Consequently, we obtain the uncovered pair of antidents, the next point p_5 is determined, and the appropriate part of polygon is cut off (see Fig. 6(b)). Finally, the last point p_6 covering the new antidents e_W and e_N is added, resulting in the output set $S = \{p_1, p_2, p_3, p_4, p_5, p_6\}$ (see Fig. 6(c-d)).

5   A 3-Approximation Algorithm for Orthogonal Polygons

In this section, using the result of Theorem 3, we provide a linear-time 3-approximation algorithm for the minimum r-star cover problem in general orthogonal polygons. Similarly as the 5-approximation algorithm discussed in Section 3, our 3-approximation algorithm starts from the partition of a polygon into histograms. Then, it groups the histograms into $[w]$-star-shaped polygons, and independently computes the minimum r-star cover for each of $[w]$-star-shaped components.

5.1 $[w]$-Star-Shaped Partition

Let P be a simple orthogonal polygon, let e be an antident of P, and assume that the interior of P is above e. Consider now the rooted histogram tree (T, v) of the histogram partition $\mathcal{HIST}(P, e)$ — discussed in Section 2 — whose vertices corresponds to histograms and two vertices are adjacent if their corresponding histograms touch each other; root v corresponds to the histogram with the base e. Clearly, (T, v) can be computed while determining $\mathcal{HIST}(P, e)$, and thus it can be computed in linear time as well by Theorem 1.

The $[w]$-star partition of (T, v) is recursively defined as follows. Let H be the histogram corresponding to root v of tree (T, v). If T is a single vertex, then $\mathcal{STAR}(T, v)$ is $\{\{v\}\}$. Otherwise, let (T', v) be the maximal rooted subtree of T such that for any two histograms H^1 and H^2 corresponding to the adjacent vertices v_1 and v_2 in tree T', where v_2 is a children of v_1, the interior of H^2 is either above H^1 or on the left to H^1. Then, the $[w]$-star partition $\mathcal{STAR}(T, v)$ is defined as

$$\{V(T')\} \cup \bigcup_{x \in N(V(T')) \setminus V(T')} \mathcal{STAR}(T_x, x),$$

where $V(T')$ is the vertex set of subtree T', T_x is a subtree of T rooted at x, and $N(V(T'))$ is the set of neighbors of vertices in $V(T')$. Note that to determine $\mathcal{STAR}(T_x, x)$, we assume that the base e_x of the histogram H_x corresponding to x is horizontal, and the interior of H_x is above e_x, that is, we rotate the relevant subpolygon if needed. Clearly, $\mathcal{STAR}(T, v)$ is uniquely defined, and it can be found in linear time just by performing the breadth-first search algorithm on (T, v).

5.2   The 3-Approximation Algorithm

Our approximation algorithm for covering an orthogonal polygon with a minimum number of r-star orthogonal polygons relies on the subroutine for finding a minimum r-star cover of a given $[w]$-star-shaped polygon (Theorem 3).

Algorithm 2

Input. An an orthogonal polygon P with n vertices.

Output. A covering of P with r-star orthogonal polygons.

1. Partition P into orthogonal histograms using the method of Levcopoulos; let (T, v) be the resulting rooted histogram tree.
2. By performing the BFS algorithm on (T, v), find the $[w]$-star partition $\mathcal{STAR}$ of T.
3. For each tree T' in $\mathcal{STAR}$, cover its corresponding $[w]$-star-shaped polygon $P_{T'}$ with the minimum number of r-star orthogonal polygons (Theorem 3).

Theorem 4. *Algorithm 2 runs in $O(n)$ time and yields a 3-approximation for the problem of covering a simple orthogonal polygon with the minimum number of r-star orthogonal polygons.*

Proof. Bearing in mind the result of Theorem 3 and [15], we only have to discuss the 3-approximation ratio (compare with the proof of Theorem 2). Consider a minimum cover OPT of P with r-star orthogonal polygons. Let $P_{T'}$ be the $[w]$-star-shaped polygon corresponding to the union of all histograms of an element (tree) of $\mathcal{STAR}$ (Step 2). On the one hand, the cardinality of a minimum cover of $P_{T'}$ by r-star orthogonal polygons is not greater than the number of r-star orthogonal polygons in OPT that have a non-empty intersection with $P_{T'}$. On the other hand, by the definition of r-star visibility and the histogram decomposition, it follows that any r-star orthogonal polygon in OPT can have a non-empty intersection with at most five histograms in the histogram partition (again see Fig. 4), say with a histogram H in the rooted histogram tree (T, v), at most two children of H on the opposite sides of H, the parent of H, and at most one sibling of H on the opposite side of the parent of H. But then, by the definition of $[w]$-star partition of tree T, exactly three of these histograms are always in one of our $[w]$-star-shaped polygons, thus any r-star orthogonal polygon in OPT can have a non-empty intersection with at most three different $[w]$-star-shaped polygons. We conclude that Algorithm 2 produces a covering of cardinality at most three times larger than that of OPT.   □

6   Extensions

Recall that the *class k* of orthogonal polygons, $k = 0, 1, 2, 3, 4$, has dents of k-different orientations {N,W,S,E}. Following this definition, in this paper we have presented our linear-time 3-approximation algorithm for the problem of covering the class 4 of orthogonal polygons with the minimum number of r-stars. However, its worth pointing out that for the class 3 of orthogonal polygons our approach leads to the linear-time 2-approximation algorithm.

Theorem 5. *Algorithm 2 yields a linear-time 2-approximation for the problem of covering the class 3 orthogonal polygon with the minimum number of r-star orthogonal polygons.*

It would be interesting to undertake modifying the above antidents pair approach for the whole class 2b of orthogonal polygons in order to provide an optimal linear-time algorithm (the class of $[w]$-star-shaped polygons is the subclass of Class 2b) — as far as we know, there are no such algorithms (again see Table 1). It is worth pointing out that our approach can be adapted to monotone polygons (class 2a), thus getting a simple linear-time algorithm for covering a monotone orthogonal polygon [16] (the first one was proposed by Gewali *et al.* [8]). Finally, we hope that our antidents pair approach will provide some intuition and structural properties that might lead to improved exact algorithms for the general orthogonal polygons. This is a future work.

References

1. Aggarwal, A.: The Art Gallery Theorem: Its Variations, Applications, and Algorithmic Aspects, Ph.D. Thesis, Johns Hopkins University (1984)
2. Carlsson, S., Nilsson, B.J., Ntafos, S.C.: Optimum guard covers and m-watchmen routes for restricted polygons. Int. J. Comput. Geometry Appl. 3(1), 85–105 (1993)
3. Chazelle, B.M., Dobkin, D.: Decomposing a polygon into convex parts. In: Proc. 11th ACM Symp. on Theory of Comput., pp. 38–48 (1979)
4. Culberson, J.C., Reckhow, R.A.: A unified approach to orthogonal polygon covering problems via dent diagrams, Technical Report TR 89-06, University of Alberta (1989)
5. Culberson, J.C., Reckhow, R.A.: Covering polygons is hard. J. Algorithms 17(1), 2–44 (1994)
6. Edelsbrunner, H., O'Rourke, J., Welzl, E.: Stationing guards in rectilinear art galleries. Computer Vision, Graphics and Image Processing 27, 167–176 (1984)
7. Franzblau, D.S., Kleitman, D.J.: An algorithm for constructing regions with rectangles: Independence and minimum generating sets for sollections of intervals. In: Proc. 16th ACM Symp. on Theory of Comput., pp. 167–174 (1984)
8. Gewali, L., Keil, J.M., Ntafos, S.C.: On covering orthogonal polygons with star-shaped polygons. Information Sciences 65, 45–63 (1992)
9. Grötschel, M., Lovász, L., Schrijver, A.: The ellipsoid method and its consequences in combinatorial optimization. Combinatorica 1, 169–197 (1981)
10. Keil, J.M.: Decomposing a polygon into simpler components. SIAM J. Computing 14, 799–817 (1985)
11. Keil, J.M.: Minimally covering a horizontally convex orthogonal polygon. In: Proc. 2nd Annual Symp. on Comput. Geometry, pp. 43–51 (1986)
12. Keil, J.M.: Polygon decomposition. In: Handbook of Computational Geometry, pp. 491–518. North-Holland Publishing, Amsterdam (2000)
13. Keil, J.M., Sack, J.-R.: Minimum decomposition of polygonal objects. In: Computational Geometry, pp. 197–216. North-Holland Publishing, Amsterdam (1985)
14. Laurent, M., Rendl, F.: Semidefinite programming and integer programming. In: Handbook on Discrete Optimization, pp. 393–514. Elsevier Science, Amsterdam (2005)
15. Levcopoulos, C.: Heuristics for Minimum Decompositions of Polygons, Ph.D. Thesis, Linköping University (1987)
16. Lingas, A., Wasylewicz, A., Żyliński, P.: Note on covering orthogonal polygons with star-shaped polygons. Inform. Process. Lett. 104(6), 220–227 (2007)

17. Lu, H.-I.: Efficient Approximation Algorithms for Some Semidefinite Programs Ph.D. Thesis, Brown Univeristy (1996)
18. Lubiw, A.: Orderings and Some Combinatorial Optimization Problems with Geometric Applications, Ph.D. thesis, University of Toronto (1986)
19. Motwani, R., Raghunathan, A., Saran, H.: Perfect graphs and orthogonally convex covers. SIAM J. Discrete Math. 2, 371–392 (1989)
20. Motwani, R., Raghunathan, A., Saran, H.: Covering orthogonal polygons with star polygons: the perfect graph approach. J. Comput. Syst. Sci. 40, 19–48 (1990)
21. Raghunathan, A.: Polygon Decomposition and Perfect Graphs, Ph.D. Thesis, University of California (1988)
22. Reckhow, R.A.: Covering orthogonally convex polygons with three orientations of dents, Technical Report TR 87-17, University of Alberta (1987)
23. O'Rourke, J., Supowit, K.J.: Some NP-hard polygon decomposition problems. IEEE Transactions on Information Theory 29(2), 181–189 (1983)
24. Preparata, F.P., Shamos, M.I.: Computational Geometry. Springer, New York (1985)
25. Worman, Ch.: Decomposing Polygons into r-stars or α-boundable Subpolygons, Ms.C. Thesis, University of Saskachewan (2004)
26. Worman, C., Keil, J.M.: Polygon decomposition and the orthogonal art gallery problem. Int. J. Comput. Geometry and Appl. 17(2), 105–138 (2007)

Multi-commodity Source Location Problems and Price of Greed

Hiro Ito[1], Mike Paterson[2], and Kenya Sugihara[1]

[1] Graduate School of Informatics, Kyoto University,
Kyoto 606-8501, Japan
{itohiro,sugihara}@kuis.kyoto-u.ac.jp
[2] Department of Computer Science, University of Warwick,
Coventry, CV4 7AL, United Kingdom
msp@dcs.warwick.ac.uk

Abstract. Given a graph $G = (V, E)$, we say that a vertex subset $S \subseteq V$ covers a vertex $v \in V$ if the edge-connectivity between S and v is at least a given integer k, and also say that S covers an edge $vw \in E$ if v and w are covered. We propose the multi-commodity source location problem, which is such that given a vertex- and edge-weighted graph G, r players each select p vertices, and obtain a profit that is the total weight of covered vertices and edges. However, vertices selected by one player cannot be selected by the other players. The goal is to maximize the total profits of all players. We show that the *price of greed*, which indicates the ratio of the total profit of cooperating players to that of selfish players, is tightly bounded by $\min\{r, p\}$. Also when $k = 2$, we obtain tight bounds for vertex-unweighted trees.

Keywords: Source location problem, price of greed.

1 Introduction

Given an undirected graph $G = (V, E)$ with vertex set V, edge set E and an edge capacity function, the *edge-connectivity* between $S \subseteq V$ and $v \in V$ is the minimum total capacity of a set of edges such that v is disconnected from S by removal of these edges. We say that a set $S \subseteq V$ of vertices (called sources) *covers* $v \in V$ if the edge-connectivity between S and v is at least k, where $k \geq 1$ is a given integer. The source location problem is to find a minimum-size source set $S \subseteq V$ covering all vertices in V. This problem has been studied widely [2,3,6,7,8,10,12,15], and such problems are important in the design of networks resistant to the failure of edges.

In real networks, there are multiple service providers, and they locate servers in networks in order to supply services. Thus we propose the *multi-commodity source location problem*. In this problem, a network $N = (G = (V, E), w, c)$, and positive integers k, r and p are given, where G is an undirected and connected graph, $c : E \to \mathbf{Z}_+$ is an edge capacity function, $w : V \cup E \to \mathbf{R}_+$ is a vertex- and edge-weight function, and r is the number of players, where $\mathbf{Z}_+$ (resp., $\mathbf{R}_+$)

S.-i. Nakano and Md. S. Rahman (Eds.): WALCOM 2008, LNCS 4921, pp. 169–179, 2008.

denotes the set of non-negative integers (resp., real numbers). Players $1, 2, \ldots, r$ each locate p sources on vertices of G. However, if a player locates a source on a vertex, then it is unavailable for the other players to locate a source on. Let player i's profit be the total weight of vertices and edges covered by the sources located by the player i, where a source set S covers an edge $e = vw$ if both v and w are covered by S. The goal of this problem is to maximize the sum of the profits of all players. When $r = 1$, the problem is the same as the maximum-cover source-location problem [12]. This problem is NP-hard [14], but it can be solved in polynomial time when $k \leq 3$ [12,13,14].

In recent years, game theory has attracted attention in certain fields of computer science. In various real problems, e.g., routing, network design and scheduling, the selfish actions of agents are obstacles to optimization for social welfare. Such phenomena are modeled as games, and the influence of selfish actions of players have been extensively analysed [1,5,11]. In this paper, we consider the influence of selfish actions of providers on network reliability. Generally the quality of services of newcomers are influenced by the services of preceding providers. For example, the location of servers becomes restricted and consequently their profits may be smaller. We analyze the phenomenon by means of a selfish model in the multi-commodity source location problem.

The selfish model is such that players $1, 2, \ldots, r$ locate p sources in this order on vertices of G so as to maximize their own profits. Note that as described previously, player i cannot locate sources on the vertices on which one of players $1, \ldots, i - 1$ has already located sources. We compare the social welfare of this model to that of the case where all players cooperate, that is, optimal solutions of the problem. Figure 1 shows an example when $k = 2$, $r = 3$ and $p = 3$. The numbers beside vertices and edges denote their weights, and the edge capacities are uniformly one. The selfish player 1 locates sources to maximize his/her profit (see Fig. 1(a)). The source set $\{a, i, j\}$ of the player 1 covers vertices a, b, c, d, h, i, j and he/she gets profit 29. The selfish player 2 then locates sources on vertices not already occupied such that his/her own profit is maximum in this situation. The source set $\{c, g, h\}$ of the player 2 covers vertices a, b, c, d, g, h and his/her profit is 25. Then selfish player 3 does similarly and gets profit 6 since his/her source set $\{b, e, f\}$ covers vertices a, b, c, d, e, f. The total profit is $29 + 25 + 6 = 60$. In contrast, the optimal players 1, 2 and 3 locate their sources as in Fig. 1(b), so that the players share as many covered vertices and edges as possible. The optimal players 1, 2 and 3 obtain profits 28, 27 and 25, respectively, and the total is 80.

As a measure of the influence of the selfish behaviours based on the ordering of players, we propose the *price of greed*, which represents the ratio of the maximum total profit of the cooperating players to the worst, i.e., minimum, total profit of the selfish players. Formally, let the price of greed for the multi-commodity source location problem be

$$POG_k(N, r, p) = \frac{\text{the optimal total profit}}{\text{the worst selfish total profit}} .$$

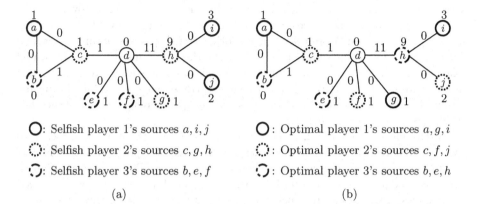

O: Selfish player 1's sources a, i, j O: Optimal player 1's sources a, g, i

◌: Selfish player 2's sources c, g, h ◌: Optimal player 2's sources c, f, j

⟁: Selfish player 3's sources b, e, f ⟁: Optimal player 3's sources b, e, h

(a)   (b)

Fig. 1. An example of behaviours of selfish and optimal players when $k = 2$, $r = 3$ and $p = 3$

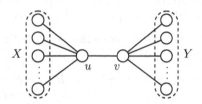

Fig. 2. An instance of $POG_k(N, r, p) = \min\{r, p\}$

The price of greed of the example in Fig. 1 is $POG_2(N, 3, 3) = (28 + 27 + 25)/(29 + 25 + 6) = 4/3$.

A well-known similar measure, the price of anarchy is the ratio of the worst cost of Nash equilibria of selfish players to the optimal cost [11]. The locations of sources derived by our selfish model are Nash equilibria, since no player can gain profit by changing their locations. However, the locations are only a component of all Nash equilibria, and we analyze influence of the greedy behaviour based on the ordered strategy in our model. Thus we use the different name.

Our Results. Our goal is to analyze the maximum value of the price of greed $POG_k(r, p) = \max_N POG_k(N, r, p)$. When $k = 1$, it is clear that $POG_1(r, p) = 1$ for any $r, p \geq 1$, since all vertices and edges are covered, wherever the sources are located. Hence we assume $k \geq 2$.

First, we show the following tight bounds for the general case.

Theorem 1. $POG_k(r, p) = \min\{r, p\}$ for any $k \geq 2$, $r \geq 1$ and $p \geq 1$.

The upper bound, $POG_k(r, p) \leq \min\{r, p\}$, is shown in Sect. 2 as Lemma 2. The lower bound is easily proved by showing an instance with $POG_k(N, r, p) = \min\{r, p\}$ in Fig. 2 where $r \geq 2$ and $p \geq 2$. Let $|X| = \min\{r, p\} - 1$ and $|Y| = rp - \min\{r, p\} - 1$. The weight of u is 1, the other vertices and edges have weight

0, and the capacities of all edges are $\lceil k/2 \rceil$. In this case, a vertex is covered when it is on a path between sources. If the selfish player 1 obtains profit 1 by locating $\min\{r, p\}$ sources on X and u, then the other selfish players obtain no profits. Hence the worst total selfish profit is 1. On the other hand, each of optimal players $1, \ldots, \min\{r, p\}$ obtains profit 1 by locating one source on $X \cup \{u\}$ and $p - 1$ sources on $Y \cup \{v\}$. Then the other optimal players cannot obtain any profit. Hence the optimal total profit is $\min\{r, p\}$. Therefore, this instance has $POG_k(N, r, p) = \min\{r, p\}$.

Furthermore, when $k = 2$, we consider the case where the input graph G is restricted to a vertex-$unweighted$ tree with every edge of capacity 1 and weight at least 0, and where sources are located only on the leaves. This vertex-unweighted tree case is equivalent to the problem that r players find the r subtrees induced by p leaves of the input tree such that the total edge-weight of these r subtrees is a maximum. Maximum edge-weight trees have many applications, e.g., communication networks [4,9]. This is a quite special case of the original problem. However, $POG_2(r, p)$ is at most only one less than that for the vertex- and edge-weighted case, as we show in the following theorem. Note that if $p = 1$, then any optimal and selfish player obtains no profit and hence we assume $p \geq 2$.

Theorem 2. *For vertex-unweighted trees and any* $r \geq 1$, $POG_2(r, 2) = \min\{r, 2\}$ *and* $POG_2(r, p) = \min\{r, p - 1\}$ *for* $p \geq 3$.

At first, we tried to construct an instance having $POG_2(N, r, p) = p$ for $r \geq p \geq 3$ in a similar way to the vertex-weighted case. However, in Fig. 2, if we let the weight of the edge uv be 1 instead of the vertex-weight of u, and let that of the other edges be 0, then X must satisfy $|X| \leq p - 1$ so that selfish player 1's sources can occupy X. In this case, the total profit of optimal players is at most $p - 1$. Hence even if $|X| = p - 1$, $POG_2(N, r, p) = p - 1$ for this instance. In fact, it can be shown that this instance has the worst price of greed, since $POG_2(r, p) \leq \min\{r, p - 1\}$ for $p \geq 3$ (Lemma 5).

2   Analysis of $POG_k(r, p)$ for the General Case

Given a network N, let W_i $(1 \leq i \leq r)$ be the profit of an optimal player i. We assume $W_1 \geq W_2 \geq \cdots \geq W_r$ without loss of generality. Let W_i' $(1 \leq i \leq r)$ be the profit of a selfish player i for the worst case, i.e., where the total profit is least. From the definition, $W_1' \geq W_2' \geq \cdots \geq W_r'$ holds, and $POG_k(N, r, p) = (\sum_{i=1}^r W_i)/(\sum_{i=1}^r W_i')$. Let S_i $(1 \leq i \leq r)$ be the set of sources located by the optimal player i and S_i' $(1 \leq i \leq r)$ be the set of sources located by the selfish player i. For a source set $S \subseteq V$, let $w_k(S)$ denote the total weight of vertices and edges covered by S. Note that $w_k(S_i) = W_i$ and $w_k(S_i') = W_i'$ for $i = 1, \ldots, r$. For a vertex set $\{v_1, \ldots, v_q\}$, we may write $w_k(\{v_1, \ldots, v_q\})$ as $w_k(v_1, \ldots, v_q)$ for notational simplicity.

We show the upper bounds for $POG_k(r, p)$ in Lemma 2, and show that the bounds are tight in Lemma 3. Lemma 2 depends on the following lemma.

Lemma 1. *For any* i $(1 \leq i \leq \lfloor \frac{r-1}{p} \rfloor + 1)$, *we have* $W_i' \geq W_{(i-1)p+1}$.

Proof. When $i = 1$, the inequality $W_1' \geq W_1$ evidently holds. Then we consider $i \geq 2$. Since $|\bigcup_{j=1}^{i-1} S_j'| = (i-1)p$, at least one, S_q say, of $S_1, \ldots, S_{(i-1)p+1}$ has no common source with any of $S_1', \ldots, S_{i-1}'$. The profit W_i' of the selfish player i is the largest profit when he/she locates sources on vertices in $V \setminus \bigcup_{j=1}^{i-1} S_j'$. From the above discussion, we obtain $W_i' \geq W_q \geq W_{(i-1)p+1}$. □

Lemma 2. *For any $k \geq 2$, $r \geq 1$ and $p \geq 1$, we have $POG_k(r,p) \leq \min\{r,p\}$.*

Proof. From $W_1' \geq W_i$ for any i, it is clear that $POG_k(N,r,p) \leq (\sum_{i=1}^r W_i)/W_1'$ $\leq r$ for any N, r and p. Then we show $POG_k(r,p) \leq p$ for $p < r$ as follows. From Lemma 1, for $1 \leq i \leq \lfloor \frac{r-1}{p} \rfloor + 1$, we have $pW_i' \geq W_{(i-1)p+1} + \cdots + W_{ip}$ (where we take $W_v = 0$ when $v > r$), and hence $p\sum_{i=1}^r W_i' \geq \sum_{i=1}^r W_i$. Therefore we have $POG_k(r,p) \leq p$ for $p < r$. □

The next lemma shows that the upper bound is tight.

Lemma 3. *$POG_k(r,p) \geq \min\{r,p\}$ for any $k \geq 2$, $r \geq 1$ and $p \geq 1$.*

Proof. This is proved by showing an instance (N,k,r,p) that has $POG_k(N,r,p)$ $= \min\{r,p\}$. Clearly, when $r = 1$ or $p = 1$, $POG_k(N,r,p) = 1$ for any network N. For $r \geq 2$ and $p \geq 2$, we have already shown such an instance in Sect. 1 (Fig. 2). □

Proof (of Theorem 1). This follows immediately from Lemmas 2 and 3. □

3 Analyses of $POG_2(r,p)$ for Vertex-Unweighted Trees

In this section, we deal with the case of $k = 2$, vertex-unweighted trees with every edge of capacity 1 and weight at least 0, and where the players locate sources only on the leaves of the tree. Locating sources on leaves does not make the problem weak, since the case where sources can be located on any vertices can be reduced to this case by adding a leaf to every non-leaf vertex. This problem is equivalent to the problem for finding r subtrees induced by p leaves of the input tree such that the r subtrees have maximum total edge-weight, and it is a basic and important problem in network optimization problems [12]. Note that the upper bound in Lemma 5 is available for the case where sources can be located on any vertices from the above reduction involving adding leaves.

First, we show the following lemma, which is important for showing the upper bound (Lemma 5). Recall that in this case $w_2(s,v)$ is the total weight of edges in the path from s to v.

Lemma 4. *Assume that $r \geq 2$ and $p \geq 3$. Let i be an integer with $1 \leq i \leq r-1$. Let $S_i' = \{s_1', \ldots, s_p'\}$, and $v_0 \notin S_1' \cup \cdots \cup S_i'$ be a vertex. Then $\sum_{j=1}^p w_2(s_j', v_0) \leq (p-1)W_i'$.*

Proof. Let P_j be the set of edges on the path between s_j' and v_0. Let $X = \bigcap_{j=1}^p P_j$ be the set of edges contained in all of $P_1, \ldots, P_p$, and let $Y = \bigcup_{1 \leq j < \ell \leq p}(P_j \cap P_\ell) \setminus X$

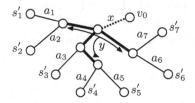

Fig. 3. The edges on the dotted line are in X and those on the bold lines are in Y

be the set of edges contained in at least two of $P_1, \ldots, P_p$ and not contained in X (see Fig. 3). Let x (resp., y) be the total weight of the edges in X (resp., Y). Moreover, let a_j be the total weight of the edges in $P_j \setminus (X \cup Y)$ (See Fig. 3).

From the definition, the following inequality holds.

$$\sum_{1 \leq j \leq p} w_2(s'_j, v_0) \leq px + (p-1)y + \sum_{1 \leq j \leq p} a_j .$$

Clearly, $W'_i = y + \sum_{j=1}^{p} a_j$. Here we can observe that $x \leq \min_j\{a_j\}$, since S'_i has the largest profit obtained by p sources not contained in any of $S'_1, \ldots, S'_{i-1}$, and hence $a_j - x = W'_i - w_2(S'_i \cup \{v_0\} \setminus \{s'_j\}) \geq 0$. Thus for $p \geq 3$,

$$\sum_{1 \leq j \leq p} w_2(s'_j, v_0) \leq px + (p-1)y + \sum_{1 \leq j \leq p} a_j$$

$$\leq (p-1)y + 2 \sum_{1 \leq j \leq p} a_j$$

$$\leq (p-1)W'_i . \qquad \square$$

Lemma 5. *For vertex-unweighted trees and any $r \geq 1$, $POG_2(r,2) \leq \min\{r,2\}$ and $POG_2(r,p) \leq \min\{r, p-1\}$ for $p \geq 3$.*

Proof. From Lemma 2, $POG_2(r,p) \leq \min\{r,p\}$ for any $r \geq 1$ and $p \geq 2$. We show $POG_2(r,p) \leq p-1$ for $r \geq p \geq 3$ in the following part.

Let $S_i = \{s^i_1, \ldots, s^i_p\}$ for $1 \leq i \leq r$. We consider a source $v_0 = s^t_1 \in S_t$ of an optimal player t $(1 \leq t \leq r)$. For any pair of vertices $u, s \in V$ and a vertex subset $X \subseteq V$, $w_2(X \cup \{u, s\}) \leq w_2(X \cup \{u\}) + w_2(u, s)$. Hence

$$W_t = w_2(v_0, s^t_2, \ldots, s^t_p) \leq \sum_{2 \leq j \leq p} w_2(s^t_j, v_0) ,$$

and

$$W_i = w_2(s^i_1, \ldots, s^i_p)$$

$$\leq w_2(s^i_1, \ldots, s^i_p, v_0)$$

$$\leq \sum_{1 \leq j \leq p} w_2(s^i_j, v_0) \quad \text{for } i \neq t \text{ and } 1 \leq i \leq r .$$

If we sum each side of the inequalities, then for any $v_0 \in \bigcup_{i=1}^{r} S_i$,

$$\sum_{1 \leq i \leq r} W_i \leq \sum_{s \in \left(\bigcup_{i=1}^{r} S_i\right) \setminus \{v_0\}} w_2(s, v_0) . \qquad (1)$$

In the rest of the proof, we consider two cases (i) $\bigcup_{i=1}^{r} S_i' = \bigcup_{i=1}^{r} S_i$ and (ii) $\bigcup_{i=1}^{r} S_i' \neq \bigcup_{i=1}^{r} S_i$.

Case (i). ($\bigcup_{i=1}^{r} S_i' = \bigcup_{i=1}^{r} S_i$). Let $v_0 \in S_r'$. Note that v_0 is also contained in $\bigcup_{i=1}^{r} S_i$. Since $v_0 \notin S_1' \cup \cdots \cup S_i'$ for $1 \le i \le r-1$, from Lemma 4,

$$\sum_{s' \in S_i'} w_2(s', v_0) \le (p-1)W_i' \tag{2}$$

for each i with $1 \le i \le r-1$. On the other hand, for any $s' \in S_r'$, $w_2(s', v_0) \le W_r'$, since $v_0 \in S_r'$. Hence

$$\sum_{s' \in S_r' \setminus \{v_0\}} w_2(s', v_0) \le (p-1)W_r' . \tag{3}$$

Consequently, if we sum up (2) for $1 \le i \le r-1$ and (3), we get

$$\sum_{s' \in (\bigcup_{i=1}^{r} S_i') \setminus \{v_0\}} w_2(s', v_0) \le (p-1) \sum_{1 \le i \le r} W_i' . \tag{4}$$

Since $\bigcup_{i=1}^{r} S_i' = \bigcup_{i=1}^{r} S_i$, the left side of (4) is equal to the right side of (1), so

$$\sum_{1 \le i \le r} W_i \le \sum_{s \in (\bigcup_{i=1}^{r} S_i) \setminus \{v_0\}} w_2(s, v_0)$$

$$= \sum_{s' \in (\bigcup_{i=1}^{r} S_i') \setminus \{v_0\}} w_2(s', v_0)$$

$$\le (p-1) \sum_{1 \le i \le r} W_i' .$$

Therefore, $POG_2(r, p) \le p - 1$, for any r, p with $r \ge p \ge 3$.

Case (ii). ($\bigcup_{i=1}^{r} S_i' \neq \bigcup_{i=1}^{r} S_i$). In this case, there exist selfish source sets S_i' with $S_i' \nsubseteq \bigcup_{j=1}^{r} S_j$. Let $S_{\ell(1)}', S_{\ell(2)}', \ldots, S_{\ell(h)}'$ with $\ell(1) \le \ell(2) \le \cdots \le \ell(h)$ be such source sets.

For each $S_{\ell(i)}'$, we make a new source set $S_{\ell(i)}'' \subseteq \bigcup_{j=1}^{r} S_j$ with $|S_{\ell(i)}''| = p$ in the following way. Let $T_{\ell(i)} = S_{\ell(i)}' \cap \bigcup_{j=1}^{r} S_j$ be a source set that consists of the sources contained in not only $S_{\ell(i)}'$ but also contained in some of the optimal sources $S_1, \ldots, S_r$. For $T_{\ell(1)}$, we select q ($= p - |T_{\ell(1)}|$) sources $s_1, \ldots, s_q$ and let $S_{\ell(1)}'' = T_{\ell(1)} \cup \{s_1, \ldots, s_q\}$. The selection of $s_1, \ldots, s_q$ is such that $s_1, \ldots, s_q$ are selected from the vertices contained in the optimal source sets $S_1, \ldots, S_r$ but not contained in any of the selfish source sets $S_1', \ldots, S_r'$, i.e., $\bigcup_{i=1}^{r} S_i \setminus \bigcup_{i=1}^{r} S_i'$, and maximize the total weight $W_{\ell(1)}'' := w_2(S_{\ell(1)}'')$ of edges covered by $S_{\ell(1)}''$. Similarly, for $T_{\ell(2)}$, we select $p - |T_{\ell(2)}|$ sources from not yet selected vertices in $\bigcup_{i=1}^{r} S_i \setminus \bigcup_{i=1}^{r} S_i'$ so as to maximize the total weight $W_{\ell(2)}'' := w_2(S_{\ell(2)}'')$, where $S_{\ell(2)}''$ is a set of the selected sources and the sources in $T_{\ell(2)}$. By repeating the above operations, $S''_{\ell(1)}, \ldots, S''_{\ell(h)}$ are obtained. Note that $W''_{\ell(i)} \le W'_{\ell(i)}$ for $1 \le i \le h$, from the behaviour of selfish players.

Now, let $v_0 \in S''_{\ell(h)} \setminus T_{\ell(h)}$ be a source of the selfish player $\ell(h)$ which is selected in the above operation. Note that $v_0 \in \bigcup_{i=1}^{r} S_i$. Let $S''_{\ell(i)} = \{s''_{\ell(i),1}, \ldots, s''_{\ell(i),p}\}$ for $1 \le i \le h$. We show the following inequality similar to that of Lemma 4. For $1 \le i \le h-1$,

$$\sum_{1 \le j \le p} w_2(s''_{\ell(i),j}, v_0) \le (p-1) W'_{\ell(i)} . \tag{5}$$

Similarly to the notation of the proof of Lemma 4 (Fig. 3), let P_j with $1 \le j \le p$ be the set of edges on the path between v_0 and $s''_{\ell(i),j} \in S''_{\ell(i)}$. Let $X = \bigcap_{j=1}^{p} P_j$, and let $Y = \bigcup_{1 \le j < \ell \le p}(P_j \cap P_\ell) \setminus X$ be the set of edges contained in at least two of $P_1, \ldots, P_p$ and not contained in X. Let x (resp., y) be the total weight of the edges in X (resp., Y). Moreover, let a_j with $1 \le j \le p$ be the total weight of the edges in $P_j \setminus (X \cup Y)$, and we abbreviate $\sum_{1 \le j \le p} a_j$ by A. Clearly,

$$\sum_{1 \le j \le p} w_2(s''_{\ell(i),j}, v_0) \le px + (p-1)y + \sum_{1 \le j \le p} a_j = px + (p-1)y + A . \tag{6}$$

From the definition of $S''_{\ell(i)}$, there exists a_q such that $x \le a_q$, (because if such a_q did not exist, then by selecting v_0 instead of a source $s''_{\ell(i),t} \in S''_{\ell(i)}$ in the above operation, the total weight of edges covered by $S''_{\ell(i)} \cup \{v_0\} \setminus \{s''_{\ell(i),t}\}$ would be larger than $W''_{\ell(i)}$.) Moreover, from the behaviour of the selfish player i, for an arbitrary t with $1 \le t \le p$,

$$x + y + \sum_{1 \le j \le p : j \ne t} a_j = x + y + A - a_t \le W'_{\ell(i)} . \tag{7}$$

Summing (7) for all $t \ne q, 1 \le t \le p$, yields

$$(p-1)(x+y+A) - (A - a_q) \le (p-1)W'_{\ell(i)} ,$$

and hence

$$px + (p-1)y + (p-2)A \le (p-1)W'_{\ell(i)} , \tag{8}$$

since $x \le a_q$. From the above discussion, (5) can be shown as follows.

$$\sum_{1 \le j \le p} w_2(s''_{\ell(i),j}, v_0) \le px + (p-1)y + A \qquad \text{(from (6))}$$

$$\le (p-1)W'_{\ell(i)} - (p-3)A \qquad \text{(from (8))}$$

$$\le (p-1)W'_{\ell(i)} . \qquad \text{(since } p \ge 3\text{)}$$

On the other hand, for any $s'' \in S''_{\ell(h)}$, $w_2(s'', v_0) \le W''_{\ell(h)}$ since $v_0 \in S''_{\ell(h)}$. Hence

$$\sum_{s'' \in S''_{\ell(h)} \setminus \{v_0\}} w_2(s'', v_0) \le (p-1)W''_{\ell(h)} \le (p-1)W'_{\ell(h)} . \tag{9}$$

In addition, we consider the selfish source sets S_i' other than $S_{\ell(1)}', S_{\ell(2)}', \ldots, S_{\ell(h)}'$. For such S_i',

$$\sum_{s' \in S_i'} w_2(s', v_0) \le (p-1)W_i' \tag{10}$$

from Lemma 4, since no S_i' contains v_0.

Let S' be the union of every S_i' except $S_{\ell(t)}'$ for all $1 \le t \le h$, and $S'' = \bigcup_{1 \le i \le h} S_{\ell(i)}''$. Now $S' \cup S''$ is equal to the union $\bigcup_{i=1}^r S_i$ of the optimal source sets. Thus, the summation of the left side of inequalities (5), (9) and (10) is equal to the right side of (1), i.e.,

$$\sum_{1 \le i \le r} W_i \le \sum_{s \in (\bigcup_{i=1}^r S_i) \setminus \{v_0\}} w_2(s, v_0)$$

$$= \sum_{s'' \in S'' \setminus \{v_0\}} w_2(s'', v_0) + \sum_{s' \in S'} w_2(s', v_0)$$

$$= \sum_{i=1}^{h-1} \sum_{j=1}^{p} w_2(s_{\ell(i),j}'', v_0) + \sum_{s'' \in S_{\ell(h)}'' \setminus \{v_0\}} w_2(s'', v_0) + \sum_{s' \in S'} w_2(s', v_0)$$

$$\le (p-1) \sum_{1 \le i \le r} W_i' .$$

Therefore, $POG_2(r, p) \le p - 1$, for any r, p with $r \ge p \ge 3$.   □

The following lemma shows that the upper bounds are tight.

Lemma 6. *For vertex-unweighted trees and any $r \ge 1$, $POG_2(r, 2) \ge \min\{r, 2\}$ and $POG_2(r, p) \ge \min\{r, p-1\}$ for any $p \ge 3$.*

Proof. This is proved by showing an instance (N, k, r, p) that has $POG_2(N, r, 2) = \min\{r, 2\}$ and $POG_2(N, r, p) = \min\{r, p-1\}$ for $p \ge 3$.

When $r = 1$, clearly $POG_2(N, 1, p) = 1$ for any network N and p.

For $p = 2$, $POG_2(N, r, 2) = \min\{r, 2\}$ for the network N in Fig. 4. The numbers beside edges denote their weights, and $|Y| = 2r - 2$. The selfish player 1 may locate two sources on X and obtains profit 1. The other selfish players cannot obtain any profits. On the other hand, the optimal players 1 and 2 each obtain profit 1, by locating one source on X and the other one on Y. The other optimal players do not obtain profits. Hence the least selfish total profit and the optimal one are 2 and 4, respectively. Hence $POG_2(N, r, 2) = 2$ for this network N.

For $r \ge p \ge 3$, in Fig. 2, let the weight of the edge uv be 1 and that of the other edges be 0. Let $|X| = p - 1$ and $|Y| = rp - p + 1$. The selfish player 1 may locate $p - 1$ sources on X and one source on Y, and obtains profit 1. Then the other selfish players cannot obtain profits. Each of the optimal players $1, 2, \ldots, p-1$ ($\le r$) obtains profit 1 by locating one source on X and $p-1$ sources on Y. Since the other optimal players obtain no profit, the total optimal profit is $p - 1$. Hence this network N has $POG_2(N, r, p) = p - 1$ for $r \ge p \ge 3$.

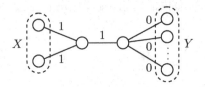

Fig. 4. An instance of $POG_2(N, r, 2) = \min\{r, 2\}$

In the case of $p > r$ (≥ 2), let $|X| = r$, $|Y| = rp - r$, and the weight of the edge uv be 1 in Fig. 2. The rest is similar. This network N has $POG_2(N, r, p) = r$ for $p > r$.   □

Note that even if we do not assume that sources are located only on the leaves, there exist instances equivalent to those for $r \geq p \geq 3$ and for $p > r \geq 2$ in the proof of Lemma 6 by removing an arbitrary leaf from X and also one from Y.

Proof (of Theorem 2). This follows immediately from Lemmas 5 and 6.   □

4   Conclusion and Future Work

In this paper we presented a new problem, the multi-commodity source location problem and analyzed the value of the price of greed. We showed the tight bound $POG_k(r, p) = \min\{r, p\}$ in the general case for any $k \geq 2$, $r \geq 1$ and $p \geq 1$. In addition, for a vertex-unweighted tree and $k = 2$, we showed $POG_2(r, p) = \min\{r, p - 1\}$ for $p \geq 3$ and $POG_2(r, 2) = \min\{r, 2\}$ for any $r \geq 1$ if the players locate sources only on the leaves.

Further work is to analyze the value of the price of greed and the behaviours of selfish players when each player in turn locates sources one at a time. It would also be interesting to consider the problem when all players simultaneously locate sources, with possibly several players choosing the same vertices. The profit of vertices and edges covered by several players' sources would be divided among those players in some appropriate way.

Acknowledgements

We are grateful to Professor Kazuo Iwama for his very helpful discussions of the problem.

References

1. Anshelevich, E., Dasgupta, A., Kleinberg, J., Tardos, É., Wexler, T., Roughgarden, T.: The price of stability for network design with fair cost allocation. In: Proc. of the 45th Annual IEEE Symposium on Foundations of Computer Science, pp. 295–304 (2004)

2. Arata, K., Iwata, S., Makino, K., Fujishige, S.: Locating sources to meet flow demands in undirected networks. J. of Algorithms 42, 54–68 (2002)
3. Bárász, M., Becker, J., Frank, A.: An algorithm for source location in directed graphs. EGRES Technical Report, 2004-06 (2004)
4. Farley, A.M., Fragopoulou, P., Krumme, D., Proskurowski, A., Richards, D.: Multi-source spanning tree problems. J. Interconnection Networks 1, 61–71 (2000)
5. Hayrapetyan, A., Tardos, É., Wexler, T.: The effect of collusion in congestion games. In: Proc. of the 38th annual ACM symposium on Theory of computing, pp. 89–98 (2006)
6. Heuvel, J.V.D., Johnson, M.: Transversals of subtree hypergraphs and the source location problem in digraphs, CDAM Research Report, LSE-CDAM-2004-10 (2004)
7. Ito, H., Ito, M., Itatsu, Y., Nakai, K., Uehara, H., Yokoyama, M.: Source location problems considering vertex-connectivity and edge-connectivity simultaneously. Networks 40(2), 63–70 (2002)
8. Ito, H., Uehara, H., Yokoyama, M.: A faster and flexible algorithm for a location problem on undirected flow networks. IEICE Trans. Fundamentals E83-A, 704–712 (2000)
9. Lee, W.C.: Spanning tree method for link state aggregation in large communication networks. In: Proc. of the 14th Annual Joint Conference of the IEEE Computer and Communication Societies, vol. 1, pp. 297–302 (1995)
10. Nagamochi, H., Ishii, T., Ito, H.: Minimum cost source location problem with vertex-connectivity requirements in digraphs. Information Processing Letters 80, 287–294 (2001)
11. Roughgarden, T.: Selfish routing and the price of anarchy. MIT Press, Cambridge (2005)
12. Sugihara, K., Ito, H.: Maximum-cover source-location problems. IEICE Trans. Fundamentals E89-A, 1370–1377 (2006)
13. Sugihara, K., Ito, H.: Maximum-cover source-location problem with objective edge-connectivity three. In: Proc. of CTW 2006, University of Duisburg-Essen, Germany, pp. 131–136 (2006), Electronic Notes in Discrete Mathematics 25, 165–171 (2006)
14. Sugihara, K., Ito, H.: Maximum-cover source-location problem with objective edge-connectivity three (journal version submitted)
15. Tamura, H., Sengoku, M., Shinoda, S., Abe, T.: Location problems on undirected flow networks. IEICE Trans. E73, 1989–1993 (1990)

Inverse Booking Problem: Inverse Chromatic Number Problem in Interval Graphs

Yerim Chung[1], Jean-François Culus[2], and Marc Demange[3]

[1] Paris School of Economics, Paris I University, France
Yerim.Chung@malix.univ-paris1.fr
[2] LIPN, Paris XIII University, France
Jean-francois.Culus@lipn.univ-paris13.fr
[3] ESSEC Business School, DIS department, France
demange@essec.fr

Abstract. We consider inverse chromatic number problems in interval graphs having the following form: we are given an integer K and an interval graph $G = (V, E)$, associated with $n = |V|$ intervals $I_i =]a_i, b_i[$ ($1 \leq i \leq n$), each having a specified length $s(I_i) = b_i - a_i$, a (preferred) starting time a_i and a completion time b_i. The intervals are to be newly positioned with the least possible discrepancies from the original positions in such a way that the related interval graph can be colorable with at most K colors. We propose a model involving this problem called *inverse booking problem.* We show that inverse booking problems are hard to approximate within $O(n^{1-\epsilon}), \epsilon > 0$ in the general case with no constraints on lengths of intervals, even though a ratio of n can be achieved by using a result of [13]. This result answers a question recently formulated in [12] about the approximation behavior of the unweighted case of single machine just-in-time scheduling problem with earliness and tardiness costs. Moreover, this result holds for some restrictive cases.

Keywords: Inverse combinatorial optimization, Inverse chromatic number problem, Interval graphs, Machine(s)-scheduling with earliness and/ or tardiness costs, NP-hardness, Approximation.

1 Introduction

Inverse combinatorial optimization problems have been extensively studied for several weighted problems during the last two decades [1,8,16]. Given an instance of a weighted combinatorial optimization problem Π and its feasible solution, the corresponding inverse problem, denoted by $I\Pi$, consists in modifying as little as possible (with respect to a fixed norm) the weight system in such a way that the given solution becomes an optimal solution of Π in the new instance, defined by the new weight system. Here, we consider the L_1 norm. In [4,5], we studied several variants of such problems: the case where the original problem is non-weighted for which we are supposed to modify the structure of the given instance rather than the weight system, the case where the fixed solution is supposed to be selected not by any optimal algorithm but by a fixed (optimal or

S.-i. Nakano and Md. S. Rahman (Eds.): WALCOM 2008, LNCS 4921, pp. 180–187, 2008.

approximated) algorithm and also the case where we are given a target solution value K instead of a feasible solution. In this last case, if Π is a minimization problem, we wish to find a minimal modification of the given instance so that the optimal solution value of the new instance does not exceed K. For instance, the inverse chromatic number problem consists, given any graph $G = (V, E)$ and an integer K, in modifying as little as possible the given instance in such a way that the chromatic number of the new instance is not greater than K.

In this paper, we address the inverse booking problem, denoted by IBP_K. Although we describe it in the case of a hotel booking system, it may arise in the framework of any booking system. We are given an integer K and $n = |V|$ intervals $I_i =]a_i, b_i[$, $(1 \leq i \leq n)$ with integral endpoints a_i, $b_i \in \mathbb{N}$, each corresponding to a room reservation in a hotel for $s(I_i) = b_i - a_i$ days without interruption (no change of room during the stay). We will refer to a_i as the *preferred starting time* of interval I_i. We assume that the hotel has only K identical rooms for which the reservations are interchangeable and that there are at the beginning more than K reservations overlapping in their staying periods, implying that there is no legal assignment to K rooms. With the goal of satisfying all demands, the hotel manager proposes the clients to change their holiday plans: to come stay in the hotel a little bit earlier or later than their preferred starting dates (movement of holiday period) or to stay less than their preferred duration (reduction of holiday duration), etc. In this paper, we only consider the first type of modification. Of course, these negotiations incur some compensating costs such as discount on price for which the clients are willing to accept the proposition. So, the hotel manager has to schedule n reservations using K rooms in such a way that the total changes are minimized. The objective function to minimize is the total movements (total discrepancies) which are calculated by $\sum_i |b_i' - b_i|$ (or equivalently $\sum_i |a_i' - a_i|$) where b_i and b_i' (a_i and a_i') denote respectively the original and new right (left) endpoints of the interval I_i. If $(b_i' - b_i) > 0$ $((b_i' - b_i) < 0)$, then it means that the interval I_i is shifted to the right (left) side, respectively.

This problem is exactly a version of the inverse chromatic number problem in interval graphs under L_1-norm. Since interval graphs are perfect [7], it equivalently consists in shifting some intervals so that the clique number of the corresponding interval graph does not exceed K. Consequently, it is closely related to the inverse stability number problem, introduced in [4], in co-interval graphs. Nevertheless, the main difference arises in the way of modifying an instance. Instead of modifying the related graph (as studied in [4]), it is more relevant in the inverse booking problem to modify the interval system.

For $K = 1$, the inverse booking problem can be seen as a particular case of the so-called *total discrepancy problem* [13]. An interval $I_i =]a_i, b_i[$ corresponds to a non-preemptive job J_i with a processing time $p(J_i) = b_i - a_i = s(I_i)$, a due date $d(J_i) = b_i$ and a completion time $c(J_i) = b_i'$. Translation of an interval to the right side corresponds to the tardiness $T_i = max\{0, b_i' - b_i\}$ of the related job, and translation movement to the left side means its earliness $E_i = max\{0, b_i - b_i'\}$. Using the common scheduling notations, introduced by Graham et al. [14], we can also denote $IBP_{K=1}$ by $1||\sum_i(E_i + T_i)$. Garey, Tarjan and Wilfong [13] showed its hardness in the ordinary sense. We also consider a variant of $IBP_{K=1}$ where the intervals can be shifted only to the right side; let us denote it by $IBP_{K=1,\rightarrow}$. In this case, the inverse booking problem can be regarded as the problem to minimize total tardiness on one machine with arbitrary release dates and no preemption, which is known to be strongly NP-hard [9]. Indeed, the left endpoint a_i of an interval I_i can be seen as a release date r_i of a job J_i: translations are permitted only to the right side, meaning that a job cannot start before its release date. We can denote this variant by $1|r_i|\sum_i T_i$. Obviously, for any bounded integer K, IBP_K and $IBP_{K,\rightarrow}$ respectively correspond to the multiprocessor scheduling problems, $K||\sum_i(E_i + T_i)$ and $K|r_i|\sum_i T_i$.

Notice that $IBP_{K=1}$ can be viewed as the unweighted version of the so-called *single machine just-in-time scheduling problem with earliness and tardiness costs* [12] for which it is shown that no polynomial approximation ratio can be guaranteed in polynomial time (see [2] for basic definitions in approximation theory). Since weights play a crucial role in the proof, the approximation behavior of the unweighted case is stated as an open question. On the other hand, it is shown in [13] that minimizing the maximum discrepancy is polynomially solvable, meaning that the version of $IBP_{K=1}$ where the objective function is defined with the L_∞ norm is polynomially solvable; moreover, it is easy to see that the same holds for $IBP_{K=1,\rightarrow}$. Since, for every vector v of dimension n, we have $|v|_\infty \leq |v|_1 \leq n|v|_\infty$, the following holds:

Proposition 1. $IBP_{K=1}$ and $IBP_{K=1,\rightarrow}$ are n-approximable.

We show that $IBP_{K=1}$ and $IBP_{K=1,\rightarrow}$ cannot be approximated in polynomial time within a ratio of $O(n^{1-\epsilon})$ for any $\epsilon > 0$. This result also holds for the case where the interval sizes are polynomially bounded, implying that both problems are NP-hard in the strong sense; to our knowledge, it was not known for the former. From scheduling problems' point of view, it is worth noting that if processing times are polynomially bounded then Lawler's algorithm [10] for $1||\sum_i T_i$ becomes polynomial. For this case, it points out a huge gap of complexity between $1||\sum_i T_i$ and $1||\sum_i(E_i + T_i)$ or $1|r_i|\sum_i T_i$.

2   Hardness of Approximating $IBP_{K=1}$ and $IBP_{K=1,\rightarrow}$

We show that $IBP_{K=1}$ and $IBP_{K=1,\rightarrow}$ are both hard to approximate. This result also holds if the original interval graph is supposed to be bipartite.

Theorem 1. If $\mathbf{P} \neq \mathbf{NP}$, then there is no polynomial time approximation algorithm for $IBP_{K=1}$ or for $IBP_{K=1,\rightarrow}$ guaranteeing an approximation ratio of $O(n^{1-\epsilon}), \epsilon > 0$, even if all intervals have polynomially bounded lengths and the related interval graph is bipartite.

Proof. We use a polynomial time reduction from 3-*Partition*, known to be strongly NP-complete [6]. An instance of 3-*Partition* is given by a bound $B \in \mathbb{N}^+$ and a finite set $X = \{x_1, \cdots, x_{3m}\}$ of $3m$ elements, each having size $s(x_i) =$

$d_i \in \mathbb{N}^+$ for $i \in \{1, \cdots, 3m\}$ such that $B/4 < d_i < B/2$ and $\sum_{i=1}^{3m} d_i = mB$. It is said to be positive if there is a partition of the d_i's into m groups of three elements, each summing exactly to B. Since 3-*Partition* is strongly NP-complete, we assume that the d_i's (and consequently also B) are polynomially bounded. The proof is done for $IBP_{K=1}$, the case of $IBP_{K=1,\rightarrow}$ being very similar. Let us suppose that a ratio $\rho(n) \leq cn^{1-\epsilon}$, with a fixed $\epsilon > 0$ and a constant c, can be guaranteed for $IBP_{K=1}$, where n is the number of intervals of the instance (the order of the related interval graph).

1. For the conciseness of the proof, we first omit the constraint that the original interval graph is bipartite, and we assume that some intervals have rational lengths. Given an instance of 3-*Partition* with polynomially bounded d_i's, we construct an instance of $IBP_{K=1}$ as follows (see figure 1):

- $3m$ intervals: $I_i =]a_i, b_i[$ ($i \in \{1, \ldots, 3m\}$) of length $s(I_i) = b_i - a_i = d_i$.
- $m + 1$ interval blocks: $L^j = \cup_{i=1}^{T_j} L_i^j$ ($j \in \{0, \ldots, m\}$) where $T_j = 3Q(m)m^2 B$, $j \in \{0, \cdots, m\}$, $Q(m)$ being a fixed polynomial function (to be defined later), and $L_i^j =]l_i^j, r_i^j[$ is such that $\forall i$, $r_i^j = l_{i+1}^j$. For $j \in \{0, \cdots, m\}$, blocks L^j are composed of T_j contiguous intervals L_i^j ($i \in \{1, \ldots, T_j\}$) of length $s(L_i^j) = 1/(3Q(m)m^2)$ for $j \in \{1, \ldots, m-1\}$ and of length 1 for $j \in \{0, m\}$. The blocks L^j have size $s(L^j) = B$ for $j \in \{1, \ldots, m-1\}$ and $s(L^0) = s(L^m) = 3Q(m)m^2 B$ for $j \in \{0, m\}$.
- Interval I_i ($i \in \{1, \ldots, 3m\}$) has the same right endpoint as the interval $L_{T_0}^0$; we set explicitly $r_{T_0}^0 = b_1 = b_2 = \cdots = b_{3m} = 0$
- Blocks L^j ($j \in \{0, \cdots, m\}$) are distributed on a line in order of increasing indices and with a gap of length B between L^{j-1} and L^j, $j \in \{1, \cdots, m\}$. Explicitly, we have:
for $j = 0$ and $i \in \{1, \cdots, T_0\}$, $L_i^0 =] - 3Q(m)m^2 B + (i-1), -3Q(m)m^2 B + i[$,
for $j \in \{1, \cdots, m-1\}$ and $i \in \{1, \cdots, T_j\}$, $L_i^j =](2j-1)B + \frac{i-1}{3Q(m)m^2}, (2j-1)B + \frac{i}{3Q(m)m^2}[$,
for $j = m$ and $i \in \{1, \cdots, T_m\}$, $L_i^m =](2m-1)B + (i-1), (2m-1)B + i[$.
- Since B is polynomially bounded, the total number N of intervals clearly satisfies $N \leq P(m)Q(m)$ for a fixed polynomial function P. We then choose $Q > c^{1/\epsilon} P^{(1-\epsilon)/\epsilon}$, implying $cN^{1-\epsilon} < Q(m)$.

Lemma 1. *Let us consider a block L^j, composed of BQ' intervals each having size $\frac{1}{Q'}$ where $Q' = 3Q(m)m^2$ and $B = s(L^j)$. It will cost at least dBQ' to shift L^j by distance $d \in \mathbb{N}$ (assume that $d \leq B$) either to the right side or to the left side.*

Indeed, there are two ways of moving L^j: either by shifting every interval of L^j by distance d to the right (or left) side or by shifting the dQ' first (last) intervals of L^j by distance B to the right (left) side. It is straightforward to verify that both cases yield the same cost of dBQ'. □

We first show that, if the instance of 3-*Partition* is positive, then the optimal value β of the instance of $IBP_{K=1}$ satisfies $\beta \leq 3m^2 B$, else $\beta \geq 3Q(m)m^2 B$.

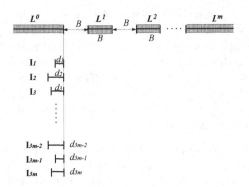

Fig. 1. Construction of an instance of $IBP_{K=1}$

We then conclude that no polynomial time algorithm guarantees the ratio $\rho(n)$ where n is the number of intervals.

Let us first consider the case where there exists a 3-partition for the original instance: the set X can be partitioned into m disjoint sets $X_1, \cdots, X_m$ such that for $1 \leq j \leq m$, $\sum_{x \in X_j} s(x) = B$. By renumbering the intervals I_i's ($1 \leq i \leq 3m$), we may assume without loss of generality that for $1 \leq j \leq m$, the three elements of the j-th group X_j correspond to the size of intervals I_{3j-2}, I_{3j-1} and I_{3j}, respectively. Let us denote by $B_j =]r_{T_j}^{j-1}, l_1^j[$ the interval (blank) between L^{j-1} and L^j, $j \in \{1, \cdots, m\}$, then by filling up each B_j (of length B) with three intervals I_{3j-2}, I_{3j-1} and I_{3j}, we can obtain a feasible solution of $IBP_{K=1}$ with value not greater than $3m^2B$. Indeed, we need to move I_{3j-2}, I_{3j-1} and I_{3j} to the right side at most by $(2j-1)B$. So, the total distance of these movements is bounded by $\sum_{j=1}^{m} 3B(2j-1) = 3B(\sum_{j=1}^{m} j + \sum_{j=1}^{m-1} j) = 3m^2B$. Hence, we deduce: $\beta \leq 3m^2B$.

Suppose now that there is no 3-partition for the original instance. By construction, to solve the instance of $IBP_{K=1}$, one has to perform at least one of the two following actions: **(i)** passing over L^0 or L^m to shift an interval I_i either to the left side of L^0 or to the right side of L^m, **(ii)** enlarging sizes of some (at least 1) intervals B_j's at least by 1 (since $d_i \in \mathbb{N}$ for $i \in \{1, \cdots, 3m\}$) in such a way that all intervals I_i's can be positioned within B_j's without intersecting each other. Since $s(L^0) = s(L^m) = 3Q(m)m^2B$, action **(i)** costs at least $3Q(m)m^2B$. For action **(ii)**, one needs either to shift a block L^j, $j \in \{0, \ldots, m\}$, at least by distance 1 (to the left or the right side) or to split up a bloc L^j, $j \in \{1, \ldots, m-1\}$, into two blocks and to shift them - the first one to the left side and the second one to the right side - to produce a blank of size at least $B+1$ between. By lemma 1, such an action induces a cost of at least $3Q(m)m^2B$. Hence, we have $\beta \geq 3Q(m)m^2B$.

To complete the proof of case **1**, let us consider a polynomial time approximation algorithm for $IBP_{K=1}$ guaranteeing the ratio $\rho(n) \leq cn^{1-\epsilon}$. If there exists a 3-partition ($\beta \leq 3m^2B$), then it delivers a solution of value λ for the instance of $IBP_{K=1}$ having size N, such that $\lambda \leq 3m^2B\rho(N) \leq 3cN^{1-\epsilon}m^2B < 3Q(m)m^2B$

(recall that $cN^{1-\epsilon} < Q(m)$). On the other hand, if there is no 3-partition, then we have $\lambda \geq \beta \geq 3Q(m)m^2B$. So, it would provide a polynomial time algorithm solving 3-*Partition* with polynomially bounded d_i's, which contradicts the strong NP-hardness of 3-*Partition* unless **P=NP**.

2. Note that the above interval graph is initially $(3m+1)$-colorable (see figure 1). To prove the same result with the constraint that the original graph is bipartite, we reposition intervals $I_i =]a_i, b_i[$ $(1 \leq i \leq 3m)$ in such a way that $b_{i+1} = a_i$, $i \in \{1, \cdots, 3m-1\}$ and $b_1 = 0$. We may assume without loss of generality that $d_1 \leq d_2 \leq \cdots \leq d_{3m}$ where $d_i = s(I_i) = b_i - a_i$ for $i \in \{1, \cdots, 3m\}$. In comparison with the $IBP_{K=1}$ instance having no constraint on the chromatic number, this instance needs, by construction, some additional movements in order to be 1-colorable. The quantity, $X = d_1 + (d_1+d_2) + \cdots + (d_1 + \cdots + d_{3m-1}) = (3m-1)d_1 + (3m-2)d_2 + \cdots + 2d_{3m-2} + d_{3m-1}$ is exactly the supplementary distance to move; to make $b_i = 0$ for $i \in \{2, \cdots, 3m\}$, we need to shift I_i to the right side by distance $d_1 + d_2 + \cdots + d_{i-1}$. Since $d_1 \leq d_2 \leq \cdots \leq d_{3m}$, X can be upper bounded by a constant $X_0 = \frac{3}{2}m^2B$, which is obtained when $d_1 = d_2 = \cdots = d_{3m} = B/3$. The same argument as in case **1** works if we suppose that blocks L^j $(1 \leq j \leq m-1)$ and L^m are composed of $5Q(m)m^2B$ intervals of length $1/(5Q(m)m^2)$ and of length 1, respectively. In addition, we need to enlarge the size of L^0 to $8Q(m)m^2B$ in order to prevent I_i's from going before L^0. As in case **1**, the polynomial function Q can be chosen such that the number N of intervals satisfies $cN^{1-\epsilon} < Q(m)$. If there exists a polynomial time $O(n^{1-\epsilon})$-approximation algorithm for $IBP_{K=1}$, then it will deliver a solution of value $\lambda < 5Q(m)m^2B$ for the case where a 3-partition exists ($\beta \leq 3m^2B + X_0 \leq 5m^2B$), and a solution of value $\lambda \geq 5Q(m)m^2B$ for the opposite case. Such an algorithm would solve 3-*Partition*, which is impossible unless **P=NP**.

To conclude the proof, note that some interval lengths are defined as a rational number. By multiplying all interval lengths by a same number (e.g. $3Q(m)m^2$ and $5Q(m)m^2$ for cases **1** and **2**, respectively), we can easily obtain an instance with integer data, which is equivalent to our initial instance. The value of each solution will be also multiplied by this common value.

Finally, an instance of $IBP_{K=1,\rightarrow}$ can be constructed in a same way as above but without the block L^0. The same arguments hold in this case, which concludes the proof of the theorem. ∎

Since all numbers involved in the above instances of $IBP_{K=1}$ and $IBP_{K=1,\rightarrow}$ are polynomially bounded, we have:

Corollary 1. $IBP_{K=1}$ and $IBP_{K=1,\rightarrow}$ are strongly NP-hard.

3   Discussion on Further Work

In this work, we introduced an inverse chromatic number problem in interval graphs, motivated by a hotel booking problem. We considered two variants whether intervals can be shifted to the left and the right side or only to the right side. We pointed out that these problems are closely related to scheduling problems, namely $K||\sum_i(E_i + T_i)$ and $K|r_i|\sum_i T_i$, respectively. For $K = 1$, these problems can also be seen as unweighted versions of a just-in-time scheduling problem, shown to be hard to approximate within a polynomial ratio in [12]. We answered a question about the hardness of approximating the unweighted case (see [12]). We showed that this case cannot be approximated within $O(n^{1-\epsilon})$ although it is n-approximated. Moreover, this result holds even if the original interval graph is bipartite - such a restriction is natural in our booking framework - and if the interval lengths are polynomially bounded. Using scheduling notations, it shows that $1||\sum_i(E_i + T_i)$ and $1|r_i|\sum_i T_i$ with polynomially bounded processing times are hard to approximate while $1||\sum_i T_i$ is known to be polynomially solvable for such processing times [10].

This pessimistic result motivates us to look for polynomial cases of the inverse booking problem. If all intervals (jobs) have the same length, these problems are known to be solvable in polynomial time [10,13] and even with rational starting dates [15]. So, for this case, $IBP_{K=1}$ and $IBP_{K=1,\rightarrow}$ can be solved in polynomial time. In particular, if all interval lengths are equal to 1, then $IBP_{K=1,\rightarrow}$ and $IBP_{K=1}$ can be easily formulated as an assignment problem [10] and as a minimum cost flow problem, respectively. In [11] Lawler devised a pseudo-polynomial algorithm for sequencing jobs to minimize total weighted tardiness $1||\sum_i w_i T_i$ where the weights have the so-called *agreeable* property, including the unweighted case.

In [12] some approximation results are devised for the case where the ratio between the maximum and minimum lengths is bounded by a constant. However, to our knowledge, the complexity status of this case is not known. So, this question and even the hardness of the cases where processing times are supposed to belong to a finite set are both interesting. As a first attempt to handle it, we consider in [3] the case where the interval lengths are either 1 or 2 and show that it can be solved in linear time by a greedy algorithm if intervals can be shifted only to the right side. Finally, it would be interesting to study inverse chromatic number problems in other classes of graphs. In particular, in [3,5] we consider it for permutation graphs and show that it is polynomially solved if the target chromatic number is fixed to a constant.

References

1. Ahuja, R.K., Orlin, J.B.: Inverse optimization. Operations Research 49, 771–783 (2001)
2. Ausiello, G., Crescenzi, P., Gambosi, G., Kann, V., Marchetti-Spaccamela, A., Protasi, M.: Complexity and approximation (Combinatorial optimization problems and their approximability properties). Springer, Heidelberg (1999)
3. Chung, Y.: On some inverse combinatorial optimization problems. PhD thesis, Paris School of Economics, Paris I University (preparation)
4. Chung, Y., Demange, M.: The 0-1 inverse maximum stable set problem (to be published)
5. Chung, Y., Demange, M.: Some inverse traveling salesman problems. In: IV Latin-American Algorithms, Graphs and Optimization Symposium (to appear, 2007)

6. Garey, M.R., Johnson, D.S.: Computers and Intractability – A Guide to the Theory of NP-Completeness. Freeman, San Francisco (1979)

7. Golumbic, M.: Algorithmic Graph Theory and Perfect Graphs. Academic press, New York (1980)

8. Heuberger, C.: Inverse combinatorial optimization: A survey on problems, methods, and results. J. Comb. Optim. 8(3), 329–361 (2004)

9. Kan, A.H.G.R.: Machine scheduling problem: Classification, Complexity and Computation, Nijhoff, The Hague (1976)

10. Lawler, E.L.: On scheduling problems with deferral costs. Management Sci. 11, 280–288 (1964)

11. Lawler, E.L.: A "pseudopolynomial" algorithm for scheduling jobs to minimize total tardiness. Annals of Discrete Mathematics 1, 331–342 (1977)

12. Müller-Hannemann, M., Sonnikow, A.: Non-approximability of just-in-time scheduling. Technical report, TU Darmstadt, Extended abstract in MAPSP 2007, Istanbul (2006)

13. Tarjan, R.E., Garey, M.R., Wilfong, G.T.: One-processor scheduling with symmetric earliness and tardiness penalties. Mathematics of Operations Research 13, 330–348 (1988)

14. Lenstra, J.K., Graham, R.L., Lawler, E.L., Kan, A.H.G.R.: Optimization and approximation in deterministic sequencing and scheduling: A survey. Annals of Discrete Mathematics 5, 287–326 (1979)

15. Dessouky, D., Verma, S.: Single-machine scheduling of unit-time jobs with earliness and tardiness penalties. Mathematics of Operations Research 23, 930–943 (1988)

16. Zhang, J., Yang, X., Cai, M.-C.: The complexity analysis of the inverse center location problem. J. of Global optimization 15(1), 213–218 (1999)

Optimal Algorithms for Detecting Network Stability

Dimitrios Koukopoulos[1], Stavros D. Nikolopoulos[2], Leonidas Palios[2], and Paul G. Spirakis[3]

[1] Department of Cultural Heritage Management and New Technologies, University of Ioannina, GR-30100 Agrinio, Greece
koukopou@ceid.upatras.gr
[2] Department of Computer Science, University of Ioannina, GR-45110 Ioannina, Greece
{stavros,palios}@cs.uoi.gr
[3] Department of Computer Engineering and Informatics, University of Patras, GR-26500 Patras, Greece, and Research Academic Computer Tech. Institute (RACTI), N. Kazantzaki str., GR-26500, Patras, Greece
spirakis@cti.gr

Abstract. A *packet-switched* network is *universally stable* if, for any greedy protocol and any adversary of injection rate less than 1, the number of packets in the network remains bounded at all times. A natural question that arises is whether there is a fast way to detect if a network is universally stable based on the network's structure. In this work, we study this question in the context of *Adversarial Queueing Theory* which assumes that an adversary controls the locations and rates of packet injections and determines packet paths. Within this framework, we present optimal algorithms for detecting the universal stability (packet paths do not contain repeated edges but may contain repeated vertices) and the simple-path universal stability (paths contain neither repeated vertices nor repeated edges) of a network. Additionally, we describe an algorithm which decides in constant time whether the addition of a link in a universally stable network leads it to instability; such an algorithm could be useful in detecting intrusion attacks.

Keywords: Packet-switched communication networks, network stability, linear algorithms, graph theory, intrusion detection, adversarial queueing theory.

1 Introduction

In a *packet-switched* communication network, packets arrive dynamically at the nodes with predetermined paths, and they are routed at discrete time steps across the edges (links). In each node, there is a buffer (queue) associated with each outgoing link. A *packet* is an atomic entity that resides at a buffer at the end of any step. It must travel along a *directed path* in the network from its *source* to its *destination*, both of which are network nodes. When a packet is

S.-i. Nakano and Md. S. Rahman (Eds.): WALCOM 2008, LNCS 4921, pp. 188–199, 2008.
© Springer-Verlag Berlin Heidelberg 2008

injected, it is placed in the buffer of the first link on its route. At every time step, at most one packet proceeds along each edge; if more than one packets are waiting in the buffer of an edge e, a *protocol* is employed to resolve the conflict and to determine the packet that is to be routed along e and the remaining ones wait in the buffer of e. When a packet reaches its destination, it is *absorbed*.

In the field of packet-switched networks, a great deal of research has been devoted to the specification of their behavior. A major issue that arises in such a setting is that of stability: will the number of packets in the network remain bounded at all times under any traffic pattern and any protocol? We focus on a basic adversarial model for packet injection and path determination that has been introduced in a pioneering work by Borodin *et al.* [5] under the name *Adversarial Queueing Theory*. At each time step, an *adversary* may inject a set of packets into some nodes; for each such packet, the adversary specifies a path that the packet must traverse. A crucial parameter of the adversary is its *injection rate* ρ, where $0 < \rho < 1$. Among the packets that the adversary injects in any time interval I, at most $\lceil \rho|I| \rceil$ can have paths that contain any particular edge. Within this framework, a protocol is *stable* [5] on a network G against an adversary $\mathcal{A}$ of rate ρ if there is a constant B (which may depend on G and $\mathcal{A}$) such that the number of packets in the system is bounded at all times by B. We also say that *a network G is universally stable* [5] if every greedy protocol is stable against every adversary of rate less than 1 on G. In general, the paths that the packets follow are not necessarily simple (i.e., they do not contain repeated edges but they may contain repeated vertices) [2]. If the paths are restricted to be simple (no repeated vertices or edges), then we deal with *simple-path universal stability*.

The answer to the question of determining the universal stability of a network is non-trivial. Since the property of universal stability is a predicate quantified over all protocols and adversaries, it might seem that it is not a decidable property. It turns out that this is not the case; Alvarez *et al.* in [2] showed that determining the universal stability of a network can be done in polynomial time. Yet, from a practical standpoint, it is very important to have algorithms for determining network stability which are as efficient as possible. This stems from the need in contemporary large-scale platforms for distributed communication and computation (as is the *Internet*) to be able to quickly detect whether a structural change in a network (possibly performed by a malicious intruder) leads to instability. The quick detection will guarrantee the *survivability* of the network i.e., the ability of the network and the interconnected computers to be resilient in the face of an attack [13]. Intrusion detection [8,9] has emerged as a key technique for network survivability. An algorithm that performs intrusion detection should be fast enough in order to eliminate the damage a network can suffer by an intruder.

Adversarial Queueing Theory has received a lot of attention in the study of stability issues [3,2,7]. The issue of proposing a characterization for universally stable networks was first addressed in [3] in which it was proved that there exists a finite set A of basic undirected graphs such that a network is stable if and only if it does not contain any of the graphs in A as a minor. This result was improved in [2]: a network in which packets may follow non-simple paths is universally stable if and only if it does not contain as a subgraph any of the extensions of the graphs $\mathcal{U}_1$ or $\mathcal{U}_2$ (see Figures 1, 2) [2, Lemma 7]. In [2], it was also shown that a network in which packets follow simple paths is universally stable if and only if it does not contain as a subgraph any of the extensions of the graphs $\mathcal{S}_1$ or $\mathcal{S}_2$ or $\mathcal{S}_3$ or $\mathcal{S}_4$ (see Figures 3, 4) [2, Theorem 12]. Based on the above, it was then shown that detecting universal stability of networks requires polynomial time [2].

Our work focuses on the issue of detecting the universal stability of networks within the Adversarial Queueing model. Based on the characterization in [2], we present two algorithms: one for detecting whether a network is universally stable (packets may follow non-simple paths) and another for detecting whether a network is simple-path universally stable (packets follow simple paths). Both algorithms run in optimal $O(m + n)$ time where n is the number of network vertices and m is the number of network edges. Additionally, we describe an algorithm which decides in constant time whether the addition of a link in a universally stable network leads it to instability.

2   Theoretical Framework

We use the same network model as the one in [5, Section 3]: we consider that a routing network is modelled by a finite *multi-digraph* on n vertices and m edges where a multi-digraph is a directed graph (or digraph, for short) in which multiple edges may exist from a vertex to another vertex; each vertex $x \in V(G)$ represents a communication switch, and each edge $e \in E(G)$ represents a link between two switches.

We consider finite directed and undirected simple graphs and multi-digraphs with no loops. A directed (undirected, resp.) edge is an ordered (unordered, resp.) pair of distinct vertices $x, y \in V(G)$, and is denoted by xy. The *multiplicity* of a vertex-pair xy of a digraph G, denoted by $\lambda(xy)$, is the number of edges joining the vertex x to y in G. For a set $C \subseteq V(G)$ of vertices of the graph G, the subgraph of G *induced* by C is denoted $G[C]$; for a set $S \subseteq E(G)$ of edges, the subgraph of G *spanned* by S is denoted $G\langle S \rangle$.

A *path* in G is a sequence of vertices $(v_0, v_1, \ldots, v_k)$ such that $v_i v_{i+1} \in E(G)$ for $i = 0, 1, \ldots, k - 1$; we say that this is a path from v_0 to v_k (a v_0-v_k path, for short) of *length* k. A path is *undirected* or *directed* depending on whether G is undirected or directed. It is called *simple* if none of its vertices occurs more than once; it is called *trivial* if its length is equal to 0. A path is *closed* (resp. *open*) if $v_0 = v_k$ (resp. $v_0 \neq v_k$). A closed path $(v_0, v_1, \ldots, v_{k-1}, v_0)$ is a *cycle* of length k; the directed closed path (v_0, v_1, v_0) is called *2-cycle*.

A *connected component* (or component) of an undirected graph G is a maximal set of vertices, say, $C \subseteq V(G)$, such that for every pair of vertices $x, y \in C$, there exists an x-y path in the subgraph $G[C]$ of G induced by the vertices in C. A component is called *non-trivial* if it contains two or more vertices; otherwise, it is called *trivial*. A *biconnected component* (or bi-component) of an undirected graph G is a maximal set of edges such that any two edges in the set lie on a simple cycle of G [6]; G is called *biconnected* if it is connected and contains only one biconnected component.

A *strongly connected component* (or strong component) of a directed graph G is a maximal set of vertices $C \subseteq V(G)$ such that for every pair of vertices $x, y \in C$, there exists both a (directed) x-y path and a (directed) y-x path in the subgraph of G induced by the vertices in C; the graph G is called *strongly connected* if it consists of only one strong component. The *acyclic component graph* G_{scc} of a digraph G is an acyclic digraph obtained by contracting all edges within each strongly connected component of G so that only a single vertex remains in each component. The *underlying graph* $G_{u\ell}$ of a digraph G is an undirected graph which results after making all the edges of G undirected and consolidating any duplicate edges.

In our work, we will need a combination of a strong component of a digraph and the bi-components of its underlying graph. Therefore, we define a *strongly biconnected component* (a strong bi-component or a bi-scc for short) of a directed graph G to be a maximal set of edges $S \subseteq E(G)$ such that the graph $G\langle S \rangle$ spanned by S is strongly connected and its underlying graph $G\langle S \rangle_{u\ell}$ is biconnected; the graph G is called *strongly biconnected* if it is strongly connected and contains only one strong bi-component. The graph $\mathcal{U}_1$ of Fig. 1 is strongly biconnected, whereas the graph $\mathcal{U}_2$ contains two strong bi-components.

The *subdivision* operation on an edge xy of a digraph G consists of the addition of a new vertex w and the replacement of xy by the two edges xw and wy; hereafter, we shall call it *edge-subdivision* operation.

Given a digraph G on n vertices and m edges, $\mathcal{E}(G)$ denotes the family of digraphs which contains the digraph G and all the digraphs obtained from G by successive edge-subdivisions.

It has been proved that the digraphs $\mathcal{U}_1$ and $\mathcal{U}_2$ of Fig. 1 are not universally stable; they are the minimum forbidden subgraphs characterizing universal stability. Moreover, the family of digraphs obtained from $\mathcal{U}_1$ and $\mathcal{U}_2$ by successive edge-subdivisions are also not universally stable, i.e., the digraphs in $\mathcal{E}(\mathcal{U}_1) \cup \mathcal{E}(\mathcal{U}_2)$ [2] (see Fig. 2; a dashed edge with label i denotes a directed path of length i). The following result holds:

Lemma 1. *(Alvarez, Blesa, and Serna [2]): A digraph G is universally stable if and only if G does not contain as a subgraph any of the digraphs in $\mathcal{E}(\mathcal{U}_1) \cup \mathcal{E}(\mathcal{U}_2)$.*

Observation 1. *Let G be a directed graph of the family $\mathcal{E}(\mathcal{U}_1) \cup \mathcal{E}(\mathcal{U}_2)$. Then, the graph G has the following structure:*

(a) it consists of a simple cycle $C = (x_0, x_1, x_2, \ldots, x_\ell, x_0)$, $\ell \geq 1$, and
(b) a path $P = (x_i, y_1, y_2, \ldots, y_k, x_j)$ such that, $y_1, y_2, \ldots, y_k \notin C$, $x_i, x_j \in C$ and $k \geq 0$.

It is easy to see that, if P is an open path, i.e, $x_i \neq x_j$, then $G \in \mathcal{E}(\mathcal{U}_1)$, whereas if P is a closed path, i.e, $x_i = x_j$, then $G \in \mathcal{E}(\mathcal{U}_2)$.

Fig. 1. Not universally stable digraphs

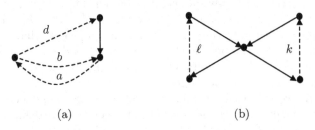

(a)   (b)

Fig. 2. The digraphs formed by extensions of $\mathcal{U}_1$ and $\mathcal{U}_2$: (a) the digraphs in $\mathcal{E}(\mathcal{U}_1)$, $a \geq 1$, $b \geq 1$, and $d \geq 0$; (b) the digraphs in $\mathcal{E}(\mathcal{U}_2)$, $\ell \geq 0$ and $k \geq 0$

Let G be a digraph on n vertices and m edges. We denote by G^* the element of $\mathcal{E}(G)$ which is obtained from G by applying an edge-subdivision operation on each edge $uv \in E(G)$. Obviously, G^* has $n+m$ vertices and $2m$ edges. Moreover, G^* does not contain 2-cycles; in particular, every cycle in G^* has length greater than or equal to 4. We call G^* the *one-subdivided* graph of G.

Below we present some results on which our algorithms for detecting universal stability and simple-path universal stability rely.

Lemma 2. *Let G be a directed graph and let G^* be its one-subdivided graph. The graph G is not universally stable if and only if G^* contains a subgraph $H \in \mathcal{E}(\mathcal{U}_1) \cup \mathcal{E}(\mathcal{U}_2)$.*

Lemma 3. *Let G be a directed graph and let G^* be its one-subdivided graph. Let $S_1, S_2, \ldots, S_k$ be the strongly connected components of G^* and let n_i and m_i be the number of vertices and edges of the strong component S_i, respectively. Then, G is not universally stable if and only if G^* has a strong component S_i such that $m_i > n_i$, $1 \leq i \leq k$.*

Regarding the *simple-path universal stability*, in which packets follow simple paths (i.e, paths not containing repeated edges or vertices), it has been shown that all the digraphs in $\mathcal{E}(\mathcal{S}_1) \cup \mathcal{E}(\mathcal{S}_2) \cup \mathcal{E}(\mathcal{S}_3) \cup \mathcal{E}(\mathcal{S}_4)$ (see Figures 3 and 4) are not simple-path universally stable [2].

Lemma 4. *(Alvarez, Blesa, and Serna [2]): A digraph G is simple-path universally stable if and only if G does not contain as a subgraph any of the digraphs in $\mathcal{E}(\mathcal{S}_1) \cup \mathcal{E}(\mathcal{S}_2) \cup \mathcal{E}(\mathcal{S}_3) \cup \mathcal{E}(\mathcal{S}_4)$.*

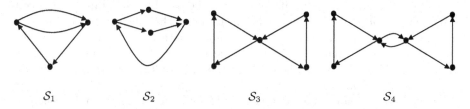

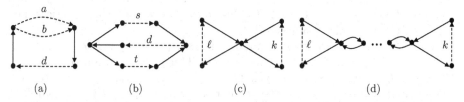

Fig. 3. Digraphs which are not simple-path universally stable

Fig. 4. The digraphs formed by extensions of S_1, S_2, S_3, and S_4: (a) the digraphs in $\mathcal{E}(S_1)$, $a \geq 1$, $b \geq 1$, and $d \geq 0$; (b) the digraphs in $\mathcal{E}(S_2)$, $s \geq 0$, $d \geq 0$, and $t \geq 0$; (c) the digraphs in $\mathcal{E}(S_3)$, $\ell \geq 1$ and $k \geq 1$; (d) the digraphs in $\mathcal{E}(S_4)$, $\ell \geq 1$ and $k \geq 1$

For the detection of simple-path universal stability, we will need what we call the *reduced* graph $\widehat{G}$ of a digraph G: the reduced graph $\widehat{G}$ is obtained from G by setting the multiplicity of each edge of G to 1. Obviously, the reduced graph $\widehat{G}$ of a digraph G on n vertices and m edges is a simple digraph and has n vertices and $m' \leq m$ edges.

Lemma 5. *Let G be a directed graph, $\widehat{G}$ the reduced graph of G, and $S_1, S_2, \ldots, S_k$ the strong components of $\widehat{G}$. Let $S_{i1}, S_{i2}, \ldots, S_{ik_i}$ be the bi-scc of the strong component S_i and let n_{ij} and m_{ij} be the number of vertices and edges of the bi-scc S_{ij}, respectively. Then, G is not simple-path universally stable if and only if $\widehat{G}$ has a strong component S_i which satisfies one of the following conditions:*

   (i)   *S_i contains a bi-scc S_{ij} such that: $n_{ij} \geq 3$, the underlying $G\langle S_{ij}\rangle_{u\ell}$ is a cycle, and there exists an edge xy in $G\langle S_{ij}\rangle$ with multiplicity $\lambda(xy) \geq 2$;*
  (ii) *S_i contains a bi-scc S_{ij} such that: $n_{ij} \geq 3$ and $G\langle S_{ij}\rangle_{u\ell}$ is not a cycle;*
 (iii) *S_i contains two bi-scc S_{ip} and S_{iq} such that: $n_{ip} \geq 3$, $n_{iq} \geq 3$, and both graphs $G\langle S_{ip}\rangle_{u\ell}$ and $G\langle S_{iq}\rangle_{u\ell}$ are cycles;*

where $1 \leq i \leq k$ and $1 \leq j, p, q \leq k_i$.

Sketch of Proof: The definition of the reduced graph $\widehat{G}$ of G implies that a bi-scc S_{ij} of the strong component S_i of $\widehat{G}$ consists of either a cycle $C = (x_0, x_1, x_2, \ldots, x_r, x_0)$, $r \geq 2$, or a cycle $C = (x_0, x_1, x_2, \ldots, x_r, x_0)$, $r \geq 2$, and a path $P = (x_i, y_1, y_2, \ldots, y_{r'}, x_j)$ such that $y_1, y_2, \ldots, y_{r'} \notin C$, $x_i \neq x_j$ and $r' \geq 1$.

($\Longleftarrow$) It is easy to see that if condition (i) holds, then the graph $G\langle S_{ij}\rangle$ contains a subgraph $H \in \mathcal{E}(\mathcal{S}_1)$. If condition (ii) holds, then the bi-scc S_{ij} consists of a cycle $C = (x_0, x_1, x_2, \ldots, x_r, x_0)$, $r \geq 2$, and a path $P = (x_i, y_1, y_2, \ldots, y_{r'}, x_j)$ such that $x_i \neq x_j$ and $r' \geq 1$. Thus, the graph $G\langle S_{ij}\rangle$ contains a subgraph $H \in \mathcal{E}(\mathcal{S}_2)$. If condition (iii) holds, then the graph $G\langle S_{ij}\rangle$ contains a subgraph $H \in \mathcal{E}(\mathcal{S}_2) \cup \mathcal{E}(\mathcal{S}_4)$.

($\Longrightarrow$) Suppose now that G is not simple-path universally stable. Then, G contains a subgraph $H \in \mathcal{E}(\mathcal{S}_1) \cup \mathcal{E}(\mathcal{S}_2) \cup \mathcal{E}(\mathcal{S}_3) \cup \mathcal{E}(\mathcal{S}_4)$, and, thus, H contains a cycle $C = (x_0, x_1, x_2, \ldots, x_r, x_0)$, $r \geq 2$. It follows that C belongs to a strong component S_i of the graph $\widehat{G}$, $1 \leq i \leq k$. Let S_{ij} be the bi-scc of S_i which contains the cycle C. Since $r \geq 2$, the bi-scc S_{ij} has at least three vertices, i.e., $n_{ij} \geq 3$. We distinguish two cases:

Case (a): S_{ij} contains the cycle C and a path $P = (x_i, y_1, y_2, \ldots, y_{r'}, x_j)$, $r' \geq 1$. Then, $H \in \mathcal{E}(\mathcal{S}_2)$ and $G\langle S_{ij}\rangle_{u\ell}$ is not a cycle. Thus, the condition (ii) holds.

Case (b): S_{ij} contains only the cycle $C = (x_0, x_1, x_2, \ldots, x_r, x_0)$, $r \geq 2$. If there exists an edge $x_i x_{i+1 \bmod r}$ in C such that $\lambda(x_i x_{i+1 \bmod r}) \geq 2$ in G, then $H \in \mathcal{E}(\mathcal{S}_1)$ and, since $G\langle S_{ij}\rangle_{u\ell}$ is a cycle, the condition (i) holds. If there exists no such edge in C, then $H \in \mathcal{E}(\mathcal{S}_3) \cup \mathcal{E}(\mathcal{S}_4)$. Thus, H contains another cycle $C' = (x'_0, x'_1, x'_2, \ldots, x'_r, x'_0)$, $r \geq 2$, which belongs to a bi-scc, say, S_{iq}, of S_i. If the conditions (i) and (ii) do not hold for the bi-scc S_{iq}, then the graph $G\langle S_{iq}\rangle_{u\ell}$ is a cycle and $n_{iq} \geq 3$. Thus, the condition (iii) holds. $\square$

3   Detecting Universal Stability

In this section we present optimal algorithms for detecting universal and simple-path universal stability on a digraph G. Both algorithms take as input a digraph G on n vertices and m edges, and decide if G is universally stable or simple-path universally stable, resp., in $O(n + m)$ time using $O(n + m)$ space.

3.1   Universal Stability

Our algorithm for detecting universal stability on a digraph G applies Lemma 3; it works as follows:

Algorithm Univ_Stability

Input: a digraph G on n vertices and m edges;
Output: yes, if G is universally stable; otherwise, no.

1. Construct the one-subdivided graph G^* of the input digraph G;
2. Compute the strongly connected components $S_1, S_2, \ldots, S_k$ of the digraph G^*, and the number of vertices n_i and edges m_i of each strong component S_i, $1 \leq i \leq k$;
3. for $i = 1$ to k do
        if $m_i > n_i$ then return no; exit;
4. return yes;

The correctness of the algorithm Univ_Stability follows from Lemma 3. We next compute its time and space complexity. The one-subdivided graph G^* of the input graph G can be constructed in $O(n + m)$ time, where n is the number of vertices and m is the number of edges of the graph G. The graph G^* has $n + m$ vertices and $2m$ edges, and, thus, the computation of the strong components of G^* can be done in $O(n + m)$ time. Thus, the whole algorithm runs in $O(n + m)$ time; the space needed is $O(n + m)$.

From the above description we conclude that the detection algorithm Univ_Stability runs in linear time and requires linear space. Hence, we have:

Theorem 1. *Let G be a digraph on n vertices and m edges. The algorithm Univ_Stability decides whether G is universal stable in $O(n+m)$ time and space.*

Theorem 2. *Let G be a digraph on n vertices and m edges. The universal stability of G can be decided in $O(n + m)$ time and space.*

3.2   Simple-Path Universal Stability

Our simple-path universal stability detection algorithm applies Lemma 5; it works as follows:

Algorithm Simple-Path_Univ_Stability

Input: a digraph G on n vertices and m edges;
Output: yes, if G is simple-path universally stable; otherwise, no.

1. Construct the reduced graph $\widehat{G}$ of the input digraph G;
2. Compute the strong components $S_1, S_2, \ldots, S_k$ of the graph $\widehat{G}$, $1 \le i \le k$;
3. Compute the bi-scc $S_{i1}, S_{i2}, \ldots, S_{ik_i}$ of each strong component S_i, $1 \le i \le k$, and the number of vertices n_{ij} and edges m_{ij} of the bi-scc S_{ij}, $1 \le j \le k_i$;
4. for $i = 1$ to k do
     for $j = 1$ to k_i do
          if $n_{ij} \ge 3$ and $G\langle S_{ij}\rangle_{u\ell}$ is not a cycle, then return no; exit;
          if $n_{ij} \ge 3$ and $G\langle S_{ij}\rangle_{u\ell}$ is a cycle, and
               there exists an edge xy in $G\langle S_{ij}\rangle$ such that $\lambda(xy) \ge 2$,
          then return no; exit;
          if $n_{ij} \ge 3$ and $G\langle S_{ij}\rangle_{u\ell}$ is a cycle, then mark the bi-scc S_{ij};
     end-for
     if S_i contains at least two marked bi-scc then return no; exit;
5. return yes;

The correctness of the algorithm Simple-Path_Univ_Stability follows from Lemma 5. The construction of the reduced graph $\widehat{G}$ of the input graph G can be done in $O(n+m)$ time. The graph $\widehat{G}$ has n vertices and $m' \le m$ edges, and, thus, the computation of the strong components of $\widehat{G}$ can be completed in $O(n + m)$ time. Then the bi-scc $S_{i1}, S_{i2}, \ldots, S_{ik_i}$ of each strong component S_i, $1 \le i \le k$, can be computed in $O(n + m)$ time because $\sum_{j=1,k} m_{ij} \le m_i$ and $n_{ij} \le m_{ij}$

since the bi-scc are biconnected and do not share edges. It is easy to see that all the operation of step 4 are executed in linear time. Thus, the algorithm runs in $O(n+m)$ time; the space needed is $O(n+m)$.

Thus, we can state the following results.

Theorem 3. *Let G be a digraph on n vertices and m edges. The algorithm Simple-Path_Univ_Stability decides whether G is simple-path universal stable in $O(n+m)$ time and space.*

Theorem 4. *Let G be a digraph on n vertices and m edges. The simple-path universal stability of G can be decided in $O(n+m)$ time and space.*

4   Detecting Intrusion Attacks

In this section, we prove that the malicious intentions of an adversary/intruder to lead a stable network in instability by adding links in specific network parts can be detected in constant time after a preprocessing phase in which we compute path information. In particular, given a universally stable digraph G and a pair of distinct vertices $x, y \in V(G)$, we want to decide whether the graph $G + xy$ is also universally stable, where xy is the directed edge from x to y.

Based on the results of Section 3, we can decide whether $G + xy$ is universally stable in linear time without any preprocessing by executing algorithm Univ_Stability. However, we are interested in being able to answer queries of the form "is the graph $G + xy$ universally stable?" in $O(1)$ time.

Let G be a universally stable digraph on n vertices and m edges, and let $\widehat{G}$ be the reduced graph of G; recall that $\widehat{G}$ has n vertices and $m' \leq m$ edges. As in Lemma 3, each non-trivial strong component of $\widehat{G}$ forms a cycle; a trivial strong component consists of only one vertex. Thus, we have the following observation.

Observation 2. *Let G be a universally stable digraph, $\widehat{G}$ be its reduced graph, and let $S_1, S_2, \ldots, S_k$ be the strong components of $\widehat{G}$, $1 \leq i \leq k$. The acyclic component graph $\widehat{G}_{scc}$ of the digraph $\widehat{G}$ has the following property: it consists of k vertices $v_1, v_2, \ldots, v_k$ (the vertex v_i corresponds to the strong component S_i of $\widehat{G}$), and each strong component S_i is either a cycle, i.e., $m_i = n_i$, or a trivial component, i.e., $n_i = 1$ and $m_i = 0$.*

Let G be a multi-digraph and let $x, y \in V(G)$ be a pair of distinct vertices. We say that the vertices x and y form an *xy-pair* if there exists a directed path from x to y in G, and we say that they form an *xy-multi-pair* if there exist more than one directed path from x to y in G, or a directed path $(x = u_0, u_1, \ldots, u_k = y)$ containing an edge $u_i u_{i+1}$ with $\lambda(u_i u_{i+1}) > 1$, $k \geq 1$. Then, we can prove the following lemma.

Lemma 6. *Let G be a universally stable digraph and let $x, y \in V(G)$ be a pair of distinct vertices. Let $\widehat{G}$ be the reduced graph of G, $S_1, S_2, \ldots, S_k$ be the strongly connected components of $\widehat{G}$ and let n_i and m_i be the number of vertices and edges* *of the strong component S_i, respectively. The graph $G + xy$ is not universally stable if and only if one of the following conditions holds:*

(i) $x, y \in S_i$, $1 \le i \le k$;
(ii) $x \in S_i$ *and* $y \in S_j$ *where* $i \ne j$ *and at least one of* n_i, n_j *is larger than* 1, *and* x, y *form a* yx-*pair in* G;
(iii) $x \in S_i$ *and* $y \in S_j$ *where* $i \ne j$ *and* $n_i = n_j = 1$, *and* x, y *form an* yx-*multi-path in* G.

Based on this lemma, we next present an algorithm for detecting whether a graph G preserves its universal stability after the addition of a link xy into the topology of G. Our algorithm works on the acyclic component graph $\widehat{G}_{scc}$ of the digraph $\widehat{G}$ and uses path information of $\widehat{G}_{scc}$ which we have computed in a preprocessing stage. Let $v_1, v_2, \ldots, v_k$ be the vertices of $\widehat{G}_{scc}$. We say that an edge $v_i v_j$ in $\widehat{G}_{scc}$ is *thick* if there exist more than one edge in G with their start-points in S_i and their end-points in S_j.

We can keep all the information regarding the types of directed paths connecting each vertex v_i of $\widehat{G}_{scc}$ to all the other vertices v_j using a two-dimensional array of $O(n^2)$ space; we call this information *path-information*. Our algorithm takes as input the path-information of a universally stable digraph G on n vertices and m edges and a pair of vertices $x, y \in V(G)$, and detects in $O(1)$ time whether the graph $G + xy$ is universally stable.

Algorithm Test_Link

Input: the path-information of a universally stable digraph G on n vertices and m edges and an (ordered) pair of vertices $x, y \in V(G)$;
Output: yes, if $G + xy$ is universally stable; otherwise, no.

1. if x, y belong to the same strong component of the digraph $\widehat{G}$, then return no; exit;
2. $\{x, y$ belong to different strong components; let $x \in S_i$ and $y \in S_j$, $i \ne j\}$
   if at least one of n_i, n_j is larger than 1 then
      if there exists a $v_j v_i$-path in $\widehat{G}_{scc}$, then return no; exit;
   else $\{n_i = n_j = 1\}$
      if there exist more than one $v_j v_i$-path in $\widehat{G}_{scc}$, then return no; exit;
      if there exists a $v_j v_i$-path in $\widehat{G}_{scc}$ with a thick edge, then return no; exit;
3. return yes;

The correctness of the algorithm follows from Lemma 6. Moreover, the time taken by the algorithm to test the addition of a (directed) link xy takes $O(1)$ time thanks to the information on the strong components of $\widehat{G}$ (their sizes and the vertices participating in them) and the array storing the path-information. Hence, we have the following theorem.

Theorem 5. *Let G be a universally stable digraph on n vertices and m edges. Given the path-information of the acyclic component graph $\widehat{G}_{scc}$ of the digraph $\widehat{G}$,* *the preservation or not of universal stability of G after the addition of a link into the topology of G can be decided in $O(1)$ time.*

Algorithm Test_Link only tells us whether the addition of a link on the same base network will preserve its universal stability; this has the advantage of guaranteeing constant time complexity but also the limitation that the much more interesting approach of dynamically maintaining the network under the addition of links that would not lead to instability is not supported (note that such a dynamic maintenance does not seem possible in constant time). We are currently working on exploiting ideas and results for the poly-logarithmic dynamic maintenance of the biconnected components of an undirected graph [11] in order to achieve dynamic maintenance of a universally stable network under the addition of links in poly-logarithmic time.

5   Directions for Further Research

In this work, we propose optimal algorithms for detecting network universal stability in the context of Adversarial Queueing Theory both when packets may follow non-simple paths and when they follow simple paths. A lot of interesting problems related to the stability of networks under a fixed protocol remain open. Known results include the study of network stability under the FFS [1], the NTG-LIS [2] and the FIFO [4] protocols, which has been shown to be decided in polynomial time. Another interesting problem is whether there are upper bounds on the injection rate that guarantee stability for the forbidden subgraphs (networks) $\mathcal{U}_1$ and $\mathcal{U}_2$.

Acknowledgment. The research of the first three authors is co-funded by the EU in the framework of the projects "Pythagoras II" and "Support of Computer Science Studies in the University of Ioannina" of the Operational Program for Education and Initial Vocational Training. The research of the fourth author is partially funded by the EU FET Proactive IP project "Aeolus".

References

1. Alvarez, C., Blesa, M., Diaz, J., Fernandez, A., Serna, M.: The Complexity of Deciding Stability under FFS in the Adversarial Model. Information Processing Letters 90, 261–266 (2004)
2. Alvarez, C., Blesa, M., Serna, M.: A Characterization of Universal Stability in the Adversarial Queuing Model. SIAM Journal on Computing 34, 41–66 (2004)
3. Andrews, M., Awerbuch, B., Fernandez, A., Kleinberg, J., Leighton, T., Liu, Z.: Universal Stability Results for Greedy Contention-Resolution Protocols. Journal of the ACM 48, 39–69 (2001)
4. Blesa, M.: Deciding Stability in Packet-Switched FIFO Networks Under the Adversarial Queuing Model in Polynomial Time. In: Fraigniaud, P. (ed.) DISC 2005. LNCS, vol. 3724, pp. 429–441. Springer, Heidelberg (2005)

5. Borodin, A., Kleinberg, J., Raghavan, P., Sudan, M., Williamson, D.: Adversarial Queueing Theory. Journal of the ACM 48, 13–38 (2001)
6. Cormen, T.H., Leiserson, C.E., Rivest, R.L., Stein, C.: Introduction to Algorithms, 2nd edn. MIT Press, Cambridge (2001)
7. Diaz, J., Koukopoulos, D., Nikoletseas, S., Serna, M., Spirakis, P., Thilikos, D.: Stability and Non-Stability of the FIFO Protocol. In: Proc. 13th Annual ACM Symposium on Parallel Algorithms and Architectures, Crete, Greece, pp. 48–52 (2001)
8. Dong, Y., Rajput, S., Hsu, S.: Application Level Intrusion Detection using Graph-based Sequence Learning Algorithm. In: Proc. IASTED Conference on Modelling and Simulation, Cancun, Mexico (2005)
9. Garfinkel, S., Spafford, G.: Practical UNIX and Internet Security, 2nd edn. O'Reilly & Associates, Inc. (1996)
10. Golumbic, M.C.: Algorithmic Graph Theory and Perfect Graphs. Academic Press, New York (1980)
11. Holm, J., de Lichtenberg, K., Thorup, M.: Poly-logarithmic deterministic fully-dynamic algorithms for connectivity, minimum spanning tree, 2-edge, and biconnectivity. In: Proc. Symposium on Theory of Computing 1998, pp. 79–89 (1998)
12. Koukopoulos, D., Mavronicolas, M., Nikoletseas, S., Spirakis, P.: The Impact of Network Structure on the Stability of Greedy Protocols. Theory of Computing Systems 38, 425–460 (2005)
13. Stallings, W.: Network and Internetwork Security Principles and Practice. Prentice-Hall, Englewood Cliffs,NJ (1995)

On Certain New Models for Paging with Locality of Reference

Reza Dorrigiv and Alejandro López-Ortiz

Cheriton School of Computer Science,
University of Waterloo,
Waterloo, ON, N2L 3G1, Canada
{rdorrigiv,alopez-o}@uwaterloo.ca

Abstract. The competitive ratio is the most common metric in online algorithm analysis. Unfortunately, it produces pessimistic measures and often fails to distinguish between paging algorithms that have vastly differing performance in practice. An apparent reason for this is that the model does not take into account the locality of reference evidenced by actual input sequences. Therefore many alternative measures have been proposed to overcome the observed shortcomings of competitive analysis in the context of paging algorithms. While a definitive answer to all the concerns has yet to be found, clear progress has been made in identifying specific flaws and possible fixes for them. In this paper we consider two previously proposed models of locality of reference and observe that even if we restrict the input to sequences with high locality of reference in them the performance of every on-line algorithm in terms of the competitive ratio does not improve. Then we prove that locality of reference is useful under some other cost models, which suggests that a new model combining aspects of both proposed models can be preferable. We also propose a new model for locality of reference and prove that the randomized marking algorithm has better fault rate on sequences with high locality of reference. Finally we generalize the existing models to several variants of the caching problem.

1 Introduction

The competitive ratio is the most common metric in on-line algorithm analysis. Formally introduced by Sleator and Tarjan, it has served as a practical framework for the study of algorithms that must make irrevocable decisions in the presence of only partial information [11]. On-line algorithms are more often than not amenable to analysis under this framework; that is, computing the competitive ratio has proven to be effective—even in cases where the exact shape of the optimal solution is unknown. On the other hand, there are known applications in which the competitive ratio produces somewhat unsatisfactory results. In some cases it results in unrealistically pessimistic measures; in others, it fails to distinguish between algorithms that have vastly differing performance under any practical characterization.

S.-i. Nakano and Md. S. Rahman (Eds.): WALCOM 2008, LNCS 4921, pp. 200–209, 2008.

Paging. The paging problem is an important on-line problem both in theory and in practice. In this problem we have a small but fast memory (cache) of size k and a larger slow memory. The input is a sequence of page requests. The on-line paging algorithm should serve the requests one after another. For each request, if the requested page is in the cache, a hit occurs and the algorithm can serve the request without incurring any cost. Otherwise a fault occurs and the algorithm should bring the requested page to the cache. If the cache is already full, the algorithm should evict at least one page in order to make room for the new page. The objective is to design efficient on-line algorithms in the sense that on a given request sequence the total cost, namely the total number of faults, is kept low. Three well known paging algorithms are *Least-Recently-Used* (LRU), *First-In-First-Out* (FIFO), and *Flush-When-Full* (FWF). On a fault, if the cache is full, LRU evicts the page that is least recently requested, FIFO evicts the page that is first brought to the cache, and FWF empties the cache.

A paging algorithm is called *conservative* if it incurs at most k page faults on any page sequence that contains at most k distinct pages. LRU and FIFO are conservative while FWF is not [3]. Another important class of on-line paging algorithms is *marking* algorithms. A marking algorithm $\mathcal{A}$ works in phases. Each phase consists of the maximal sequence of requests that contain at most k distinct pages. All the pages in the cache are unmarked at the beginning of each phase. We mark any page just after the first request to it. When an eviction is necessary, $\mathcal{A}$ should evict an unmarked page until none exists, which marks the end of the phase. LRU and FWF are marking algorithms while FIFO is not [3]. We only consider *demand paging* algorithms, i.e., algorithms that do not evict any pages on a hit. Any paging algorithm can be modified into a demand paging algorithm that has no more faults [3].

The cost of a paging algorithm $\mathcal{A}$ on an input sequence I is the number of faults it incurs to serve I. In competitive analysis, we compare on-line algorithms to the off-line optimal algorithm OPT which knows the entire sequence in advance. An on-line algorithm $\mathcal{A}$ is said to have competitive ratio c if $\mathcal{A}(I) \leq c \times OPT(I)$ for all input sequences I.

Alternative Measures. It is known that the competitive ratio of any deterministic on-line paging algorithm is at least k and the competitive ratio of any conservative or marking paging algorithm is at most k (see e.g. [3]). This means that all conservative and marking algorithms are optimal with respect to the competitive ratio. Therefore competitive analysis cannot distinguish among the algorithms LRU, FIFO, and FWF. However, LRU is preferable in practice and furthermore its practical performance ratio is much better than its competitive ratio. These drawbacks of the competitive ratio have led to several proposals for better performance measures. For a survey on alternative measures see the monograph by Dorrigiv and López-Ortiz [6].

One reason that LRU has good experimental behaviour is that in practice page requests show *locality of reference*. This means that when a page is requested it is more likely to be requested in the near future. Most measures for the analysis of the paging algorithms try to use this fact by restricting their legal inputs to those with high locality. Several models have been suggested for paging with locality of reference (e.g. [1,2,4,9,10,12]).

In this paper we consider the models proposed by Torng [12] and Albers et al. [1]. These models are based on the Denning's working set model [5] and we term them the *k-phase model* and the *working set model*, respectively. In the k-phase model, we consider decompositions of input sequences to phases in the same way as marking algorithms. For an input sequence I, let $D(I,k)$ be the decomposition of I into phases of a marking algorithm with cache of size k, $|D(I,k)|$ be the number of phases of $D(I,k)$, and $L(I,k) = |I|/|D(I,k)|$ be the average length of phases. We call I a-local if $L(I,k) \geq a \cdot k$. Torng models locality by restricting the input to a-local sequences for some constant a. In the working set model, a request sequence has high locality of reference if the number of distinct pages in a window of size n is small. For a concave function f, we say that a request sequence is consistent with f if the number of distinct pages in any window of size n is at most $f(n)$, for any $n \in \mathcal{N}$. In this way, Albers et al. model locality by restricting the input to sequences that are consistent with a concave function f.

Torng shows that marking algorithms perform better on sequences with high locality under the *full access cost model*. In this model, the cost of a hit is one and the cost of a fault is $p+1$, where p is a parameter of model that is called the *miss penalty*. We denote the cost of an algorithm $\mathcal{A}$ on a sequence I in the full access cost model by $\mathcal{A}_{FA}(I)$. Albers et al. use the *fault rate* as their performance measure. The fault rate of an algorithm $\mathcal{A}$ on an input sequence I, denoted by $F_{\mathcal{A}}(I)$, is defined as $\mathcal{A}(I)/|I|$, where $|I|$ is the length of I.

Our Results. First we formally show that under the standard cost model the competitive ratio of every on-line paging algorithm remains the same under the Albers et al. and the Torng locality of reference models. Hence, new results about paging algorithms necessitate a change to the cost model. We apply the fault rate cost model to the k-phase model and the full access cost model to the working set model. These two had been previously studied under the alternate combination. The full access cost model compares on-line algorithms to the optimal off-line algorithm, while the fault rate cost model does not. Therefore there is not a direct relationship between the two cost models and our results cannot be directly concluded from the results of Torng [12] and Albers et al. [1]. Recently, Angelopoulos et al. [2] considered the standard cost model (number of faults) together with a new comparison model, called bijective analysis, and proved that LRU is the best on-line paging algorithm on sequences that show locality of reference in the working set model. Furthermore, we propose a new model for locality of reference and show that the randomized marking algorithm of [7] benefits from the locality of reference assumption. Finally we apply the locality of reference assumption to the caching problem, a generalization of the paging problem in which pages have different sizes and retrieval costs. We extend the existing models of locality of reference and show that certain caching algorithms perform better on sequences with good locality of reference under this model.

This paper is part of a systematic study of alternative models for paging. It contrasts two previously proposed models under locality of reference assumptions. Then using lessons learned it introduces a new model and shows that this new model is able to reflect locality of reference assumptions for the randomized marking algorithm. While the ultimate model for paging analysis remains yet to be discovered, the lessons learned (both positive and negative) from the results in this paper are of value to the field, and to judge from the past, likely to be of use in further future refinements in the quest for the ultimate model for paging analysis.

2   Limitations of the Competitive Ratio Model

We prove that restricting input sequences to those with high locality of reference is not reflected as an improvement on the competitive ratio.

Observation 1. *If we restrict the input to sequences that are consistent with a concave function f, the competitive ratio of deterministic on-line paging algorithms does not improve.*

Proof. The proof idea is the same as the one used to show that finite lookahead does not improve competitive ratio of on-line paging algorithms [3]. Let $\mathcal{A}$ be a deterministic paging algorithm, we can obtain a worst case sequence by requesting always a page which is not in the cache. To ensure that requesting such a page is consistent with f, we repeat the last request of I a sufficient number of times. Since we only consider demand paging algorithms, this has no effect on the contents of the cache.   □

A similar result for the k-phase model can be proven, as we can make any sequence a-local by repeating each request a times.

Observation 2. *If we restrict the input to a-local sequences, the competitive ratio of deterministic on-line paging algorithms does not change.*

3   The Fault Rate of the k-phase Model

We obtain new results based on the k-phase model by considering the fault rate as the cost model. The fault rate of an algorithm $\mathcal{A}$ on a-local sequences is defined as $F_{\mathcal{A}}(a) = \inf\{r \mid \exists n \in \mathcal{N} : \forall I, L(I,k) \geq ak, |I| \geq n : F_{\mathcal{A}}(I) \leq r\}$. We also denote by $F_{\mathcal{A}}(I)$ the fault rate of $\mathcal{A}$ on an arbitrary sequence I. We obtain the following bound for the fault rate of marking algorithms.

Theorem 1. *Let $\mathcal{A}$ be an arbitrary marking algorithm and $a > 0$ be a constant. Then $F_{\mathcal{A}}(a) \leq 1/a$.*

Proof. Consider an arbitrary a-local sequence I of size at least n. We show that $F_{\mathcal{A}}(I) \leq 1/a$. Consider the decomposition $D(I,k)$ of I. $\mathcal{A}$ does not fault more than once on a page P in a phase ϕ because after the first fault, it marks P

and does not evict it in the reminder of ϕ. Since each phase contains at most k distinct pages, $\mathcal{A}$ does not fault more than k times in a phase.

Thus $\mathcal{A}$ incurs at most $k \cdot |D(I,k)|$ faults on I and we have $F_{\mathcal{A}}(I) \leq \frac{k \cdot |D(I,k)|}{|I|}$. Using $L(I,k) = |I|/|D(I,k)|$, we get $F_{\mathcal{A}}(I) \leq \frac{k}{L(I,k)} \leq \frac{k}{ak} = 1/a$. $\qquad\square$

This theorem shows that the fault rate of any marking algorithm decreases as the locality of reference of the input increases. Note that this holds for every algorithm $\mathcal{A}$ that incurs at most k faults in each phase. Since any phase contains at most k distinct pages, we obtain the following result.

Corollary 1. *Let $\mathcal{A}$ be an arbitrary conservative algorithm and $a > 0$ be a constant. Then $F_{\mathcal{A}}(a) \leq 1/a$.*

4   The Working Set Model Under Full Access Cost Model

In this section we apply the full access cost model to the working set model. Earlier we proved that the standard competitive ratio does not improve for sequences with high locality of reference in this model. Now we show that the competitive ratio of the classical algorithms in the full access cost model improves for such sequences.

First we use some results of Albers et al. [1] about the fault rate of paging algorithms. These results are expressed in term of f^{-1}, the inverse function of f, defined as

$$f^{-1}(m) = \min\{n \in \mathcal{N} \mid f(n) \geq m\}.$$

In other words, $f^{-1}(m)$ denotes the minimum size of a window that contains at least m distinct pages.

Theorem 2. *The competitive ratio of LRU with respect to a concave function f in the full access cost model is at most $\frac{p \cdot k \cdot (k-1) + k(f^{-1}(k+1)-2)}{p \cdot (k-1) + k(f^{-1}(k+1)-2)}$.*

Proof. Albers et al. proved that the fault rate of LRU is at most $\frac{k-1}{f^{-1}(k+1)-2}$ [1]. Consider an arbitrary sequence I that is consistent with f. Suppose that LRU (OPT) incurs m (m') faults on I. We have $\frac{m}{|I|} \leq \frac{k-1}{f^{-1}(k+1)-2} \implies |I| \geq \frac{m \cdot (f^{-1}(k+1)-2)}{k-1}$. Since $m' \geq m/k$, we have $OPT_{FA}(I) = p \cdot m' + |I| \geq p \cdot m/k + |I|$. We also have $LRU_{FA}(I) = p \cdot m + |I|$, and therefore $\frac{LRU_{FA}(I)}{OPT_{FA}(I)} \leq \frac{p \cdot m + \frac{m \cdot (f^{-1}(k+1)-2)}{k-1}}{p \cdot m/k + \frac{m \cdot (f^{-1}(k+1)-2)}{k-1}} = \frac{p \cdot k \cdot (k-1) + k(f^{-1}(k+1)-2)}{p \cdot (k-1) + k(f^{-1}(k+1)-2)}$. Since I was an arbitrary sequence, this proves the theorem. $\qquad\square$

As the cost of a fault p increases, the upper bound of Theorem 2 approaches k. When p is not too large, the term $(f^{-1}(k+1)-2)$ becomes important. For a fixed p, the larger the value of $f^{-1}(k+1)$, the better the upper bound of the theorem. This supports our intuition that LRU has better performance on sequences with more locality of reference.

It is also known that $F_{FIFO}(f) \leq \frac{k}{f^{-1}(k+1)-1}$ and $F_{\mathcal{A}}(f) \leq \frac{k}{f^{-1}(k+1)-1}$ for any marking algorithm $\mathcal{A}$ [1]. We can use these results to prove the following theorem in an analogous way to Theorem 2.

Theorem 3. *Let $\mathcal{A}$ be a marking algorithm or FIFO. The competitive ratio of $\mathcal{A}$ with respect to a concave function f in the full access cost model is at most $\frac{p \cdot k + (f^{-1}(k+1)-1)}{p+(f^{-1}(k+1)-1)}$.*

Finally we prove a result for all marking and conservative algorithms.

Theorem 4. *Let $\mathcal{A}$ be a marking or conservative algorithm. The competitive ratio of $\mathcal{A}$ with respect to a concave function f in the full access cost model is at most $\frac{p \cdot k + f^{-1}(k)}{p+f^{-1}(k)}$.*

Proof. Let I be a sequence consistent with f and consider the decomposition $D(I,k)$ of I. We know that $\mathcal{A}$ incurs at most k faults in each phase. Let m denote the number of faults $\mathcal{A}$ incurs on I. We have $m \leq k \cdot |D(I,k)| \Rightarrow |D(I,k)| \geq m/k$. Each phase has length at least $f^{-1}(k)$ because I is consistent with f. Therefore $|I| \geq |D(I,k)| \cdot f^{-1}(k) \geq m \cdot f^{-1}(k)/k$, and $\frac{A_{FA}(I)}{OPT_{FA}(I)} \leq \frac{p \cdot m + |I|}{p \cdot m/k + |I|} \leq \frac{p \cdot m + m \cdot f^{-1}(k)/k}{p \cdot m/k + m \cdot f^{-1}(k)/k} = \frac{p \cdot k + f^{-1}(k)}{p+f^{-1}(k)}$. $\square$

5   A New Model for Locality of Reference

In this section we introduce a new model for locality of reference that can be used to show that the *randomized marking* algorithm [7] benefits from locality of reference. We can generalize the definitions of the competitive ratio and the fault rate to the randomized algorithms by considering the expected number of page faults. Let H_k denote the kth harmonic number: $H_k = 1 + 1/2 + \ldots + 1/k$. Fiat et al. introduced the randomized marking algorithm, $\mathcal{RM}$, and showed that it is $2H_k$-competitive [7]. The phases of $\mathcal{RM}$ are defined as deterministic marking algorithms. On a fault, $\mathcal{RM}$ evicts a page chosen uniformly at random from among the unmarked pages. A page is called *clean* if it was not requested in the previous phase and *stale* otherwise. Intuitively, a sequence with locality of reference does not have many clean pages in a phase. In order to formalize this intuition we generalize the k-phase model as follows.

Definition 1. *Let I be a sequence and consider its k-decomposition. For constants $a > 1$ and $b < 1$, I is called (a,b)-local if $L(I,k) \geq ak$ and each phase of $D(I,k)$ has at most bk clean pages.*

Now we can define the fault rate of an algorithm $\mathcal{A}$ on (a,b)-local sequences, $F_{\mathcal{A}}(a,b)$, by restricting the input sequences to (a,b)-local sequences.

The following theorem shows that $\mathcal{RM}$ works better on sequences that are "more" local.

Theorem 5. *For any constants $a > 1$ and $b < 1$,*

$$F_{\mathcal{RM}}(a,b) \leq \frac{b \cdot (H_k - H_{bk} + 1)}{a} \sim \frac{b(1 - \ln b)}{a}.$$

Proof. Consider an arbitrary (a, b)-local sequence I such that $|I| \geq n$. We should show that $F_{\mathcal{RM}}(I) \leq \frac{b \cdot (H_k - H_{bk} + 1)}{a}$. Consider the k-decomposition of I. Let l_i denote the number of clean pages of phase i. Fiat et al. proved that the expected number of faults of $\mathcal{RM}$ in phase i, f_i, is at most $B_i = l_i \cdot (H_k - H_{l_i} + 1)$ [7]. Note that $1 \leq l_i \leq bk$; the first page of each phase is clean and I is an (a, b)-local sequence. Since B_i is strictly increasing for $l_i \leq k$, we get $f_i \leq b \cdot k \cdot (H_k - H_{bk} + 1)$. Therefore the expected number of faults that $\mathcal{RM}$ incurs on I is at most $bk \cdot (H_k - H_{bk} + 1) \cdot |D(I, k)|$. On the other hand we have $|I| \geq ak \cdot |D(I, k)|$. Therefore

$$F_{\mathcal{RM}}(I) \leq \frac{bk \cdot (H_k - H_{bk} + 1) \cdot |D(I, k)|}{ak \cdot |D(I, k)|} = \frac{b \cdot (H_k - H_{bk} + 1)}{a}. \qquad \square$$

Since $H_n \approx \ln n$, we can get an upper bound of $b \cdot (1 - \ln b)/a$ for $F_{\mathcal{RM}}(a, b)$. Thus the fault rate of $\mathcal{RM}$ decreases as a increases and b decreases. Note that several other results can be obtained by imposing more restrictions on input sequences. For example if I is an a-local sequence that contains only $k + 1$ distinct pages we have $l_i = 1$ for each phase i and therefore

$$F_{\mathcal{RM}}(I) \leq \frac{1 \cdot (H_k - H_1 + 1)|D(I, k)|}{ak|D(I, k)|} = \frac{H_k}{ak}.$$

6   Caching with Locality of Reference

In the paging problem, all pages have the same size and the same retrieval cost on a fault. However, in some applications such as caching files on the Web, pages have different sizes and the cost of bringing a page to the cache varies for different pages. We can generalize the paging problem in different ways. These generalized variants of the problem are usually called *caching* problems. There are various models for the caching problem [13,8]:

- **General Model.** In this model pages can have arbitrary sizes and arbitrary retrieval costs.
- **Weighted Caching.** Pages have uniform sizes, but they can have arbitrary retrieval costs (weights).
- **Fault Model.** Pages have arbitrary weights, however, they have uniform retrieval costs.
- **Bit Model.** Pages have arbitrary sizes and the retrieval cost of a page is proportional to its size.

Each of these models is appropriate for certain applications. Irani describes some applications in Web caching that are best modeled using the Fault/Bit model [8]. In this section we study the behaviour of marking caching algorithms on sequences with high locality of reference.

6.1 Weighted Caching

For weighted caching, we introduce some new notation. Consider an on-line paging algorithm $\mathcal{A}$. Each page π has a weight $w(\pi)$. In the full access cost model, the cost of a hit is 1 and the cost of a fault on a page π is $p \cdot w(\pi) + 1$ for some parameter p. Let $W_{\mathcal{A}}(I)$ be the total weight of pages on which $\mathcal{A}$ incurs a fault when it serves a sequence I and $W_{OPT}(I)$ be the same value for the optimal off-line algorithm. Define the average weight of faults in a phase as $AW_{\mathcal{A}}(I) = W_{\mathcal{A}}(I)/|D(I,k)|$ and $AW_{OPT}(I) = W_{OPT}(I)/|D(I,k)|$. Note that the full access cost of $\mathcal{A}$ and OPT on I is $|I| + p \cdot W_{\mathcal{A}}(I)$ and $|I| + p \cdot W_{OPT}(I)$, respectively. Let $C_{FA}(\mathcal{A}, I) = \frac{|I| + p \cdot W_{\mathcal{A}}(I)}{|I| + p \cdot W_{OPT}(I)}$; then we have $C_{FA}(\mathcal{A}) = \sup_I C_{FA}(\mathcal{A}, I)$.

Now assume that I is an a-local sequence, i.e. $L(I,k) \geq ak$ for some constant $a > 1$. We have

$$C_{FA}(\mathcal{A}, I) = \frac{L(I,k) + p \cdot AW_{\mathcal{A}}(I)}{L(I,k) + p \cdot AW_{OPT}(I)} \leq \frac{ak + p \cdot AW_{\mathcal{A}}(I)}{ak + p \cdot AW_{OPT}(I)}.$$

Note that the standard competitive ratio of $\mathcal{A}$ is

$$C(\mathcal{A}) = \sup_I \frac{AW_{\mathcal{A}}(I)}{AW_{OPT}(I)}.$$

Therefore when p is large, $C_{FA}(\mathcal{A})$ approaches the standard competitive ratio. For smaller values of p, $C_{FA}(\mathcal{A})$ improves as the locality of reference increases.

6.2 Bit Model

There is a close connection between the Bit model and the full access cost model. Let $s(\pi)$ denote the size of a page π and k be the size of cache. In the bit model, the retrieval cost of π is $r \cdot s(\pi)$ for some fixed constant r. In the full access cost model, the cost of a hit is 1 and the cost of a fault is $p + 1$ for some parameter p. Therefore we can have a generalization of the full access cost model for the Bit model as follows. The cost of a hit is 1 and the cost of a fault on a page π is $q \cdot s(\pi) + 1$ for some parameter q.

We obtain results in this model using the idea of k-decomposition. However since pages can have arbitrary sizes, we modify the definition of the decomposition. We upper bound the total size of distinct pages in a phase, rather than the number of distinct pages. For an input sequence I and an integer $m > 1$, the m-decomposition $D(I,m)$ is defined as partitioning I into consecutive phases so that each phase is a maximal subsequence that contains a set Π of distinct pages such that the total of pages in Π adds up to at most m units of information.[1] Note that each phase may contain a set of distinct pages whose total size is strictly less than m. $|D(I,m)|$ and $L(I,m)$ are as before.

A marking algorithm in this model works in phases. At the beginning of each phase all pages in the cache are unmarked. A page is marked when it is requested. On a fault, the algorithm brings the requested page to the cache and evicts as

---

[1] Depending on the application, the unit of information can be: bit, byte, word, etc.

many (unmarked) pages as necessary from the cache to make room for this page. If all pages in the cache are marked, the phase ends and all pages are unmarked. As before, we call a sequence I a-local if $L(I, k) \geq ak$ for some constant $a > 1$. Also assume that we have normalized the sizes of pages so that the smallest pages have unit size.

Theorem 6. *Let $\mathcal{A}$ be an arbitrary marking algorithm on an a-local input sequence. Then under the Bit model we have $C_{FA}(\mathcal{A}) \leq 1 + q/a$.*

Proof. Consider an arbitrary a-local sequence I. $\mathcal{A}$ incurs at most one fault on any page π in any phase ϕ because π is marked after the first fault and will not be evicted in the remaining steps of ϕ. Since the total size of distinct pages in a phase is at most k, the full access cost of $\mathcal{A}$ on I is at most $|I| + |D(I, k)| \cdot (q \cdot k)$. On the other hand, according to the definition of the decomposition, the optimal off-line algorithm should incur at lease one fault in each phase and therefore its full access cost is at least $|I| + |D(I, k)| \cdot (q \cdot 1)$. Therefore we get

$$C_{FA}(\mathcal{A}, I) \leq \frac{|I| + |D(I, k)| \cdot (q \cdot k)}{|I| + |D(I, k)| \cdot (q \cdot 1)} = \frac{L(I, k) + q \cdot k}{L(I, k) + q \cdot 1}.$$

Now since I is an $a$local sequence, $L(I, k) \geq ak$ and

$$C_{FA}(\mathcal{A}, I) \leq \frac{ak + q \cdot k}{ak + q \cdot 1} = 1 + \frac{(k - 1)}{ak/q + 1} < 1 + q/a.$$

This completes the proof as I is an arbitrary a-local sequence.   □

7   Conclusions

In this paper we studied some models for paging with locality of reference. In particular we proved that in general the competitive ratio does not improve on input sequences with high locality of reference under the models of Torng [12] and Albers et al. [1]. We also proposed a new model for locality of reference and proved that the randomized marking algorithm has better fault rate on sequences with high locality of reference. Finally we generalized the existing models to several variants of the caching problem.

Acknowledgements. We thank Spyros Angelopoulos for useful discussions on this subject.

References

1. Albers, S., Favrholdt, L.M., Giel, O.: On paging with locality of reference. JCSS: Journal of Computer and System Sciences 70 (2005)
2. Angelopoulos, S., Dorrigiv, R., López-Ortiz, A.: On the separation and equivalence of paging strategies. In: SODA 2007. Proceedings of the 18th ACM-SIAM Symposium on Discrete Algorithms, pp. 29–237 (2007)

3. Borodin, A., El-Yaniv, R.: Online Computation and Competitive Analysis. Cambridge University Press, Cambridge (1998)
4. Borodin, A., Irani, S., Raghavan, P., Schieber, B.: Competitive paging with locality of reference. Journal of Computer and System Sciences 50, 244–258 (1995)
5. Denning, P.J.: The working set model for program behaviour. Communications of the ACM 11(5) (May 1968)
6. Dorrigiv, R., López-Ortiz, A.: A survey of performance measures for on-line algorithms. SIGACTN: SIGACT News (ACM Special Interest Group on Automata and Computability Theory) 36(3), 67–81 (2005)
7. Fiat, A., Karp, R.M., Luby, M., McGeoch, L.A., Sleator, D.D., Young, N.E.: Competitive paging algorithms. Journal of Algorithms 12(4), 685–699 (1991)
8. Irani, S.: Page replacement with multi-size pages and applications to web caching. Algorithmica 33(3), 384–409 (2002)
9. Irani, S., Karlin, A.R., Phillips, S.: Strongly competitive algorithms for paging with locality of reference. SIAM J. Comput. 25, 477–497 (1996)
10. Panagiotou, K., Souza, A.: On adequate performance measures for paging. In: STOC 2006. Proceedings of the 38th Annual ACM Symposium on Theory of Computing, pp. 487–496 (2006)
11. Sleator, D.D., Tarjan, R.E.: Amortized Efficiency of List Update and Paging Rules. Communications of the ACM 28, 202–208 (1985)
12. Torng, E.: A unified analysis of paging and caching. Algorithmica 20(2), 175–200 (1998)
13. Young, N.E.: On-line file caching. Algorithmica 33(3), 371–383 (2002)

Listing All Plane Graphs

(Extended Abstract)

Katsuhisa Yamanaka and Shin-ichi Nakano

Department of Computer Science, Gunma University,
1-5-1 Tenjin-cho Kiryu, Gunma, 376-8515 Japan
yamanaka@nakano-lab.cs.gunma-u.ac.jp,nakano@cs.gunma-u.ac.jp

Abstract. In this paper we give a simple algorithm to generate all connected rooted plane graphs with at most m edges. A "rooted" plane graph is a plane graph with one designated (directed) edge on the outer face. The algorithm uses $O(m)$ space and generates such graphs in $O(1)$ time per graph on average without duplications. The algorithm does not output the entire graph but the difference from the previous graph. By modifying the algorithm we can generate all connected (non-rooted) plane graphs with at most m edges in $O(m^3)$ time per graph.

1 Introduction

Generating all graphs with some property without duplications has many applications, including unbiased statistical analysis [9]. A lot of algorithms to solve these problems are known [1,2,8,9,10,14, etc]. See textbooks [5,6,7,12,13].

In this paper we wish to generate all connected "rooted" plane graphs, which will be defined precisely in Section 2, with at most m edges. Such graphs play an important role in many algorithms, including graph drawing algorithms [3,4,11, etc].

To solve these all-graph-generating problems some types of algorithms are known.

Classical method algorithms [5, p57] first generate all the graphs with a given property allowing duplications, but output only if the graph has not been output yet. Thus this method requires quite a huge space to store a list of graphs that have already been output. Furthermore, checking whether each graph has already been output requires a lot of time.

Orderly method algorithms [5, p57] need not store the list, since they output a graph only if it is a "canonical" representative of each isomorphism class.

Reverse search method algorithms [1] also need not store the list. The idea is to implicitly define a connected graph H such that the vertices of H correspond to the graphs with the given property, and the edges of H correspond to some relation between the graphs. By traversing an implicitly defined spanning tree of H, one can find all the vertices of H, which correspond to all the graphs with the given property.

The main idea of our algorithms is that for some problems(biconnected triangulations [8], and triconnected triangulations [10]) we can define a tree (not

S.-i. Nakano and Md. S. Rahman (Eds.): WALCOM 2008, LNCS 4921, pp. 210–221, 2008.

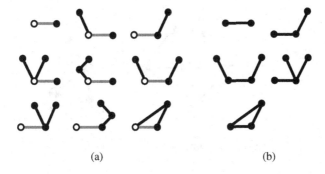

(a)   (b)

Fig. 1. (a) Connected rooted plane graphs, and (b) Connected (non-rooted) plane graphs

a general graph) as the graph H of the reverse search method. Thus our algorithms do not need to find a spanning tree of H, since H itself is a tree. With some other ideas we give the following two simple but efficient algorithms.

Our first algorithm generates all simple connected rooted plane graphs with at most $m(m > 0)$ edges. *Simple* means there is neither self loops nor multiple edges. A *rooted* plane graph means a plane graph with one designated "root" edge on the outer face. For instance there are nine simple connected rooted plane graphs with at most three edges, as shown in Fig. 1(a). The root edges are depicted by thick grey lines. However, there are only five simple connected (non-rooted) plane graphs with at most three edges. See Fig. 1(b). The algorithm uses $O(m)$ space and runs in $O(g(m))$ time, where $g(m)$ is the number of nonisomorphic connected rooted plane graphs with at most m edges. The algorithm generates each graph in $O(1)$ time on average without duplications. The algorithm does not output the entire graph but the difference from the previous graph.

By modifying the algorithm we can generate all connected (non-rooted) plane graphs with at most m edges in $O(m^3)$ time per graph.

The rest of the paper is organized as follows. Section 2 gives some definitions. Section 3 shows a tree structure among connected rooted plane graphs. Section 4 presents our first algorithm to generate all connected rooted plane graphs. Then, by modifying the algorithm we give an algorithm to generate all connected (non-rooted) plane graphs. Section 5 analyzes the running time of our algorithm. Finally Section 6 is a conclusion.

2   Preliminaries

In this section we give some definitions.

Let G be a connected graph with m edges. In this paper all graphs are simple, so there is neither self loops nor multiple edges. An edge connecting vertices u and w is denoted by (u, w). The *degree* of a vertex v is the number of neighbors of v in G.

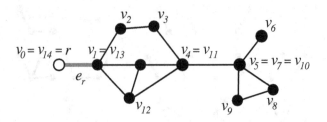

Fig. 2. A connected rooted plane graph

A graph is *planar* if it can be embedded in the plane so that no two edges intersect geometrically except at a vertex to which they are both incident. A *plane* graph is a planar graph with a fixed planar embedding. A plane graph divides the plane into connected regions called *faces*. The unbounded face is called *the outer face*, and other faces are called *inner faces*. We regard *the contour* of a face as the clockwise cycle formed by the vertices on the boundary of the face. We denote the contour of the outer face of plane graph G by $C_o(G)$. For instance, in Fig. 2, $C_o(G) = v_0, v_1, v_2, v_3, v_4, v_5, v_6, v_7 = v_5, v_8, v_9, v_{10} = v_5, v_{11} = v_4, v_{12}, v_{13} = v_1, v_{14} = v_0$. Note that a vertex may appear several times on $C_o(G)$. We say each v_i on $C_o(G)$ is an *appearance* of a vertex. For instance v_5, v_7 and v_{10} are the appearances of the same vertex $v_5 = v_7 = v_{10}$. A *rooted* plane graph is a plane graph with one designated edge $e_r = (v_r, v_l)$ on $C_o(G)$. We assume v_l succeeds v_r on $C_o(G)$. The designated edge is called *the root edge*, and vertex v_l is called *the root vertex*. Note that a rooted plane graph has one or more edges. From now on we write r for the root vertex.

3 The Removing Sequence and the Family Tree

Let S_m be the set of all connected rooted plane graphs with at most m edges. In this section we explain a tree structure relating the graphs in S_m.

Let G be a connected rooted plane graph with two or more edges. Let $e_r = (v_{k-1}, v_0)$ be the root edge of G and $C_o(G) = v_0(= r), v_1, v_2, \ldots, v_{k-1}, v_0(= r)$. Note that v_0 succeeds v_{k-1} on $C_o(G)$.

We classify the edges on $C_o(G)$ into three types as follows. If e on $C_o(G)$ is included in a cycle of G then e is *a cycle edge*. Otherwise, if at least one vertex of e has degree 1 then e is *a pendant*. Otherwise e is *a bridge*. We can observe if we remove a bridge from G then the resulting graph is disconnected. For instance, in Fig. 2, edge (v_2, v_3) is a cycle edge, edge (v_5, v_6) is a pendant, and edge (v_4, v_5) is a bridge.

An edge $e \neq e_r$ on $C_o(G)$ is *removable* if after removing e from G the remaining edges induce a connected graph. Thus each edge $e \neq e_r$ is removable if and only if e is either a pendant or a cycle edge.

Since G is a rooted plane graph, the resulting graph after removing a removable edge is also a rooted plane graph with the same root edge.

We have the following lemma.

Lemma 1. *Every connected rooted plane graph with two or more edges has at least one removable edge.*

Proof. Let G be a connected rooted plane graph with two or more edges, with the root edge (v_{k-1}, v_0), and $C_o(G) = v_0(= r), v_1, v_2, \ldots, v_{k-1}, v_0$. Let $e_1 = (v_0, v_1) \neq (v_{k-1}, v_0) = e_r$ be any edge on $C_o(G)$. Now e_1 must be one of the three types, that is, a bridge, a pendant or a cycle edge. If e_1 is a pendant or a cycle edge, it is removable, and we are done. Otherwise e_1 is a bridge, then on $C_o(G)$ the next edge of e_1 is either a pendant, a bridge or a cycle edge. By repeating this procedure we can find at least one pendant or cycle edge, which is removable. □

If $e_a = (v_{a-1}, v_a)$ is removable but none of $(v_0, v_1), (v_1, v_2), \ldots, (v_{a-2}, v_{a-1})$ is removable, then e_a is called *the first removable edge* of G. We can observe that if e_a is the first removable edge then each of $(v_0, v_1), (v_1, v_2), \ldots, (v_{a-2}, v_{a-1})$ is a bridge or the root edge. (So they are not removable.)

For each graph G in S_m except K_2, if we remove the first removable edge then the resulting edge-induced graph, denoted by $P(G)$, is also a graph in S_m having one less edge. Thus we can define the unique graph $P(G)$ in S_m for each G in S_m except K_2. We say G is a *child* graph of $P(G)$.

Given a graph G in S_m, by repeatedly removing the first removable edge, we can have the unique sequence $G, P(G), P(P(G)), \ldots$ of graphs in S_m which eventually ends with K_2. By merging those sequences we can have *the family tree* T_m of S_m such that the vertices of T_m correspond to the graphs in S_m, and each edge corresponds to each relation between some G and $P(G)$. For instance T_4 is shown in Fig. 3, in which each first removable edge is depicted by a thick black line. We call the vertex in T_m corresponding to K_2 *the root* of T_m.

4   Algorithms

Given S_m we can construct T_m by the definition, possibly with huge space and much running time. However, how can we construct T_m efficiently only given an integer m? Our idea [8,10] is by reversing the removing procedure as follows.

Given a connected rooted plane graph G in S_m with at most $m - 1$ edges, we wish to find all child graphs of G. Let e_r be the root edge. Let $C_o(G) = v_0(= r), v_1, \ldots, v_{k-1}, v_0(= r)$, and (v_{a-1}, v_a) be the first removable edge of G. Note that k is the number of appearances of the vertices on the contour of the outer face. Since K_2 has no removable edge, for convenience, we regard $e_1 = (v_0, v_1)$ as the first removable edge for K_2. We denote by $G(i)$, $0 \leq i < k$, the rooted plane graph obtained from G by adding a new pendant at v_i, and by $G(i, j)$, $0 \leq i < j < k$, the rooted plane graph obtained from G by adding a new cycle edge connecting v_i and v_j on the outer face of G, as shown in Fig. 4. We can observe that each child of G is either $G(i)$ or

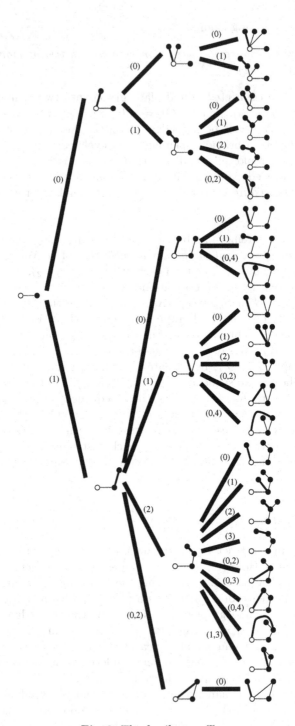

Fig. 3. The family tree T_4

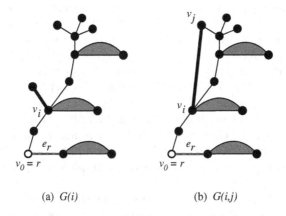

(a) *G(i)*   (b) *G(i,j)*

Fig. 4. Illustration for (a) $G(i)$ and (b) $G(i,j)$

$G(i,j)$ for some i and j, and $G(i)$ or $G(i,j)$ is a child graph of G if and only if the newly added edge of $G(i)$ or $G(i,j)$ is the first removable edge of $G(i)$ or $G(i,j)$.

If (v_{a-1}, v_a) is the first removable edge of G, then edges $(v_0, v_1), (v_1, v_2), \ldots,$ (v_{a-2}, v_{a-1}) are bridges or the root edge, and vertices $v_0, v_1, v_2, \ldots, v_a$ form a path on $C_o(G)$. We call this path *the critical path* of G and denote it $P_c(G)$. For instance, in Fig. 2, $P_c(G)=(v_0, v_1, v_2)$.

Now we are going to find all child graphs of G. We have the following two cases to consider. Let $b(i)$ be the largest integer satisfying $v_i = v_{b(i)}$. Thus $v_{b(i)}$ is the last appearance of v_i on $C_o(G)$.

Case 1: The first removable edge (v_{a-1}, v_a) of G is a pendant. (including the special case when G is K_2)

Consider graphs $G(i)$, $0 \leq i \leq k$. For each i, $0 \leq i \leq a$, the newly added edge in $G(i)$ is the first removable edge of $G(i)$, thus $P(G(i)) = G$. For each i, $a < i < k$, (v_{a-1}, v_a) is still the first removable edge of $G(i)$, so $P(G(i)) \neq G$.

Then consider graphs $G(i,j)$, $0 \leq i < j < k$. For each i and $j, (i < j)$ such that (1) $v_i \neq v_j$, (2) $0 \leq i \leq a - 1$, (3) (v_i, v_j) is not an edge of G, and (4) $j < b(i)$, the newly added edge in $G(i,j)$ is the first removable edge of $G(i,j)$, thus $P(G(i,j)) = G$. Note that if $v_i = v_j$ edge (v_i, v_j) is a self loop, and so $G(i,j)$ is not simple. Also if G has edge (v_i, v_j) then $G(i,j)$ has a multiple edge, and so $G(i,j)$ is not simple. If $i \geq a$, then the newly added edge in $G(i,j)$ is not the first removable edge of $G(i,j)$, since (v_{a-1}, v_a) is still removable, thus $P(G(i,j)) \neq G$. Otherwise, $0 \leq i \leq a - 1$ and $j > b(i)$ holds. Now edge (v_{i-1}, v_i) becomes removable in $G(i,j)$, so $P(G(i,j)) \neq G$.

Case 2: The first removable edge (v_{a-1}, v_a) of G is a cycle edge.

Consider graphs $G(i)$, $0 \leq i < k$. For each i, $0 \leq i \leq a - 1$, the newly added edge in $G(i)$ is the first removable edge of $G(i)$, so $P(G(i)) = G$. For each i, $a \leq i < k$, (v_{a-1}, v_a) is still the first removable edge of $G(i)$, so $P(G(i)) \neq G$.

Then consider graphs $G(i,j)$, $0 \le i < j < k$. For each i and $j, (i < j)$ such that (1) $v_i \ne v_j$, (2) $0 \le i \le a - 1$, (3) (v_i, v_j) is not an edge of G, and (4) $j < b(i)$, the newly added edge in $G(i,j)$ is the first removable edge of $G(i,j)$, thus $P(G(i,j)) = G$. If $i \ge a$, then the newly added edge in $G(i,j)$ never becomes the first removable edge of $G(i,j)$, so $P(G(i,j)) \ne G$. Otherwise, $0 \le i \le a - 1$ and $j > b(i)$ holds. Now edge (v_{i-1}, v_i) becomes removable in $G(i,j)$, so $P(G(i,j)) \ne G$.

Based on the case analysis above we can find all child graphs of any given graph in S_m. If G has l child graphs, then we can find them in $O(l)$ time with a suitable data structure, which will be described in Section 5. This is an intuitive reason why our algorithm generates each graph in $O(1)$ time per graph on average.

And recursively repeating this process from the root of T_m corresponding to K_2 we can traverse T_m without constructing the whole part of T_m at once. During the traversal of T_m, we assign a label (i) or (i,j) to each edge connecting G and either $G(i)$ or $G(i,j)$ in T_m, as shown in Fig. 3. Each label denotes how to add a new edge to G to generate a child graph $G(i)$ or $G(i,j)$, and each sequence of labels on a path starting from the root specifies a graph in S_m. For instance, the sequence $(1)(0,2)(0)$ specifies the right-bottom graph in Fig. 3. During our algorithm we will maintain these labels only on the path from the root to the "current" vertex of T_m, because those labels carry enough information to generate the "current" graph. To generate the next graph, we need to maintain more information only for the graphs on the "current" path, which has length at most m, and each graph can be represented as a constant size of difference from the preceding one. This is an intuitive reason why our algorithm uses only $O(m)$ space, while the number of graphs may not be bounded by a polynomial in m.

Our algorithm is as follows.

Procedure find-all-child-graphs(G)
begin
01   Output G {Output the difference from the previous graph.}
02   Assume (v_{a-1}, v_a) is the first removable edge of G.
03   **if** G has exactly m edges **then return**
04   **for** $i = 0$ **to** $a - 1$ {Case 1 and 2}
05   **find-all-child-graphs($G(i)$)**
06   **if** (v_{a-1}, v_a) is a pendant **then**   {Case 1}
07   **find-all-child-graphs($G(a)$)**
08   **for** $i = 0$ **to** $a - 1$ {Case 1 and 2}
09   **for** $j = i + 2$ **to** $b(i) - 1$
10   **if** $v_i \ne v_j$ and (v_i, v_j) is not an edge of G **then**
11   **find-all-child-graphs($G(i,j)$)**
    end

Algorithm find-all-graphs(T_m)
begin
1   Output K_2
2    $G = K_2$
3   **find-all-child-graphs**$(G(0))$
4   **find-all-child-graphs**$(G(1))$
end

We have the following theorem. The proof is given in Section 5.

Theorem 1. *The algorithm uses $O(m)$ space and runs in $O(g(m))$ time, where $g(m)$ is the number of nonisomorphic connected rooted plane graphs with at most m edges.*

We can modify our algorithm so that it outputs all connected (non-rooted) plane graphs with at most m edges, as follows. At each vertex v of the family tree T_m, the graph G corresponding to v is checked whether the sequence of labels of G (with the root edge) is the lexicographically first one among the k sequences of labels of G for the k choices of the root edge on $C_o(G)$, and only if so G is output. Thus we can output only the canonical representative of each isomorphism class. A similar method is appeared in [8,10].

Lemma 2. *The algorithm uses $O(m)$ space and runs in $O(m^3 \cdot h(m))$ time, where $h(m)$ is the number of nonisomorphic connected (non-rooted) plane graphs with at most m edges.*

Proof. For each graph corresponding to a vertex of T_m we construct $k \leq m$ of sequences of labels corresponding to the k choices for the root edge on $C_o(G)$ in $O(m)$ time for each sequence, and find the lexicographically first one in $O(km)$ time. And for each output graph, our tree may contain k of isomorphic ones corresponding to the k choices for the root edge. Thus the algorithm runs in $O(k^2 m \cdot h(m)) = O(m^3 \cdot h(m))$ time. The algorithm clearly uses $O(m)$ space. □

5   Proof of Theorem 1

In this section we give a proof of Theorem 1, that is if G has l child graph how we can find them in $O(l)$ time.

Given a connected rooted plane graph G in S_m with at most $m - 1$ edges, we are going to find all child graphs of G by algorithm **find-all-child-graphs**. Let (v_{k-1}, v_0) be the root edge of G, $C_o(G) = v_0(= r), v_1, v_2, \ldots, v_{k-1}, v_0(= r)$, and (v_{a-1}, v_a) be the first removable edge of G.

If G has l child graphs of type $G(i)$, by only maintaining the critical path $v_0, v_1, \ldots, v_a$, we can find such child graphs in $O(l)$ time. See lines 04–07 of **find-all-child-graphs**.

On the other hand, if G has l' child graphs of type $G(i, j)$, we need to maintain a slightly complicated data structure to find all such child graphs in $O(l')$ time.

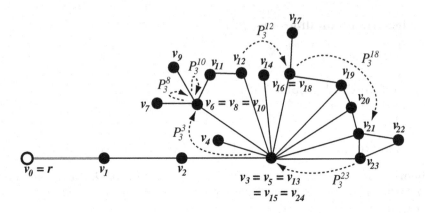

····▶ : a trace of skipping each dead path

Fig. 5. An illustration for the zombie list of v_3

Note that if either (1) $v_i = v_j$, or (2) G has an edge (v_i, v_j), then $G(i, j)$ is not simple and $G(i, j)$ is not a child graph of G, so we need to efficiently skip such j's at line 10. For each of the other j's, we need to generate $G(i, j)$, since those are child graphs of G.

Our idea is as follows. Let v_i be an appearance of a vertex on the critical path of G. We say that an appearance v_j on $C_o(G)$ is *dead* with respect to v_i if either (1) $v_i = v_j$, or (2) G has an edge (v_i, v_j). To skip dead appearances efficiently, for each vertex v_i on the critical path, we maintain a list of successive dead appearances with respect to v_i, which allow us to skip each run of successive dead appearances in $O(1)$ time. After each time skipping successive dead appearances we can always generate a child graph of G corresponding to the next "non-dead" appearance. Thus l' child graphs of type $G(i, j)$ can be generated in $O(l')$ time. The details are as follows.

Let $v_{a(i)}$ and $v_{b(i)}$ be the first and last appearance of v_i on $C_o(G)$. Let P_i be the subpath from $v_{a(i)}$ to $v_{b(i)}$ on $C_o(G)$. A maximal subpath P_i^c of P_i is called a *dead path* of v_i if all appearances $v_c, v_{c+1}, \ldots$ on P_i^c are dead with respect to v_i. For example, the graph in Fig. 5 has 6 dead paths of v_3: $P_3^3 = (v_3, v_4, v_5, v_6)$, $P_3^8 = (v_8)$, $P_3^{10} = (v_{10})$, $P_3^{12} = (v_{12}, v_{13}, v_{14}, v_{15}, v_{16})$, $P_3^{18} = (v_{18}, v_{19}, v_{20}, v_{21})$, $P_3^{23} = (v_{23}, v_{24})$. They appear on $C_o(G)$ in this order. For each $v_i(0 \le i \le a - 1)$, we maintain all dead paths as a list, and we call the list *the zombie list* of v_i. Using the zombie list we can skip each run of successive dead appearances in $O(1)$ time. After each time we skip a dead path, we can always generate at least one child graph. Thus, we can generate each child graph of type $G(i, j)$ in $O(1)$ time.

Now we show how to prepare those data structures for each child graph.

Given a connected rooted plane graph G and the zombie list of each vertex on the critical path, we are going to generate all child graphs, and for each child graph we prepare the zombie list of each vertex on the new critical path by modifying the list for G.

We have the following two cases.

Case 1: Child graphs of type $G(i)$.

We have the following two cases.

Case 1(a): $i = a$.

The first removable edge of G is a pendant, since otherwise the first removable edge of G is a cycle edge and $G(i)$ is not a child graph of G. Appending the new edge to the critical path of G generates the critical path of $G(i)$. The zombie list of each v_l, $0 \le l \le a - 2$, in $G(i)$ is identical to the ones in G.

The zombie list of v_{a-1} in $G(i)$ is derived by dividing the first dead path P of v_{a-1} in G as follows. Let $P = (v_{a-1}, v_a, v_1', v_2', \ldots)$ then we divide P into two dead paths $P_1 = (v_{a-1}, v_a)$ and $P_2 = (v_a, v_1', v_2', \ldots)$. Note that adding the new edge generates one more appearance of v_a. See an example in Fig. 6(a). The dead path P_2^2 in Fig. 6(a) is divided into P_2^2 and P_2^3. Other dead paths of v_{a-1} in $G(i)$ are identical to the ones in G.

The zombie list of v_a consists of one dead path $P = (v_a, v_x, v_a)$, where v_x is the other end vertex of the new edge.

Thus we can modify the zombie list of each vertex on the critical path in $O(1)$ time.

Case 1(b): Otherwise.

The critical path of $G(i)$ is $v_0, v_1, \ldots, v_i, v_x$, where v_x is the other end vertex of the new edge.

The zombie list of each v_l, $0 \le l \le i - 1$ in $G(i)$ is identical to the zombie list of G.

The zombie list of v_i is derived by appending (v_i, v_x) as the prefix to the first dead path of v_i, where v_x is the other end vertex of the new edge. See an example in Fig. 6(b). By appending (v_2, v_x) in to the dead path P_2^2 of v_2 in G, the dead path P_2^2 of $G(i)$ is derived. Note that the other dead path of v_i in $G(i)$ is identical to the ones in G.

Thus we can modify the zombie list of each vertex on the critical path in $O(1)$ time.

Case 2: Child graphs of type $G(i, j)$.

The critical path of $G(i, j)$ is $v_0, v_1, \ldots, v_i, v_j$.

Note that $v_{i+1}, v_{i+2}, \ldots, v_{j-1}$ are not on $C_o(G(i, j))$. So we need not maintain the zombie list of those. Also each $v_{j+1}, v_{j+2}, \ldots, v_a$ are not on the critical path of $G(i, j)$. So we need not maintain the zombie lists of those.

The zombie list of each v_l, $0 \le l \le i - 1$, is identical to the zombie list of $G(i)$.

The zombie list of v_i is derived by removing dead paths of v_i up to v_j on $C_o(G)$. If v_{j+1} is dead with respect to v_i in G, then appending (v_i, v_j) into the dead path $P_i^{j+1} = (v_{j+1}, v_{j+2}, \ldots)$ generates the zombie list of v_i in $G(i, j)$. See an example in Fig. 6(c). By appending (v_2, v_5) into the dead path P_2^6 of v_2 in G, the dead path P_2^2 of v_2 in $G(i, j)$ is derived. Otherwise if v_{j+1} is not dead then we append a new dead path $P_i^i = (v_i, v_j)$ into the zombie list of v_i. Other dead paths remain as they are.

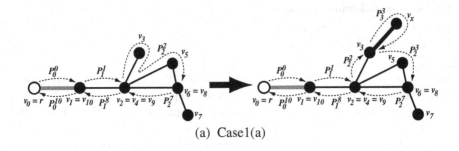

(a) Case1(a)

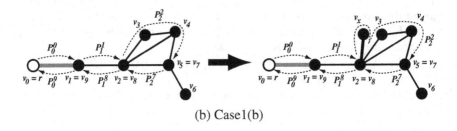

(b) Case1(b)

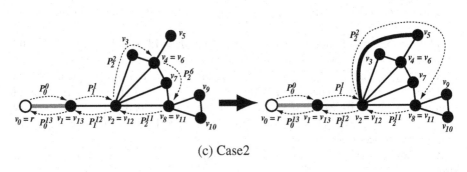

(c) Case2

Fig. 6. An update of a zombie list for (a), (b) $G(i)$ and (c) $G(i,j)$

Thus we can modify the zombie list of each vertex on the critical path in $O(1)$ time.

By the above case analysis, we can prepare the zombie list of each child graph of G in $O(1)$ time.

Next we estimate the space for zombie lists.

Since the number of dead paths of vertex v is bounded by the degree of v, the space to store the zombie lists of G is bounded by $O(m) = O(n)$.

By maintaining the zombie lists, if G has l' child graphs of type $G(i,j)$, we can find all such child graphs in $O(l')$ time. Thus, the algorithm runs in $O(g(m))$ time, where $g(m)$ is the number of nonisomorphic connected rooted plane graphs with at most m edges.

6   Conclusion

In this paper we have given a simple algorithm to generate all connected plane graphs with at most m edges. Our algorithm first defines a family tree whose vertices correspond to graphs, then outputs each graph without duplications by traversing the tree.

By implementing the algorithm one can compute the catalog of plane graphs.

References

1. Avis, D., Fukuda, K.: Reverse search for enumeration. Discrete Appl. Math. 65, 21–46 (1996)
2. Beyer, T., Hedetniemi, S.M.: Constant time generation of rooted trees. SIAM J. Comput. 9, 706–712 (1980)
3. Chrobak, M., Nakano, S.: Minimum-width grid drawings of plane graphs. Comput. Geom. Theory and Appl. 10, 29–54 (1998)
4. de Fraysseix, H., Pach, J., Pollack, R.: How to draw a planar graph on a grid. Combinatorica 10, 41–51 (1990)
5. Goldberg, L.A.: Efficient algorithms for listing combinatorial structures. Cambridge University Press, New York (1993)
6. Knuth, D.E.: The art of computer programming, vol. 4, fascicle 2, generating all tuples and permutations, Addison-Wesley (2005)
7. Kreher, D.L., Stinson, D.R.: Combinatorial algorithms. CRC Press, Boca Raton (1998)
8. Li, Z., Nakano, S.: Efficient generation of plane triangulations without repetitions. In: Orejas, F., Spirakis, P.G., van Leeuwen, J. (eds.) ICALP 2001. LNCS, vol. 2076, pp. 433–443. Springer, Heidelberg (2001)
9. McKay, B.D.: Isomorph-free exhaustive generation. J. Algorithms 26, 306–324 (1998)
10. Nakano, S.: Efficient generation of triconnected plane triangulations. Comput. Geom. Theory Appl. 27, 109–122 (2004)
11. Schnyder, W.: Embedding planar graphs on the grid. In: Proc. 1st ACM-SIAM SODA, pp. 138–148 (1990)
12. Stanley, R.P.: Enumerative combinatorics, vol. 1. Cambridge Univ. Press, Cambridge (1997)
13. Stanley, R.P.: Enumerative combinatorics, vol. 2. Cambridge Univ. Press, Cambridge (1999)
14. Wright, R.A., Richmond, B., Odlyzko, A., McKay, B.D.: Constant time generation of free trees. SIAM J. Comput. 15, 540–548 (1986)

Pairwise Compatibility Graphs

(Extended Abstract)

Muhammad Nur Yanhaona, K.S.M. Tozammel Hossain, and Md. Saidur Rahman

Department of Computer Science and Engineering
Bangladesh University of Engineering and Technology (BUET)
saidurrahman@cse.buet.ac.bd

Abstract. Given an edge weighted tree T and two non-negative real numbers d_{min} and d_{max}, a pairwise compatibility graph (PCG) of T is a graph $G = (V, E)$, where each vertex $u \in V$ corresponds to a leaf u of T and an edge $(u, v) \in E$ if and only if $d_{min} \leq distance(u, v) \leq d_{max}$ in T. In this paper we give some properties of these graphs. We establish a relationship between pairwise compatibility graphs and chordal graphs. We show that all chordless cycles and single chord cycles are pairwise compatibility graphs. We also provide a linear-time algorithm for constructing trees that can generate graphs having cycles as their maximal biconnected subgraphs as PCGs. The techniques that we used to identify various types of pairwise compatibility graphs are quite generic and may be useful to discover other properties of these graphs.

1 Introduction

Let T be an edge weighted tree and $G = (V, E)$ be a graph such that each vertex of G corresponds to a leaf of T and an edge $(u, v) \in E$ if and only if the distance between the two leaves of T corresponding to u and v is within a given range. We call G a pairwise compatibility graph (PCG) of T. Fig. 1(a) depicts an edge-weighted tree T and Fig. 1(b) depicts a pairwise compatibility graph G of T, where the given range of distance between a pair of leaves in T is from four to seven. The graph G has the edge

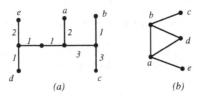

Fig. 1. (a) An edge weighted tree T and (b) a pairwise compatibility graph G of T

(a, b) since the distance between the leaves corresponding to a and b is six in T, but G does not contain the edge (a, c) since the distance between the leaves corresponding to a and c is larger than seven in T. All the remaining edges of G are drawn following the same rule. Given a PCG, the pairwise compatibility tree construction problem asks to

S.-i. Nakano and Md. S. Rahman (Eds.): WALCOM 2008, LNCS 4921, pp. 222–233, 2008.

construct an edge weighted tree that can generate the PCG. Meanwhile, the pairwise compatibility graph recognition problem asks to determine whether a given graph is a pairwise compatibility graph or not.

Pairwise compatibility graphs have applications in reconstruction of evolutionary relationships of a set of species from biological data (also called phylogeny) [JP04, L02]. Usually phylogenetic relationships are expressed as a tree, known as phylogenetic tree. Dealing with a sampling problem in a phylogenetic tree Kearney *et al.* introduced the concept of pairwise compatibility graphs [KMP03]. They showed that "the clique problem" is polynomially solvable for pairwise compatibility graphs if the pairwise compatibility tree construction problem can be solved in polynomial time. The clique problem asks to determine a maximum set of pairwise adjacent vertices in a graph [CLRS01]. It is an well known NP-complete problem that arises from many areas [PX94, BBPP99]. Thus the pairwise compatibility graph recognition problem and the pairwise compatibility tree construction problem have great potential.

There are several known specific cases of pairwise compatibility graphs, e.g., tree power graphs, phylogenetic k-root graphs [LJK00, KC98]. Some other graphs such as Steiner k-root graphs are also close in nature to pairwise compatibility graphs [LJK00]. However, to the best of our knowledge, structures of pairwise compatibility graphs and their relationships with other known graph classes have not been studied before, though Phillips showed that every graph of five vertices or less is a pairwise compatibility graph [P02].

In this paper, we determine a relationship between pairwise compatibility graphs and chordal graphs. We prove that all chordless cycles and single chord cycles are pairwise compatibility graphs. We also show that any graph having cycles as its maximal biconnected subgraphs is a pairwise compatibility graph and present a linear-time algorithm to solve the pairwise compatibility tree construction problem in such graphs. For all the graph classes that we identify as PCGs, the pairwise compatibility tree construction problem can be solved using some simple techniques.

The rest of the paper is organized as follows. Section 2 describes some of the definitions used in this paper. Section 3 presents some properties of pairwise compatibility graphs. Section 4 shows that every chordless cycle and every single chord cycle is a pairwise compatibility graph. Section 5 shows that any graph having cycles as its maximal biconnected subgraphs is a PCG. Finally, Section 6 concludes with discussion.

2 Preliminaries

In this section we define some terms used in this paper.

Let $G = (V, E)$ be a simple graph with vertex set V and edge set E. An edge between two vertices u and v is denoted by (u, v). A *path* $P_{uv} = w_0, w_1, \cdots, w_n$ is a sequence of distinct vertices in V such that $u = w_0, v = w_n$ and $(w_{i-1}, w_i) \in E$ for every $1 \le i \le n$. A *sub-path* of P_{uv} is a subsequence $P_{w_j w_k} = w_j, w_{j+1}, ..., w_k$ for some $0 \le j < k \le n$. An *internal vertex* of P_{uv} is any vertex other than u and v that is in P_{uv}. A *subgraph* of a graph $G = (V, E)$ is a graph $G' = (V', E')$ such that $V' \subseteq V$ and $E' \subseteq E$. If G' contains all the edges of G that join vertices in V', then G' is called the *subgraph induced by* V'. A graph G is *connected* if each pair of vertices belongs to a path otherwise it is disconnected. *Components* of G are its maximal connected subgraphs. A component of G is a *nontrivial component* if it has two or more vertices. A *cut vertex* of G is a vertex whose deletion increases the number of components. A *block* of G is a maximal connected subgraph of G that has no cut vertex. If G itself is connected and has no cut vertex then G is a block. An *independent set* of G is a set of pairwise nonadjacent vertices. The *block-cutpoint graph* of G is a bipartite graph H in which one partite set consists of the cut vertices of G, and the other has a vertex b_i for each block B_i of G. An edge (v, b_i) is in H if and only if $v \in B_i$.

A *cycle* $C = (w_1, w_2, \cdots, w_n = w_1)$ in a graph G is a sequence of distinct vertices starting and ending at the same vertex such that two vertices are adjacent if they appear consecutively in the list. The *length* of a cycle C is the number of its vertices. A cycle C is an *odd cycle* if the length of C is odd, otherwise C is an *even cycle*. A *chord* of a cycle C is an edge not in C whose endpoints lie on C. A *chordless cycle* is a cycle of length at least four in G that has no chord. A graph G is *chordal* if it is simple and has no chordless cycle. A *cut edge* of a graph G is an edge that belongs to no cycle. A *block-cycle graph* is a graph having cycles as its maximal biconnected subgraphs.

A *tree* T is a connected graph with no cycle. Vertices of degree one are called *leaves* and others are called *internal vertices*. A tree T is *weighted* if each edge is assigned a number as the weight of the edge. The weight of an edge (u, v) is denoted by $weight(u, v)$. A *subtree induced by a set of leaves* is entirely composed of those paths of T that connect only the leaves of that set. *The length of a path* P_{uv} in T, denoted by l_{uv}, is the sum of weights of edges of P_{uv}. A *caterpillar* is a tree in which a single path (*the spine*) is incident to or contains every edge. The *parent of a leaf* u denoted by $parent(u)$ is the internal vertex immediately adjacent to that leaf in the tree. In this paper every tree we consider is a weighted tree. We use the convention that if an edge of a tree has no number assigned to it then its default weight is one.

If T is an edge weighted tree then a pairwise compatibility graph of T with d_{min} and d_{max} is a graph $G = (V, E)$, where each vertex $v \in V$ corresponds to a leaf v of T, and there is an edge $(u, v) \in E$ if and only if the distance between leaves u and v in T satisfies the condition $d_{min} \leq distance(u, v) \leq d_{max}$. Here d_{min} and d_{max} are two nonnegative real numbers. We represent a pairwise compatibility graph of T with d_{min} and d_{max} by $PCG(T, d_{min}, d_{max})$.

3   Properties of Pairwise Compatibility Graphs

In this section we prove some properties of pairwise compatibility graphs.

Factors determining the structure of a PCG are the topology of the edge weighted tree T and values of d_{min} and d_{max}. Here we establish relationships between the structure of a PCG and the values of d_{min} and d_{max} that are used to construct it. Chordal graphs are one of the few classes of graphs for which the clique problem is solvable in polynomial time [W03]. Lin *et al.* proved that a class of graphs similar to PCG, known as "phylogenetic k-root graphs" are chordal graphs [LJK00]. However, PCGs are neither a subset of chordal graphs nor are they disjoint from that class [KMP03]. Here we show that a restricted set of PCGs that include phylogenetic k-root graphs are chordal graphs, as in the following theorem.

Theorem 1. *Let $G = (V, E)$ be a pairwise compatibility graph of a tree T with d_{min} and d_{max} where $distance(u, v) \geq d_{min}$ in T for every pair of vertices u, v that are in the same cycle of G. Then G is a chordal graph.* □

To prove Theorem 1 it is sufficient to prove that no chordless cycle can exist in G. Clearly the minimum length of a chordless cycle is four. In the following lemma we prove that no chordless cycle of length four can exist in G.

Lemma 1. *Let $G = (V, E)$ be a pairwise compatibility graph of a tree T with d_{min} and d_{max} where $distance(u, v) \geq d_{min}$ in T for every pair of vertices u, v that are in the same cycle of G. Then G does not contain any chordless cycle of length four.*

Proof. Assume for a contradiction that a chordless cycle $C = (a, b, c, d)$ of length four exists in G. Since a, b, c, d are leaves of T, unique paths P_{ab}, P_{ac}, P_{bc} and P_{ad} exist in T. Let p be the common internal vertex between P_{ac} and P_{ab} such that p is the furthest vertex from a, as illustrated in Fig. 2. Similarly, let q be the common internal vertex between P_{ac} and P_{ad} such that q is the furthest vertex from a. Since vertices a and c

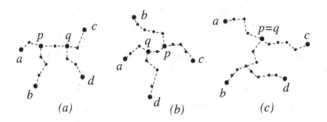

Fig. 2. Possible subtrees induced by leaves a, b, c, d

are not adjacent in G, $l_{ac} > d_{max}$. However, $l_{bc} \leq d_{max}$ since b and c are adjacent in G. Therefore, $l_{ac} > l_{bc}$. Since p is contained in P_{ac}, path P_{ac} can be decomposed into P_{ap} and P_{pc} such that $P_{ac} = P_{ap} + P_{pc}$. Similarly, $P_{bc} = P_{bp} + P_{pc}$. Since $l_{ac} > l_{bc}$, $l_{ap} + l_{pc} > l_{bp} + l_{pc}$. This implies $l_{ap} > l_{bp}$. Similarly, we can prove that $l_{cp} > l_{bp}$ since P_{ab} contains p and a and b are adjacent in G. Vertex q is either included in P_{pc} or it comes before p in the path from a to c. Without loss of generality, assume that q lies in P_{pc}. Since a and d are adjacent in G, $l_{ad} \leq d_{max}$ in T. By decomposing P_{ad} into P_{ap}, P_{pq} and P_{qd} we get $l_{ap} + l_{pq} + l_{qd} \leq d_{max}$. As $l_{ap} > l_{bp}$ from the above, we have $l_{bp} + l_{pq} + l_{qd} < d_{max}$. This implies $l_{bd} < d_{max}$. However, $l_{bd} \geq d_{min}$ since b and d are in a cycle. Therefore, b and d must be adjacent in G, a contradiction. □

We now show that G does not have any chordless cycle of length greater than four using following Lemmas 2, 3 and 4. Proofs of Lemmas 2 and 3 are omitted.

Lemma 2. *Let $G = (V, E)$ be a pairwise compatibility graph of a tree T with d_{min} and d_{max} where $distance(u, v) \geq d_{min}$ in T for every pair of vertices u, v that are in the same cycle of G. Let a, b, c be three consecutive vertices on a chordless cycle C of G, and let o be the common internal vertex of P_{ac} and P_{ab} in T such that o is the furthest vertex from a. Then $l_{ob} \leq \frac{d_{max}}{2}$ and $l_{ao} > l_{ob}$.* □

Lemma 3. *Let $G = (V, E)$ be a pairwise compatibility graph of a tree T with d_{min} and d_{max} where $distance(u, v) \geq d_{min}$ in T for every pair of vertices u, v that are in the same cycle of G. Let a, b, c, d be four vertices of a cycle in G such that $(a, d) \in E$ but $(b, d) \notin E$ and let o be the common internal vertex of P_{ac} and P_{ab} in T that is the furthest vertex from a. Then P_{ad} does not contain o and $l_{do} > l_{ob}$ if $l_{ob} \leq \frac{d_{max}}{2}$ and $l_{ao} > l_{ob}$.* □

Lemma 4. *Let $G = (V, E)$ be a pairwise compatibility graph of a tree T with d_{min} and d_{max} where $distance(u, v) \geq d_{min}$ in T for every pair of vertices u, v that are in the same cycle of G. Then G does not contain any chordless cycle of length greater than four.*

Proof. Assume for a contradiction that $C = (a_1, a_2, a_3, a_4, ..., a_n)$ is a chordless cycle of length greater than four in G. Select any three consecutive vertices a_{i-1}, a_i and a_{i+1} of C. Let o be the common internal vertex of $P_{a_{i-1}a_i}$ and $P_{a_{i-1}a_{i+1}}$ in T such that o is the furthest vertex from a_{i-1}. According to Lemma 2, $l_{oa_i} \leq \frac{d_{max}}{2}$ and $l_{a_{i-1}o} > l_{oa_i}$. According to Lemma 3, $P_{a_{i-1}a_{i-2}}$ does not contain o and $l_{a_{i-2}o} > l_{oa_i}$ since $(a_{i-2}, a_{i-1}) \in E$ and $(a_{i-2}, a_i) \notin E$. Applying Lemma 3 repeatedly for the rest of the vertices of C in the sequence $a_{i+1}, a_{i+2}, \cdots, a_{i-4}, a_{i-3}$ we can conclude that $P_{a_{i-4}a_{i-3}}$ in T does not contain o and $l_{a_{i-3}o} > l_{oa_i}$. However, $P_{a_{i-2}a_{i-3}}$ in T contains o, as illustrated in Fig. 3(c). Since a_{i-2} and a_i are not adjacent in C, $l_{a_{i-2}a_i} > d_{max}$ in T. Furthermore, $P_{a_{i-2}a_i}$ in T contains o. Therefore, $l_{a_{i-2}o} + l_{oa_i} > d_{max}$. This implies $l_{a_{i-2}o} > d_{max} - l_{oa_i}$. Since $l_{oa_i} \leq \frac{d_{max}}{2}$, $l_{a_{i-2}o} > \frac{d_{max}}{2}$. Similarly, we can prove that $l_{a_{i-3}o} > \frac{d_{max}}{2}$. Since $P_{a_{i-2}a_{i-3}}$ contains o, $l_{a_{i-2}a_{i-3}} = l_{a_{i-2}o} + l_{oa_{i-3}}$. Furthermore,

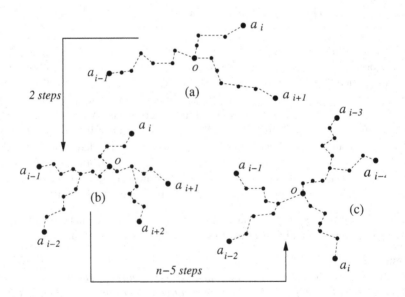

Fig. 3. (a) Subtree induced by a_{i-1}, a_i, a_{i+1}, (b) subtree induced by $a_{i-2}, a_{i-1}, a_i, a_{i+1}, a_{i+2}$, and (c) subtree induced by $a_{i-2}, a_{i-1}, a_i, a_{i-3}, a_{i-4}$

both $l_{a_{i-2}o}$ and $l_{oa_{i-3}}$ are larger than $\frac{d_{max}}{2}$. Therefore, $l_{a_{i-2}a_{i-3}} > d_{max}$. Hence, a_{i-1} and a_{i-3} cannot be adjacent in G and C cannot exist. $\qquad\square$

Lemma 1 and Lemma 4 immediately prove Theorem 1. $\qquad\square$

Let G be a PCG of a tree T with $d_{min} = x$ and $d_{max} = y$. Any other graph G' that is also constructed from T using $d_{min} \leq x$ and $d_{max} \geq y$ has all the edges of E and perhaps some additional edges. Since adding an edge in an existing graph never increases the number of components, G has at least as many components as in G'. However, the number of nontrivial components in G and G' cannot be related in such a straightforward manner. The following theorem states that the number of nontrivial components is not affected when d_{max} is set to a certain large value and only d_{min} is varied to construct different PCGs from a single tree (proof omitted).

Theorem 2. *Let $G = (V, E)$ be a pairwise compatibility graph of a tree T with $d_{min} \leq d_{max}$ where $d_{max} \geq distance(u, v)$ for every pair of leaves of T. Then G has at most one nontrivial component.* $\qquad\square$

4   Chordless Cycles and Single Chord Cycles

This section shows that every graph that is a chordless or a single chord cycle is a PCG. Kearney *et al.* posted an open problem - "Is any (or every) cycle of length greater than five a pairwise compatibility graph?" [KMP03]. We give an affirmative answer to that open problem as in the following theorem.

Theorem 3. *Let d_{min} and d_{max} be any two positive real numbers such that $d_{min} \leq d_{max}$ and let G be a cycle. Then there is a tree T such that $G = PCG(T, d_{min}, d_{max})$.*

Proof. We will show that we can construct a tree T and assign weights to the edges of T such that $G = PCG(T, d_{min}, d_{max})$. We have to consider the following two cases depending on whether G is an odd or even cycle.

Case 1: G is an odd cycle
To generate an arbitrary odd cycle $G = (a_1, a_2, \cdots, a_{2n}, a_{2n+1})$ we can construct a tree T, as illustrated in Fig. 4(a). In T the distance between any two successive odd indexed leaves a_{2i-1} and a_{2i+1} is $d_i + 2w$. We denote by d_1 and d_n the *beginning and ending intervals* respectively. We call d_i for $1 < i < n$, an *internal interval*. We call the path connecting a_1 and a_{2n+1} *the spine of T*. Since a_1 and a_{2n+1} are adjacent in G, the length of the spine of T must be at least d_{min}. Since the length of the spine of T is $\sum_{i=1}^{n} d_i + 2w$, we assign the value of d_i so that $\sum_{i=1}^{n} d_i + 2w = d_{min}$ holds. Therefore, for every pair of odd indexed leaves $\{a_i, a_j\}$ other than a_1 and a_{2n+1} in T, $distance(a_i, a_j) < d_{min}$. Every even indexed vertex a_{2i} is adjacent only to the odd indexed vertices a_{2i-1} and a_{2i+1} in G. In the tree T, $distance(a_{2i}, a_{2i-1}) = distance(a_{2i}, a_{2i+1}) = k_i + \frac{d_i}{2} + w$. In order to satisfy the adjacency relationships that are present in the cycle, we assign value of each k_i in T in such a way that the relations $d_{min} \leq k_i + \frac{d_i}{2} + w \leq d_{max}$, $k_i + \frac{d_i}{2} + d_{i+1} + w > d_{max}$ and $k_i + \frac{d_i}{2} + d_{i-1} + w > d_{max}$ hold in T. For example, we can assign $d_i = \frac{d_{min}}{n+1}$, $w = \frac{d_i}{2}$ and $k_i = d_{max} - d_i$ for $1 \leq i \leq n$ to satisfy all the conditions mentioned above. Hence, $G = PCG(T, d_{min}, d_{max})$.

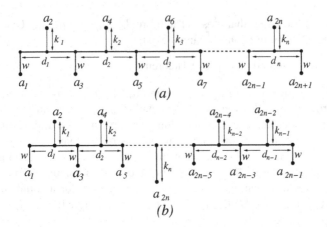

Fig. 4. (a) Generic tree organization for odd cycles, (b) generic tree organization for even cycles

Case 2: G is an even cycle

To generate an arbitrary even cycle $G = (a_1, a_2, \cdots, a_{2n-1}, a_{2n})$ we can construct a tree T, as illustrated in Fig. 4(b). Similar to the case of odd cycles, we denote by d_1 and d_{n-1} the *beginning and ending intervals* respectively and call rest of the d_i for $1 < i < n-1$ an *internal interval*. The only exception from Case 1 is the placement of the leaf a_{2n}. We call the path connecting a_1 and a_{2n-1} *the spine of T*. Since no two odd indexed vertices are adjacent to each other in G, the length of the spine of T must be less than d_{min}. Therefore, we assign values of w and each d_i so that the condition $d_{min} > \sum_{i=1}^{n-1} d_i + 2w$ is satisfied. For each even indexed vertex other than a_{2n}, we assign the value of k_i in T similarly to the case of odd cycles. Since a_{2n} is adjacent only to a_1 and a_{2n-1} in G, we connect a_{2n} through an edge of weight k_n to the midpoint of the spine of T and assign the value of k_n so that $d_{min} \leq distance(a_1, a_{2n}), distance(a_{2n-1}, a_{2n}) \leq d_{max}$ in T. For example, we can assign $d_i = \frac{d_{min}}{n+1}$, $w = \frac{d_i}{2}$, $k_i = d_{max} - d_i$ for $1 \leq i \leq n-1$ and $k_n = d_{min} - \frac{nd_i}{2}$ to satisfy all these conditions. Hence, $G = PCG(T, d_{min}, d_{max})$. $\qquad\square$

We next use the result of Theorem 3 to show that every graph that contains only a cycle with a single chord in it is a PCG, as in the following theorem (proof omitted).

Theorem 4. *Let G be a cycle with a single chord. Then G is a pairwise compatibility graph.* $\qquad\square$

5   Block-Cycle Graphs Are Pairwise Compatibility Graphs

In this section we generalize the result of Section 4 for block-cycle graphs as in the following theorem.

Theorem 5. *Every block-cycle graph is a pairwise compatibility graph.* $\qquad\square$

In the rest of this section we give a constructive proof of Theorem 5. It is sufficient to prove that for any given block-cycle graph G there is a tree T such that G is a PCG of

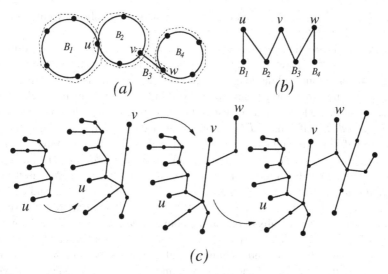

Fig. 5. (a) A block-cycle graph G, (b) block-cutpoint graph G' of G, and (c) incremental construction of a tree T

T. We first assume that G is a connected block cycle graph and later we will consider the case where G is disconnected. Let G' be the block-cutpoint graph of G. Since G is connected, G' is a tree [W03]. Fig. 5(b) illustrates such a block-cutpoint graph G' of a block-cycle graph G in Fig. 5(a). We incrementally construct a tree T that has G as a PCG. Initially we have an empty tree T and begin a depth first search (DFS) in G'. The first time we encounter a block in our depth first search, we create a tree for that block and add that tree as a branch of T. Whenever we encounter a cut vertex we do nothing. Fig. 5(c) illustrates the steps of constructing the tree T for the block-cycle graph G of Fig. 5(a). We start the depth first search from B_1 of G' and construct a tree T for the corresponding block. When we encounter B_2 of G' we add a new branch in T for corresponding block of G. Similarly, we encounter vertices B_3 and B_4 in sequence and add branches in T for the corresponding blocks respectively. Hence, after every step the tree T have a connected induced subgraph of G as its PCG. When the DFS is complete G becomes a PCG of T. For such a scheme to work correctly two constraints must be satisfied. Firstly, every branch that is added in the tree must be able to generate its corresponding block as a PCG for the globally chosen values of d_{min} and d_{max}. Secondly, the distance between the leaves of any two branches must satisfy the adjacency relationships between the corresponding vertices of the graph.

Before giving the detail of our proof we need some definitions. Let G be a block-cycle graph. Then l is the smallest integer such that l is greater than or equal to half the length of a maximal cycle in G. We assume that $d_{min} \geq 3$ if G has no cycle and always assign d_{max} the value $d_{min} + 2$. We also assume that each edge of every tree has integer weight. If T_1 and T_2 are two trees and C_1 and C_2 are two cycles in G such that C_1 is a PCG is of T_1 and C_2 is a PCG of T_2, we call T_1 and T_2 are *connection consistent* when a tree T can be constructed by connecting T_1 and T_2 that can generate the subgraph $C_1 \bigcup C_2$ as a *PCG*. Furthermore, we define the following categories of trees for generating cycles.

We construct all trees that generate cycles as *PCG*s according to the procedure described in the proof of Theorem 3. In those trees leaves are connected with their parents by an edge of weight one and the length of each internal interval is four. Furthermore, every leaf of the tree that corresponds to an even indexed vertex of the cycle is at a distance d_{max} from the leaves that correspond to its adjacent odd indexed vertices. We call a tree T an *odd tree* if T has a *PCG* which is an odd cycle. Similarly, we call a tree T an *even tree* if T has a *PCG* which is an even cycle. We call an odd tree T whose *PCG* is a cycle $C = (a_1, a_2, a_3, \cdots, a_{2k}, a_{2k+1})$ a *balanced odd tree* if its beginning interval $d_1 = 16l$ and ending interval $d_k = d_{max} - d_1 - 4(k - 2) - 2$. We call the odd tree T a *head-long odd tree* if $d_k = 4$ and $d_1 = d_{max} - 4(k - 1) - 2$, and call T a *tail-long odd tree* if $d_1 = 4$ and $d_k = d_{max} - 4(k - 1) - 2$. Every odd tree that we will use in the rest of this section falls in one of the above three categories. We call an even tree T whose *PCG* is a cycle $C = (a_1, a_2, a_3, \cdots, a_{2k-1}, a_{2k})$ a *balanced even tree* if its beginning interval $d_1 = 16l$ and ending interval $d_{k-1} = 4$. We call the even tree T a *head-long even tree* if $d_1 = 20l$ and $d_{k-1} = 4$, and call T a *tail-long even tree* when $d_1 = 4$ and $d_k = 20l$. Every even tree that we will use in the rest of this section falls in one of the above three categories.

The following lemma shows how we construct connection consistent branches for cycles that share a vertex with each other and determine an appropriate value for d_{min}.

Lemma 5. *Let* $C_1 = (a_1, a_2, \cdots, a_k)$ *and* $C_2 = (b_1, b_2, \cdots, b_l)$ *be two cycles of a connected block-cycle graph* G *such that* $b_1 = a_i$ *for some* $1 \leq i \leq k$. *Let* T_1 *be a tree such that* $C_1 = PCG(T_1, d_{min}, d_{min} + 2)$. *Then there exists a tree* T_2 *such that* $C_2 = PCG(T_2, d_{min}, d_{min} + 2)$ *and* T_1 *and* T_2 *are connection consistent if* $d_{min} = 32l - 6$ *where* l *is the smallest integer that is greater than or equal to half the length of a maximal cycle in* G.

Proof. Assume that there is a tree T_2 that can generate C_2 as a *PCG*. According to the definitions we have given before, T_1 and T_2 both fall in one of our defined categories. Since C_1 is a *PCG* of T_1 and C_2 is a *PCG* of T_2, there are leaves correspond to $a_i = b_1$ in both of the trees. Let w_1 be the parent of a_i in T_1 and w_2 be the parent of b_1 in T_2. We connect T_1 and T_2 by merging the edge (w_1, a_i) of T_1 and the edge (w_2, b_1) of T_2. To prove that T_1 and T_2 are connection consistent it is sufficient to prove that the distance between any two leaves that are taken from T_1 and T_2 respectively, is either larger than d_{max} or less than d_{min}. We now show how to construct the tree T_2 so that the above condition is satisfied. Meanwhile, we will determine the value of d_{min} that is consistent with our assumption. If a_i is not the first or last odd indexed leaf of T_1 then we will not construct T_2 as a tail long tree. It is now trivial to prove that any even indexed leaf of either of T_1 and T_2 is always at a distance larger than d_{max} from any leaf of the other one. Therefore, from now on we will only consider odd indexed leaves. We have to consider the following cases depending on the category in which T_1 may belong.

Case 1: T_1 *is a head long even or odd tree.*
If $a_i = a_1$ then we construct a head long even or odd tree T_2 for C_2 depending on whether C_2 is even or odd. Then we connect T_1 and T_2 to form a new tree T, as

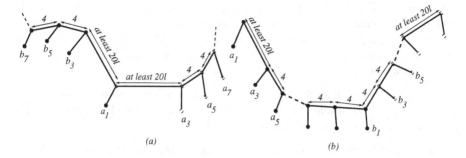

Fig. 6. (a) Adding a head long tree at the head of a head long tree, and (b) adding a tail long tree at the tail of a head long tree

illustrated in Fig. 6(a). In a head long even tree the longest distance between a pair of odd indexed leaves is $20l + 4(l-2) + 2$. Since the vertices corresponding to odd indexed leaves of a head long even tree are not adjacent in the cycle, we have to assign a value to d_{min} such that $d_{min} > 20l + 4(l-2) + 2$. The smallest distance of a leaf of T_2 from a leaf of T_1 is the distance between a_3 and b_3 where $distance(a_3, b_3) \geq 40l + 2$. Therefore, if we assign to d_{min} a value that is within the range form $20l + 4(l-2) + 4$ to $40l - 2$ then T_1 and T_2 are connection consistent. If a_i is the last odd indexed leaf of T_1 then we construct a tail long even or odd tree T_2 for C_2 depending on whether C_2 is even or odd. Then we connect T_2 and T_1 to form a new tree T, as illustrated in Fig. 6(b). In T the distance between the last odd indexed leaf of T_2 and a_1 of T_1 is larger than $40l + 2$. Since we assign a value to d_{min} within the range form $20l + 4(l-2) + 4$ to $40l - 2$ and $d_{max} = d_{min} + 2$, those two leaves are at a distance larger than d_{max}. The maximum distance of a_1 from any odd indexed leaf of T_2 other than the last odd indexed leaf is $20l + 8(l-2) + 2$. Hence, if we increase the lower limit of d_{min} from $20l + 4(l-2) + 4$ to $20l + 8(l-2) + 4$ and assign a value to d_{min} accordingly then every other odd indexed leaf of T_2 will be at a distance less than d_{min} from a_1. Again, T_1 and T_2 are connection consistent. If a_i is any leaf of T_1 other than the first and the last odd indexed leaves then we create a tree T_2 similarly. The proof of this sub case is omitted in this extended abstract.

The proofs for the remaining cases are omitted in this extended abstract.   □

We now present the detail of our tree construction procedure. For any given connected block-cycle graph we construct a tree block by block. We start from an empty tree T. Whenever we encounters a new block we construct a tree for generating that block, then this newly constructed tree is added in T as a branch. The correctness of our construction is based on the ability of constructing a branch at each step, for the corresponding block, that can be added in T without violating any condition. In a block-cycle graph every block is either a cut edge or a cycle. Furthermore, a cycle may or may not share vertices with other cycles. Therefore, we have to consider three cases depending on the type of a block. The following lemma states that we can construct a branch, that can be connected with the rest of the tree consistently in the case where the corresponding block is a cut edge (proof omitted).

Lemma 6. *Let H and H' be two connected induced subgraphs of a connected block-cycle graph G such that H' is formed by adding a cut edge B with H and $H = PCG(T, d_{min}, d_{min}+2)$ for some tree T where $d_{min} = 32l-6$ and l is the smallest integer that is greater than or equal to half the length of a maximal cycle in G. Then there is a tree T' such that $H' = PCG(T', d_{min}, d_{min}+2)$.*   □

The following lemma states that we can construct a branch that can be connected with the rest of the tree consistently in the case where the corresponding block is a cycle that shares no vertex with other cycles (proof omitted).

Lemma 7. *Let H and H' be two connected induced subgraphs of a connected block-cycle graph G such that H' is formed by adding a new cycle C in H and u be the cut vertex shared by H and C. Furthermore, assume that C shares no vertex with any cycle that is present in H and $H = PCG(T, d_{min}, d_{min} + 2)$ for some tree T where $d_{min} = 32l - 6$ and l is the smallest integer that is greater than or equal to half the length of a maximal cycle in G. Then $H' = PCG(T', d_{min}, d_{max})$ for some tree T'.* □

We now prove that we can generate a branch that can be connected with the rest of the tree consistently in case of a cycle that shares vertices with other cycles. Whenever the we encounter a new cycle that shares a vertex with another already visited cycle, we construct a tree for the new one that is connection consistent with the older one. Let two branches of the tree generating two different cycles as a PCG be isolated from each other. Within the time of visiting those two cycles we must add branches for at least one cut edge and one cycle that shares no vertex with other cycles in the tree using Lemma 6 and Lemma 7, respectively. Hence, any path in the tree connecting two leaves of these two different branches is larger than d_{max}. Therefore, to prove that a branch that can be connected with the rest of the tree consistently in the case of a cycle that shares vertices with other cycles, it is sufficient to prove the following lemma (proof omitted).

Lemma 8. *Let $C_0, C_1, C_2, \cdots, C_k$ be a sequence of distinct cycles of a block-cycle graph G such that C_i shares a vertex v_i with C_{i-1} for $1 \leq i \leq k$ and all of this vertices are distinct. Let T_i for $0 \leq i \leq k$, be a tree that generates C_i as a PCG taking v_i as the first odd indexed vertex where $d_{min} = 32l-6$ and l is the smallest integer that is greater than or equal to half the length of a maximal cycle in G. Then there is a tree T that can generate the subgraph induced by the vertices of these cycles as a PCG if T_i and T_{i+1} are connection consistent for $0 \leq i \leq k$.*   □

From Lemma 6, Lemma 7 and Lemma 8 it is evident that we can always construct a tree that has a given connected block-cycle graph as a PCG. For disconnected block-cycle graphs we can easily construct a tree by connecting those trees that generate the components of the graphs as their $PCGs$. Hence we have proved Theorem 5.

One can easily implement the tree construction procedure in linear time and hence the following theorem holds.

Theorem 6. *Let G be a block-cycle graph of n vertices. Let l be the smallest integer that is greater than or equal to half the length of a maximal cycle in G. Then one can construct a tree T in linear time such that $G = PCG(T, d_{min}, d_{min}+2)$ where $d_{min} = 32l - 6$.*   □

6 Conclusion

In this paper we viewed the structure of PCGs in different ways to identify relationships between them and some other known graph classes. We showed that some restricted subset of PCGs are chordal graphs. We also constructively proved that all chordless cycles as well as single chord cycle are PCGs. We also developed a procedure for constructing trees that can generate graphs having cycles as their maximum biconnected subgraphs. Most of our study resulted in finding new techniques for constructing trees that can generate specific classes of graphs. Those approaches that we used to develop these techniques can be used to establish relationships between PCGs and many other graph classes. We suggest interested researchers to investigate those relationships.

References

[CLRS01] Cormen, T.H., Leiserson, C.E., Rivest, R.L., Stein, C.: Introduction to Algorithms. The MIT Press, Cambridge (2001)

[BBPP99] Bomze, I.M., Budinich, M., Pardalos, P.M., Pelillo, M.: Handbook of Combinatorial Optimization, vol. 4. Kluwer Academic Publishers, Boston, MA (1999)

[HMPV00] Habib, M., McConnell, R., Paul, C., Viennot, L.: Lex-BFS and partition refinement, with applications to transitive orientation, interval graph recognition, and consecutive ones testing. Theoretical Computer Science 234, 59–84 (2000)

[JP04] Jones, N.C., Pevzner, P.A.: An Introduction to Bioinformatics Algorithms. The MIT Press, Cambridge (2004)

[KMP03] Kearney, P., Munro, J.I., Phillips, D.: Efficient generation of uniform samples from phylogenetic trees. In: Benson, G., Page, R.D.M. (eds.) WABI 2003. LNCS (LNBI), vol. 2812, pp. 177–189. Springer, Heidelberg (2003)

[KC98] Kearney, P., Corneil, D.G.: Tree powers. Journal of Algorithms, 111–131 (1998)

[L02] Lesk, A.M.: Introduction to Bioinformatics. Oxford University Press, Oxford (2002)

[LJK00] Lin, G.H., Jiang, T., Kearney, P.E.: Phylogenetic k-root and steiner k-root. In: Lee, D.T., Teng, S.-H. (eds.) ISAAC 2000. LNCS, vol. 1969, pp. 539–551. Springer, Heidelberg (2000)

[P02] Phillips, D.: Uniform sampling from phylogenetic trees, Master's thesis, University of Waterloo (August 2002)

[PX94] Pardalos, M., Xue, J.: The maximum clique problem. Journal of Global Optimization, 301–328 (1994)

[W03] West, D.B.: Introduction to Graph Theory. Prentice Hall of India, New Delhi (2003)

Multilevel Bandwidth and Radio Labelings of Graphs

Riadh Khennoufa and Olivier Togni

LE2I, UMR CNRS 5158
Université de Bourgogne, 21078 Dijon cedex, France
Riadh.Khennoufa@u-bourgogne.fr, Olivier.Togni@u-bourgogne.fr

Abstract. This paper introduces a generalization of the graph bandwidth parameter: for a graph G and an integer $k \leq \mathrm{diam}(G)$, the k-level bandwidth $B^k(G)$ of G is defined by
$B^k(G) = \min_\gamma \max\{|\gamma(x) - \gamma(y)| - d(x,y) + 1 : x, y \in V(G), d(x,y) \leq k\}$, the minimum being taken among all proper numberings γ of the vertices of G.

We present general bounds on $B^k(G)$ along with more specific results for $k = 2$ and the exact value for $k = \mathrm{diam}(G)$. We also exhibit relations between the k-level bandwidth and radio k-labelings of graphs from which we derive a upper bound for the radio number of an arbitrary graph.

Keywords: generalized graph bandwidth, radio labeling, frequency assignment.

1 Introduction

Let G be a connected graph. The distance between two vertices u and v of G is denoted by $d_G(u,v)$ or simply $d(u,v)$ and the diameter of G is denoted by $\mathrm{diam}(G)$.

The graph bandwidth problem is an old and well studied NP-complete problem (see e.g. [10,11,4,9,5]) that remains NP-complete for several simple graphs like some special types of trees. This problem arises from sparse matrix computation, coding theory, and circuit layout of VLSI design. The *bandwidth* $B(G)$ of a graph G is the minimum of the quantity $\max\{|\gamma(x) - \gamma(y)| : xy \in E(G)\}$ taken over all proper numberings γ of G.

In this paper, we introduce a generalisation of the bandwidth parameter called *k-level bandwidth* which consists, for a proper numbering of the vertices, in taking into acount not only the differences of the labels of adjacent vertices, but also of vertices at distance $i, 1 \leq i \leq k$: for a graph G and an integer $k \leq \mathrm{diam}(G)$, the *k-level bandwidth* $B^k(G, \gamma)$ of a proper numbering of the vertices of a graph G is defined by

$$B^k(G, \gamma) = \max\{|\gamma(x) - \gamma(y)| - d(x,y) + 1 : x, y \in V(G), d(x,y) \leq k\}.$$

S.-i. Nakano and Md. S. Rahman (Eds.): WALCOM 2008, LNCS 4921, pp. 234–239, 2008.

The *k-level bandwidth* $B^k(G)$ of G is

$$B^k(G) = \min\{B^k(G, \gamma) : \gamma \text{ is a proper numbering of } V(G)\}.$$

A numbering γ for which $B^k(G, \gamma) = B^k(G)$ is said to be a *k-level bandwidth numbering*.

We shall also use the notation $B^*(G) = B^{\mathrm{diam}(G)}(G)$.

Thus, the 1-level bandwidth corresponds with the graph bandwidth: $B^1(G) = B(G)$ while the 2-level bandwidth B^2 generalises the bandwidth in an analogous manner as the $\lambda_{2,1}$ parameter for the chromatic number.

Figure 1 presents three numberings of a graph G: the left one γ_1 is such that $B(G, \gamma_1) = 4, B^2(G, \gamma_1) = 5$, the one on the center γ_2 is such that $B(G, \gamma_2) = 4, B^2(G, \gamma_2) = 6$ and the right one γ_3 is such that $B(G, \gamma_3) = 3, B^2(G, \gamma_3) = 5$.

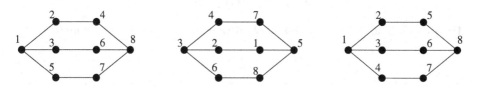

Fig. 1. Example of numberings of a graph with different bandwidths and 2-level bandwidths

Other labelings of interest since few years are radio labelings: for a graph G and an integer k, a *radio k-labeling* f of G is an assignment of non negative integers to the vertices of G such that

$$|f(u) - f(v)| \geq k + 1 - d(u, v),$$

for every two distinct vertices u and v of G. The span of the function f denoted by $\lambda^k(f)$, is $\max\{f(x) - f(y) : x, y \in V(G)\}$. The *radio k-chromatic number* $\lambda^k(G)$ of G is the minimum span of all radio k-labelings of G.

Radio k-labelings were introduced by Chartrand et al. [1], motivated by radio channel assignment problems with interference constraints. Quite few results are known concerning radio k-labelings. The radio k-chromatic number for paths was studied in [1], where lower and upper bounds were given. These bounds have been improved in [6].

Radio k-labelings generalize many other graph labelings. A radio 1-labeling is a proper vertex-colouring and $\lambda^1(G) = \chi(G) - 1$. For $k = 2$, the radio 2-labeling problem corresponds to the well studied $L(2, 1)$-labeling problem. For $k = \mathrm{diam}(G) - 1$, radio k-labelings were studied under the name of *(radio) antipodal labelings* [1,3,7,8]. A radio k-labeling with $k = \mathrm{diam}(G)$ is known as a *radio labeling*. Radio labelings of paths and of cycles have been studied in [2,12].

The aim of this paper is to introduce the k-level bandwidth parameter, to raise links between it and the radio k-labeling problem and to present bounds on the k-level bandwidth of some graphs.

2 Multilevel Bandwidth

To begin, we claim that for paths, cycles and complete bipartite graphs, the k-level bandwidth is easy to determine:

Claim 1. $-$ For any $k \leq n-1$, $B^k(P_n) = 1$;
 $-$ for any $k < \frac{n}{2}$, $B^k(C_n) = k+1$ and $B^p(C_{2p}) = p$;
 $-$ $B^2(K_{m,n}) = m+n-2$.

If G^k stands for the kth power of G (i.e. the graph with the same vertex set as G and with edges between vertices at distance at most k in G), then the following relation is easily seen:

Proposition 1. For any graph G and any $k \leq diam(G)$,

$$B(G) \leq B^k(G) \leq B(G^k).$$

Proposition 2. For any graph G of order n and any k, $1 \leq k \leq diam(G)$,

$$B^k(G) \geq \frac{k(n-1)}{diam(G)} - k + 1.$$

Proof. Let $\gamma : V(G) \longrightarrow \{1, \ldots, n\}$ be a k-level bandwidth numbering of G and let $u = \gamma^{-1}(1)$ and $v = \gamma^{-1}(n)$.

If $d(u,v) \leq k$, then we have $B^k(G, \gamma) \geq n - 1 - d(u,v) + 1 \geq n - k \geq \frac{k(n-1)}{diam(G)} - k + 1$ since $k \leq diam(G)$.

If $d(u,v) > k$, then by the pigeonhole principle, there exist two vertices x and y at mutual distance k along a shortest path between u and v such that $|\gamma(x) - \gamma(y)| \geq \frac{k(n-1)}{diam(G)} - k + 1$. $\square$

The Cartesian product $G \square H$ of graphs G and H is the graph with vertex set $V(G) \times V(H)$ and edge set $E(G \square H) = \{((a,x)(b,y)), ab \in E(G)$ and $x = y$ or $xy \in E(H)$ and $a = b\}$.

The following theorem generalize the known upper bound (see [5]) on the bandwidth of the Cartesian product of two graphs:

Theorem 1. For any graphs G and H, and for any positive integer $k \geq 1$,

$$B^k(G \square H) \leq |V(G)|B^k(H) + \max\{(|V(G)|-1)(k-1), B^k(G) + (|V(G)|-1)(k-2)-1\}.$$

Proof. The numbering of $G \square H$ which gives this upper bound is $\gamma(a,x) = \gamma_G(a) + (\gamma_H(x)-1)|V(G)|$, where γ_G and γ_H are k-level bandwidth numberings of G and of H, respectively. Due to space constraints, the rest of the proof is left to the reader.

Corollary 1. For any graph G of diameter at least 2,

$$B^2(G \square K_2) \leq 2B^2(G) + 1.$$

Theorem 2. *For any graph G of order n and any $k \leq diam(G)$,*

$$B^k(G) \leq n - diam(G).$$

Proof. Let u and v be vertices of G such that $d(u,v) = diam(G)$ and let $u_0 = u, u_1, u_2, \ldots, u_{n-1} = v$ be a distance ordering of $V(G)$ from u_0, i.e. if $i < j$ then $d(u_0, u_i) \leq d(u_0, u_j)$. Let also $V_i = \{u_j \in V(G), d(u_0, u_j) = i\}$ and $n_i = |V_i|$, $0 \leq i \leq diam(G)$.

Thus, the V_i partition $V(G)$ into $diam(G) + 1$ levels, each level V_i represent the set of vertices which are at distance i with the vertex u_0.

Now, we show that the simple numbering γ of the vertices of G given by $\gamma(u_i) = i + 1$ attains the desired bandwidth, i.e. that the the condition $|\gamma(y) - \gamma(x)| - d(x,y) + 1 \leq n - diam(G)$ is verified for any two vertices x and y, with $d(x,y) \leq k$.

Assume w.l.o.g. that $x \in V_i(G)$ and $y \in V_j(G)$ for some i and j, $0 \leq i \leq j \leq diam(G)$ and $j - i \leq k$. Then we have

$$\gamma(x) \geq 1 + \sum_{\ell=0}^{i-1} n_\ell, \text{ and}$$
$$\gamma(y) \leq \sum_{\ell=0}^{j} n_\ell.$$

Thus,

$$\gamma(y) - \gamma(x) - d(x,y) + 1 \leq n - diam(G) \Leftrightarrow$$
$$\sum_{\ell=0}^{j} n_\ell - 1 - \sum_{\ell=0}^{i-1} n_\ell - d(x,y) + 1 \leq n - diam(G) \Leftrightarrow$$
$$\sum_{\ell=i}^{j} n_\ell - d(x,y) \leq n - diam(G).$$

As $d(x,y) \geq j-i$, we have $\sum_{\ell=i}^{j} n_\ell - d(x,y) \leq \sum_{\ell=i}^{j}(n_\ell - 1) + 1 \leq \sum_{\ell=0}^{diam(G)}(n_\ell - 1) + 1 = \sum_{\ell=0}^{diam(G)} n_\ell + \sum_{\ell=0}^{diam(G)}(1) - 1 = n - diam(G)$. Hence the above inequality is verified and we obtain $B^k(G) \leq n - diam(G)$. $\square$

Combining Theorem 2 and Proposition 2 with $k = diam(G)$, we obtain the exact value of $B^*(G)$ i.e. of $B^{diam(G)}(G)$:

Corollary 2. *For any graph G of order n,*

$$B^*(G) = n - diam(G).$$

3   Relation Between the Radio k-Chromatic Number and the k-Level Bandwidth

Theorem 3. *For any graph G of order n and for any positive integer $k \geq 1$,*

$$\lambda^k(G) \leq \lambda^{k+B^k(G)-1}(P_n).$$

Proof. Let γ be a k-level bandwidth numbering of G and let $f : V(P_n) \to \{0, \ldots, \lambda^{k'}(P_n) - 1\}$ be a radio k'-labeling of P_n, with $k' = k + B^k(G) - 1$.

Consider the vertex ordering $u_1, u_2, \ldots, u_n$ of G induced by γ: $u_i = \gamma^{-1}(i)$ and label each vertex u_i using f, as if $u_1, \ldots, u_n$ was a path.

Then, we have for $i > j$,

$$|f(u_i) - f(u_j)| \geq k' + 1 - d_{P_n}(u_i, u_j) = k' + 1 - (i - j) = k + B^k(G) - (i - j).$$

As $B^k(G) \geq |\gamma(u_i) - \gamma(u_j)| - d_G(u_i, u_j) + 1 = (i - j) - d_G(u_i, u_j) + 1$, we obtain

$$|f(u_i) - f(u_j)| \geq k + (i - j) - d_G(u_i, u_j) + 1 - (i - j) = k + 1 - d_G(u_i, u_j),$$

and thus f is a radio k-labeling of G.   □

This result, along with the upper bound of Chartrand et al. for the radio k-chromatic number of the path [1] yield the following:

Corollary 3. *For any graph G of order n and for any positive integer $1 \leq k \leq diam(G)$,*

$$\lambda^k(G) \leq \frac{1}{2}(B^k(G)(B^k(G) + 2k) + k^2 - 1).$$

Corollary 4. *For any graph G of order n and diameter D,*

$$\lambda^D(G) \leq \begin{cases} \frac{1}{2}(n^2 - 2n + 2) & \text{if } n \text{ is even,} \\ \frac{1}{2}(n^2 - 2n + 5) & \text{if } n \text{ is odd.} \end{cases}$$

Proof. By using the result given in Theorem 2 and applying it in Theorem 3 with $k = diam(G) = D$, we obtain $\lambda^D(G) \leq \lambda^{D+B^*(G)-1}(P_n) = \lambda^{n-1}(P_n)$. Liu and Zhu [12], have shown that $\lambda^{n-1}(P_n) = \begin{cases} \frac{1}{2}(n^2 - 2n + 2) & \text{if } n \text{ is even,} \\ \frac{1}{2}(n^2 - 2n + 5) & \text{if } n \text{ is odd.} \end{cases}$ □

4   The 2-Level Bandwidth B^2

For the 2-level bandwidth, one can derive a better lower bound than that of Proposition 2 in some cases by refining the argument of the proof:

Proposition 3. *For any graph G of diameter at least 2,*

$$B^2(G) \geq 2 \left\lceil \frac{n-1}{diam(G)} - \frac{1}{2} \right\rceil - 1.$$

Since it is easily seen that $B^2(C_4 = H_2) = 2$, Corollary 1 gives a upper bound for the 2-level bandwidth of the hypercube:

Proposition 4. *For the n-dimensional hypercube H_n, $n \geq 2$,*

$$B^2(H_n) \leq 2^{n-1} + n - 2.$$

5   Concluding Remarks

We have introduced the k-level bandwidth of a graph and presented first results about it. Nevertheless, many questions and problems remain open. Among all:

- We have shown that the k-level bandwidth is easy to determine when $k = \mathrm{diam}(G)$ and it is known that computing the 1-level bandwidth is NP-complete. An open problem is thus to determine the algorithmic complexity of computing $B^k(G)$ for $2 \le k \le \mathrm{diam}(G) - 1$.
- Another interesting question is: given a graph G, does there exist a numbering γ of $V(G)$ which is optimal for the k-level bandwidth (i.e. such that $B^k(G, \gamma) = B^k(G)$) for any k ?

References

1. Chartrand, G., Nebeský, L., Zhang, P.: Radio k-colorings of paths. Discussiones Mathematicae Graph Theory 24, 5–21 (2004)
2. Chartrand, G., Erwin, D., Harary, F., Zhang, P.: Radio labelings of graphs. Bull. Inst. Combin. Appl. 33, 77–85 (2001)
3. Chartrand, G., Erwin, D., Zhang, P.: Radio antipodal colorings of graphs. Math. Bohem. 127(1), 57–69 (2002)
4. Chen, M.-J., Kuo, D., Yan, J.-H.: The bandwidth sum of join and composition of graphs. Discrete Math. 290(2-3), 145–163 (2005)
5. Chinn, P.Z., Chvátalová, J., Dewdney, A.K., Gibbs, N.E.: The bandwidth problem for graphs and matrices—a survey. J. Graph Theory 6(3), 223–254 (1982)
6. Kchikech, M., Khennoufa, R., Togni, O.: Linear and cyclic radio k-labelings of trees. Discussiones Mathematicae Graph Theory 27(1), 105–123 (2007)
7. Khennoufa, R., Togni, O.: A note on radio antipodal colourings of paths. Math. Bohemica 130(3), 277–282 (2005)
8. Khennoufa, R., Togni, O.: The Radio Antipodal and Radio Numbers of the Hypercube (submmited, 2007)
9. Kojima, T., Ando, K.: Bandwidth of the Cartesian product of two connected graphs. Discrete Math. 252(1-3), 227–235 (2002)
10. Lai, Y.-L., Tian, C.-S.: An extremal graph with given bandwidth. Theoret. Comput. Sci. 377(1-3), 238–242 (2007)
11. Lin, M., Lin, Z., Xu, J.: Graph bandwidth of weighted caterpillars. Theoret. Comput. Sci. 363(3), 266–277 (2006)
12. Liu, D., Zhu, X.: Multi-level distance labelings for paths and cycles. SIAM J. Discrete Mathematics 19(3), 610–621 (2005)

Author Index

Lecture Notes in Computer Science

Sublibrary 1: Theoretical Computer Science and General Issues

For information about Vols. 1– 4580
please contact your bookseller or Springer

Vol. 4706: O. Gervasi, M.L. Gavrilova (Eds.), Computational Science and Its Applications – ICCSA 2007, Part II. XXIII, 1129 pages. 2007.

Vol. 4705: O. Gervasi, M.L. Gavrilova (Eds.), Computational Science and Its Applications – ICCSA 2007, Part I. XLIV, 1169 pages. 2007.

Vol. 4703: L. Caires, V.T. Vasconcelos (Eds.), CONCUR 2007 – Concurrency Theory. XIII, 507 pages. 2007.

Vol. 4700: C.B. Jones, Z. Liu, J. Woodcock (Eds.), Formal Methods and Hybrid Real-Time Systems. XVI, 539 pages. 2007.

Vol. 4699: B. Kågström, E. Elmroth, J. Dongarra, J. Waśniewski (Eds.), Applied Parallel Computing. XXIX, 1192 pages. 2007.

Vol. 4698: L. Arge, M. Hoffmann, E. Welzl (Eds.), Algorithms – ESA 2007. XV, 769 pages. 2007.

Vol. 4697: L. Choi, Y. Paek, S. Cho (Eds.), Advances in Computer Systems Architecture. XIII, 400 pages. 2007.

Vol. 4688: K. Li, M. Fei, G.W. Irwin, S. Ma (Eds.), Bio-Inspired Computational Intelligence and Applications. XIX, 805 pages. 2007.

Vol. 4684: L. Kang, Y. Liu, S. Zeng (Eds.), Evolvable Systems: From Biology to Hardware. XIV, 446 pages. 2007.

Vol. 4683: L. Kang, Y. Liu, S. Zeng (Eds.), Advances in Computation and Intelligence. XVII, 663 pages. 2007.

Vol. 4681: D.-S. Huang, L. Heutte, M. Loog (Eds.), Advanced Intelligent Computing Theories and Applications. XXVI, 1379 pages. 2007.

Vol. 4672: K. Li, C. Jesshope, H. Jin, J.-L. Gaudiot (Eds.), Network and Parallel Computing. XVIII, 558 pages. 2007.

Vol. 4671: V.E. Malyshkin (Ed.), Parallel Computing Technologies. XIV, 635 pages. 2007.

Vol. 4669: J.M. de Sá, L.A. Alexandre, W. Duch, D. Mandic (Eds.), Artificial Neural Networks – ICANN 2007, Part II. XXXI, 990 pages. 2007.

Vol. 4668: J.M. de Sá, L.A. Alexandre, W. Duch, D. Mandic (Eds.), Artificial Neural Networks – ICANN 2007, Part I. XXXI, 978 pages. 2007.

Vol. 4666: M.E. Davies, C.J. James, S.A. Abdallah, M.D. Plumbley (Eds.), Independent Component Analysis and Blind Signal Separation. XIX, 847 pages. 2007.

Vol. 4665: J. Hromkovič, R. Královič, M. Nunkesser, P. Widmayer (Eds.), Stochastic Algorithms: Foundations and Applications. X, 167 pages. 2007.

Vol. 4664: J. Durand-Lose, M. Margenstern (Eds.), Machines, Computations, and Universality. X, 325 pages. 2007.

Vol. 4661: U. Montanari, D. Sannella, R. Bruni (Eds.), Trustworthy Global Computing. X, 339 pages. 2007.

Vol. 4649: V. Diekert, M.V. Volkov, A. Voronkov (Eds.), Computer Science – Theory and Applications. XIII, 420 pages. 2007.

Vol. 4647: R. Martin, M.A. Sabin, J.R. Winkler (Eds.), Mathematics of Surfaces XII. IX, 509 pages. 2007.

Vol. 4646: J. Duparc, T.A. Henzinger (Eds.), Computer Science Logic. XIV, 600 pages. 2007.

Vol. 4644: N. Azémard, L. Svensson (Eds.), Integrated Circuit and System Design. XIV, 583 pages. 2007.

Vol. 4641: A.-M. Kermarrec, L. Bougé, T. Priol (Eds.), Euro-Par 2007 Parallel Processing. XXVII, 974 pages. 2007.

Vol. 4639: E. Csuhaj-Varjú, Z. Ésik (Eds.), Fundamentals of Computation Theory. XIV, 508 pages. 2007.

Vol. 4638: T. Stützle, M. Birattari, H. H. Hoos (Eds.), Engineering Stochastic Local Search Algorithms. X, 223 pages. 2007.

Vol. 4630: H.J. van den Herik, P. Ciancarini, H.H.L.M.(J.) Donkers (Eds.), Computers and Games. XII, 283 pages. 2007.

Vol. 4628: L.N. de Castro, F.J. Von Zuben, H. Knidel (Eds.), Artificial Immune Systems. XII, 438 pages. 2007.

Vol. 4627: M. Charikar, K. Jansen, O. Reingold, J.D.P. Rolim (Eds.), Approximation, Randomization, and Combinatorial Optimization. XII, 626 pages. 2007.

Vol. 4624: T. Mossakowski, U. Montanari, M. Haveraaen (Eds.), Algebra and Coalgebra in Computer Science. XI, 463 pages. 2007.

Vol. 4623: M. Collard (Ed.), Ontologies-Based Databases and Information Systems. X, 153 pages. 2007.

Vol. 4621: D. Wagner, R. Wattenhofer (Eds.), Algorithms for Sensor and Ad Hoc Networks. XIII, 415 pages. 2007.

Vol. 4619: F. Dehne, J.-R. Sack, N. Zeh (Eds.), Algorithms and Data Structures. XVI, 662 pages. 2007.

Vol. 4618: S.G. Akl, C.S. Calude, M.J. Dinneen, G. Rozenberg, H.T. Wareham (Eds.), Unconventional Computation. X, 243 pages. 2007.

Vol. 4616: A.W.M. Dress, Y. Xu, B. Zhu (Eds.), Combinatorial Optimization and Applications. XI, 390 pages. 2007.

Vol. 4614: B. Chen, M. Paterson, G. Zhang (Eds.), Combinatorics, Algorithms, Probabilistic and Experimental Methodologies. XII, 530 pages. 2007.

Vol. 4613: F.P. Preparata, Q. Fang (Eds.), Frontiers in Algorithmics. XI, 348 pages. 2007.

Vol. 4600: H. Comon-Lundh, C. Kirchner, H. Kirchner (Eds.), Rewriting, Computation and Proof. XVI, 273 pages. 2007.

Vol. 4599: S. Vassiliadis, M. Bereković, T.D. Hämäläinen (Eds.), Embedded Computer Systems: Architectures, Modeling, and Simulation. XVIII, 466 pages. 2007.

Vol. 4598: G. Lin (Ed.), Computing and Combinatorics. XII, 570 pages. 2007.

Vol. 4596: L. Arge, C. Cachin, T. Jurdziński, A. Tarlecki (Eds.), Automata, Languages and Programming. XVII, 953 pages. 2007.

Vol. 4595: D. Bošnački, S. Edelkamp (Eds.), Model Checking Software. X, 285 pages. 2007.

Vol. 4590: W. Damm, H. Hermanns (Eds.), Computer Aided Verification. XV, 562 pages. 2007.

Vol. 4588: T. Harju, J. Karhumäki, A. Lepistö (Eds.), Developments in Language Theory. XI, 423 pages. 2007.

Vol. 4583: S.R. Della Rocca (Ed.), Typed Lambda Calculi and Applications. X, 397 pages. 2007.